DYNAMICS AND SIMULATION OF FLEXIBLE ROCKETS

DYNAMICS AND SIMULATION OF FLEXIBLE ROCKETS

TIMOTHY M. BARROWS

JEB S. ORR

ELSEVIER

ACADEMIC PRESS
An imprint of Elsevier

Cover photo: The Saturn IB SA-205 launch vehicle carries the first crewed Apollo spacecraft into orbit on October 11, 1968. This photograph was taken from the Airborne Lightweight Optical Tracking System (ALOTS) aboard a specially modified C-135 aircraft. (NASA)

Academic Press is an imprint of Elsevier
125 London Wall, London EC2Y 5AS, United Kingdom
525 B Street, Suite 1650, San Diego, CA 92101, United States
50 Hampshire Street, 5th Floor, Cambridge, MA 02139, United States
The Boulevard, Langford Lane, Kidlington, Oxford OX5 1GB, United Kingdom

Library of Congress Cataloging-in-Publication Data
A catalog record for this book is available from the Library of Congress

British Library Cataloguing-in-Publication Data
A catalogue record for this book is available from the British Library

ISBN: 978-0-12-819994-7

For information on all Academic Press publications
visit our website at https://www.elsevier.com/books-and-journals

Publisher: Matthew Deans
Acquisitions Editor: Carrie Bolger
Editorial Project Manager: Fernanda A. Oliveira
Production Project Manager: Kamesh Ramajogi
Designer: Mark Rogers

Typeset by VTeX

Contents

Acknowledgments

The authors are indebted to the many people that helped make this work possible. We would like to thank our present and past friends and colleagues in the dynamics and control community at NASA's Marshall Space Flight Center, Langley Research Center, Armstrong Flight Research Center, and the NASA Engineering and Safety Center. We make no attempt to list their names as they are too numerous.

The support of systems engineers and managers during the NASA Constellation and Space Launch System programs was helpful in the advancement and standardization of methods and software tools for analyzing large rockets. In addition, we would like to acknowledge the many lively discussions we enjoyed among the technical staff during our tenure at the Charles Stark Draper Laboratory.

Finally, we would like to recognize the contributions of Mr. Rekesh Ali, who as a graduate student researcher, contributed significantly to the typesetting of this book.

CHAPTER 1

Introduction

Rockets, like most things, become more complicated as they grow larger. Judging from the similarity of external appearance, it might seem that going from a small rocket to a large rocket would be a simple extrapolation according to size. However, this is not the case. Some idea of the reason for the added difficulty can be obtained from the following quote from J. B. S. Haldane:

> *... consider a giant man sixty feet high – about the height of Giant Pope and Giant Pagan in the illustrated Pilgrim's Progress of my childhood. These monsters were not only ten times as high as Christian, but ten times as wide and ten times as thick, so that their total weight was a thousand times his, or about eighty to ninety tons. Unfortunately the cross sections of their bones were only a hundred times those of Christian, so that every square inch of giant bone had to support ten times the weight borne by a square inch of human bone. As the human thigh-bone breaks under about ten times the human weight, Pope and Pagan would have broken their thighs every time they took a step. This was doubtless why they were sitting down in the picture I remember. But it lessens one's respect for Christian and the Giant Killer.*

In this example, increasing the bone cross section by a factor of a hundred is not enough – it must be increased by more than a hundred. In other words, the structural weight fraction must be increased. In the design of rockets, however, the mere suggestion of increasing the structural weight fraction will produce the most pained anguish. A good portion of this extra weight will be taken out of the payload. As a typical payload weight is less than ten percent of the total rocket weight at launch, it is easy to see how the payload can disappear entirely without a stringent effort to minimize the structural weight. The result is that the design of large rockets becomes an almost desperate effort to improve structural efficiency.

From a dynamic standpoint, as the scale increases, the rocket grows flimsier and flimsier. The natural frequencies of more and more flexible modes creep downward into a range that is within the control bandwidth. The opportunities for dynamic interaction proliferate. The control engineer must verify that all of these interactions are benign and stable. Doing this requires

Dynamics and Simulation of Flexible Rockets
https://doi.org/10.1016/B978-0-12-819994-7.00006-6

methods for constructing simulations that can efficiently deal with a large number of dynamic modes.

Perhaps the most famous large rocket ever built was the Saturn V of the Apollo space program. Since the time of that program, major advances have taken place in our ability to analyze structures using finite element methods. At the same time, modern computer tools such as MATLAB® have promoted the use of matrix techniques and made it increasingly easy to deal with large matrices. The purpose of this book is to provide a uniform foundation for modeling all these interactions that takes advantage of these developments.

The dynamics of an ascending rocket are typically presented for planar motion. That is, the resulting equations are valid for either the pitch plane or the yaw plane. This approach does not provide any insight into the possible coupling that may exist between motion in one plane and that in another. Such coupling may arise from asymmetries in either the mass distribution or the stiffness distribution.

The planar dynamics of a rocket can be found in many sources. These sources fall into two separate camps, which might called the "reduced body" approach and the "integrated" approach. The characteristic feature of the former is that the translation and rotation equations are written for a reduced body consisting of the rocket without the sloshing fuel mass. One example of this approach is the textbook by Greensite [1]. A comprehensive treatment of the planar motion of a rocket was developed for the 1960's Atlas rocket program, although the technical reports (and similarly, company reports that are cited elsewhere) are not available in the open literature. Related formulations were independently derived by Rheinfurth and Hosenthein [2]; these are presented in part in the compilation by Garner [3] and eventually appear, without reference, in the classic paper by Frosch and Vallely [4]. An early example of a derivation in the open literature is the work of Bauer [5]. He provides an analysis of a flexible rocket with sloshing fuel mass. His analysis does not include a gimbaled engine.

Rocket dynamics is essentially multibody dynamics applied to a system consisting of a rocket body, engine nozzles, and slosh masses. The multibody model must be coordinated with the structural dynamic model – they must both take either the reduced body or the integrated body approach. Thus if a finite element model already exists for the rocket, the dynamicist will have to go along with whatever approach was taken during the creation of that model. An "integrated body finite element model," as the name implies, contains all of the mass of the rocket, including the slosh masses and engines. For the creation of this structural model, the slosh masses and engines are locked to the vehicle. Thus the relative motion of the slosh masses is not included, and the engine gimbal actuators are treated as rigid. The result of the finite element analysis is a set of eigenvalues (mode frequencies) and eigenvectors (mode shapes), which become input parameters to the dynamic model (the subject of the present treatise). In a "reduced body finite element model", either the slosh masses or the engine masses, or both, are removed from the rocket, and a finite element model is created from what is left. Within the dynamic model, the effects of the relative motion of the slosh and engine masses are treated in different ways for the integrated body model and the reduced body model.

This book begins with the integrated body approach, which is derived in Chapter 2. As will be seen, the reduced body approach has the disadvantage that the results contain more terms. It turns out, however, that no guarantee can be provided that the mass matrix using the integrated approach is positive-definite. Indeed, it can be shown that if a sufficiently large number of modes are included, the mass matrix will become non-positive-definite. Thus the reduced body approach, while less convenient, is the safer of the two approaches. This is discussed in Section 2.6.

Besides the issue of the integrated body approach versus the reduced body approach, there are two other major decisions that must be made before embarking on the analysis of rocket dynamics. For preliminary studies, it is often assumed that the Thrust Vector Control (TVC) actuators are very stiff, such that the engine motion can be computed independently from the rest of the dynamics. In other words, engine motion is prescribed. Chapter 2 goes into this in some detail. For purposes of the present discussion, it is sufficient to state that one must either (a) assume a given engine motion, which acts like a disturbance to the motion equations, or (b) assume a certain actuator torque on the engine, in which case the state vector is expanded to include variables that specify the engine motion. A third decision must be made as to whether to model the slosh motion as a point mass that slides in a y-z plane at the end of a spring (the spring model), or to model it as a point mass on the end of a pendulum. Thus there are a total of eight possible outcomes from making these three binary decisions about the model formulation. For this reason, this book does not provide a "final" result for the system equations of motion, but rather attempts to present the results in such a way that the analyst can select the equations and terms for the particular formulation that is most appropriate.

Notation system

The analyses herein follow the system used by Hughes [6]. His system makes a distinction between a vector and a column matrix. A vector is a mathematical quantity with both magnitude and direction in three-dimensional space, and is independent of the system of coordinates used to express it.

Suppose there is a reference frame a defined by the orthogonal unit vectors $\hat{\mathbf{a}}_1$, $\hat{\mathbf{a}}_2$, $\hat{\mathbf{a}}_3$ and a reference frame b defined by unit vectors $\hat{\mathbf{b}}_1$, $\hat{\mathbf{b}}_2$, $\hat{\mathbf{b}}_3$. A vector may be written using its frame a components

$$\vec{\mathbf{v}} = v_1^a\hat{\mathbf{a}}_1 + v_2^a\hat{\mathbf{a}}_2 + v_3^a\hat{\mathbf{a}}_3 \tag{1.1}$$

or using its frame b components

$$\vec{\mathbf{v}} = v_1^b\hat{\mathbf{b}}_1 + v_2^b\hat{\mathbf{b}}_2 + v_3^b\hat{\mathbf{b}}_3. \tag{1.2}$$

(Symbols in italics are scalars). Both expressions represent exactly the same vector. In frames a and b, the associated column matrices are expressed as

$$\begin{aligned} \mathbf{v}^a &= \begin{bmatrix} v_1^a & v_2^a & v_3^a \end{bmatrix}^T \\ \mathbf{v}^b &= \begin{bmatrix} v_1^b & v_2^b & v_3^b \end{bmatrix}^T. \end{aligned} \tag{1.3}$$

One feature of the Hughes system is that the superscript representing the frame is dropped. Thus it may be necessary to read the text to determent the frame in which each vector is expressed. This may make it more difficult to jump into the middle of a derivation and understand what everything means. However, this drawback is more than compensated by the fact that the notation is less cluttered. Appendix A contains a glossary of symbols that may be helpful in finding where each symbol is first defined.

The symbol $\mathscr{F}$ denotes a coordinate frame. Thus $\mathscr{F}_1$ is the coordinate frame of body 1. The statement "$\mathbf{v}$ is a vector expressed in $\mathscr{F}_1$" is really a shorthand for the statement that "$\mathbf{v}$ is the vector $\vec{\mathbf{v}}$ expressed as a column matrix in $\mathscr{F}_1$."

Lower case bold represents a three- or four-element column matrix. Upper case bold represents a matrix (typically 3×3). Script is used for long vectors and large matrices. Upper case bold with an arrow represents a dyadic. The notation $\dot{\vec{\mathbf{v}}}$ indicates the time derivative of $\mathbf{v}$ with respect to an inertial frame, and $\mathring{\vec{\mathbf{v}}}$ indicates the time derivative of $\mathbf{v}$ with respect to a rotating frame. Quantities that are not bold and have no arrow are scalars, typically the scalar length of a vector. Thus $\mathbf{b} \cdot \mathbf{b} = b^2$.

The notation $\dot{\mathbf{v}}$, without the arrow, indicates the time derivative of the column matrix $\mathbf{v}$. Since a particular frame must be defined as part of the definition of $\mathbf{v}$, and each element of $\mathbf{v}$ is a scalar, from a mathematical standpoint this time derivative is uniquely defined, i.e., it can only have one meaning.

If $\mathbf{v}$ is defined in a rotating body frame, then $\dot{\mathbf{v}}$ corresponds to $\overset{\circ}{\vec{\mathbf{v}}}$. The physical meaning of this derivative may not be obvious, so $\dot{\mathbf{v}}$ might best be considered as simply a mathematical entity. In particular, if $\mathbf{v}$ is a velocity vector in the body frame, then $\dot{\mathbf{v}}$ cannot be integrated to get $\mathbf{v}$. That is what is meant by the phrase "$\mathbf{v}$ is not holonomic." For an excellent discussion of this issue, the reader is referred to Appendix B of the textbook by Hughes [5].

Hughes uses the following notation for the cross product matrix:

$$\mathbf{v}^{\times} \equiv \begin{bmatrix} 0 & -v_3 & v_2 \\ v_3 & 0 & -v_1 \\ -v_2 & v_1 & 0 \end{bmatrix} \tag{1.4}$$

Using the Hughes system, the dot product and cross product are translated into matrix form as follows:

$$\vec{\mathbf{u}} \cdot \vec{\mathbf{v}} \to \mathbf{u}^T \mathbf{v} \tag{1.5}$$

$$\vec{\mathbf{u}} \times \vec{\mathbf{v}} \to \mathbf{u}^{\times} \mathbf{v}. \tag{1.6}$$

It is important to recognize that $\mathbf{u}^{\times}$ is a *matrix*. Matrix operations such as gradients and time derivatives readily follow in this representation, whereas for the binary cross product $\vec{\mathbf{u}} \times \vec{\mathbf{v}}$, such operations are not as easily defined.

Matrix operations

With due attention to the order of operations, the dot product and the cross product can be interchanged;

$$\vec{\mathbf{u}} \cdot (\vec{\mathbf{v}} \times \vec{\mathbf{w}}) = \vec{\mathbf{w}} \cdot (\vec{\mathbf{u}} \times \vec{\mathbf{v}}) = (\vec{\mathbf{u}} \times \vec{\mathbf{v}}) \cdot \vec{\mathbf{w}}. \tag{1.7}$$

The matrix equivalent of this expression is

$$\mathbf{u}^T \mathbf{v}^{\times} \mathbf{w} = \mathbf{w}^T \mathbf{u}^{\times} \mathbf{v} = \left(\mathbf{u}^{\times} \mathbf{v}\right)^T \mathbf{w}. \tag{1.8}$$

Note that parentheses are essential for the last expression.

Sometimes it is useful to take the derivative with respect to a column matrix. Consider the scalar product

$$s = \mathbf{u}^T \mathbf{v}. \tag{1.9}$$

The partial derivative of this expression with respect to $\mathbf{v}$ is

$$\frac{\partial s}{\partial \mathbf{v}} = \mathbf{u}. \tag{1.10}$$

This is just a convenient way to write three derivatives at once. Suppose $\mathbf{u} = \begin{bmatrix} u_1 & u_2 & u_3 \end{bmatrix}^T$; this matrix equation is equivalent to the three scalar equations

$$\begin{aligned} \frac{\partial s}{\partial v_1} &= u_1 \\ \frac{\partial s}{\partial v_2} &= u_2 \\ \frac{\partial s}{\partial v_3} &= u_3. \end{aligned} \tag{1.11}$$

One slightly more complicated case will be presented here. Suppose

$$T = \frac{1}{2} \boldsymbol{\omega}^T \mathbf{I} \boldsymbol{\omega} \tag{1.12}$$

where $\mathbf{I}$ is a 3x3 symmetric inertia matrix and T is the rotational kinetic energy. The derivative of T with respect to $\boldsymbol{\omega}$ is

$$\frac{\partial T}{\partial \boldsymbol{\omega}} = \mathbf{I} \boldsymbol{\omega}. \tag{1.13}$$

The most convincing way to verify this is to write the complete expression for the scalar T in terms of the elements of $\mathbf{I}$ and $\boldsymbol{\omega}$, and then take derivatives term by term.

Organization of this book

Chapter 2 provides an introduction to finite element models, and shows how Lagrange's equation can be applied to the problem of a flexible rocket with sloshing fuel. The objective is to derive a mass matrix, i.e., the matrix $\mathcal{M}$ in the equation $\mathcal{M}\ddot{\mathbf{x}} = \mathcal{F}$. Several forms of the mass matrix are derived, depending on factors such as whether an integrated or reduced body is defined for the FEM, whether the engine motion is included in the dynamics or prescribed externally, etc.

Chapter 3, Section 3.1 provides a brief description of how a sloshing wave in a fuel tank can be represented by a suitable mechanical analog, either as a point mass on a spring or a point mass on a pendulum. Section 3.2 provides a Newton-Euler derivation of the nonlinear forces (Coriolis and centrifugal) on a slosh mass. Section 3.3 discusses various issues that arise if the FEM contains hydrodynamic elements that model the effect of sloshing fuel.

Chapter 4 contains a nonlinear Newton-Euler analysis of a pendulum on a spherical joint. The resulting model can be used to represent either a pendulum model of sloshing fuel or a gimbaled engine. This model is of particular importance if engine deflections or sloshing wave amplitudes are large enough that nonlinear effects must be included in the simulation.

Chapter 5 provides a discussion of the forces and moments that go on the right-hand side (RHS) of the equations. These include effects such as aerodynamics as well as apparent forces that arise in an accelerating reference frame. The phenomenon of rigid-body jet damping, which arises due to flowing propellant, is treated in detail. This chapter ends with summary of how to compute the forces that go with each equation.

Chapter 6 discusses the important topic of engine interactions, or more precisely the coupling that may exist between the engine motions and the rest of the dynamics. Special attention is given to the topic of inertial and thrust vector coupling of gimbaled engines with bending, which gives rise to thrust vector servoelasticity (TVSE). Recommendations for how the engine actuators should be modeled in the FEM are also provided.

Chapter 7 shows how the equations of motion can be put into state-space form that is suitable for either time-domain or frequency-domain analysis. Linear perturbation methods are used to introduce approximations for effects such as follower forces and aeroelasticity, and their influence on linear system eigenvalues and frequency response is summarized.

Chapter 8 discusses the important issue of producing the inputs that are provided to a simulation. Established practice is to run a Monte Carlo analysis in which parameters such as thrust, flex frequency, etc. are given a dispersed set of values, rather than a single value. During a simulation, the FEM must change at periodic intervals as the rocket mass changes. Section 8.3 shows how to minimize the disruption that occurs in a simulation during these changes.

Chapter 9 introduces the topic of stabilization and control of flexible boost vehicles using feedback. Linear analysis techniques developed in Chapter 7 are applied to synthesize feedback control structures that provide the desired closed-loop response of the rigid-body dynamics. A model for a typical actuation system, a pressure-stabilized hydraulic thrust vector control actuator, is introduced.

Finally, Chapter 10 incorporates material presented in previous chapters and discusses practical considerations for the development of production computer simulations. A simple constraint method using Lagrange multipliers is shown to be adequate for the modeling of launch pad interfaces. Numerical integration concepts specific to the present application are discussed. The important topic of designing Monte Carlo simulations and assessing results using binomial and order statistics, particularly for flight certification, is introduced.

CHAPTER 2

The system mass matrix

In this chapter, the fundamental dynamic equations of a flexible rocket with sloshing propellant and a gimbaled engine are derived from first principles. The detailed analysis of these features is applicable to many rocket configurations, but is particularly important for very large rockets. In the case of space launch vehicles, the motion of propellant sloshing within the fuel tanks is of great significance to the design as often more than 90% of the vehicle's liftoff mass is liquid propellant.

Sloshing propellant is usually modeled as a linearized pendulum or an equivalent spring, mass, and damper coupled to the vehicle rigid and elastic degrees of freedom such that the force and moment response of the mechanical analog matches that of test-correlated semi-empirical models of a rigid tank. The portion of the equivalent liquid mass that is not in motion is lumped into the rigid-body mass. The properties of the mechanical analog change as a function of propellant remaining and the vehicle acceleration.

Engine dynamics can also play a significant role in the global vehicle behavior. For very large booster systems, particularly space launch vehicles, the use of large thrust-vectored engines results in a total moving engine mass that is a significant fraction of the total vehicle mass. Engine position control is often provided by high-power hydraulic or electromechanical actuators. This combination of moving mass, high actuator loads, and the lightweight, flexible stage structure leads to a variety of coupling effects between the engines and vehicle that must be accounted for explicitly in the design.

In the following development, the equations of motion will be developed initially for a rocket with a single fuel tank and a single engine. Generalization of these techniques to the case of multiple tanks and engines is straightforward.

2.1 Problem formulation

Consider a rocket with one fuel tank and one gimbaled engine, as shown in Fig. 2.1. Thus three bodies are considered, one of which is modeled as a point mass. There may also be one or more non-gimbaled engines, not

Dynamics and Simulation of Flexible Rockets
https://doi.org/10.1016/B978-0-12-819994-7.00007-8

shown. The origin of the coordinate system is placed on the undeformed centerline, at any convenient location.

The rocket body includes the non-sloshing fuel and the non-gimbaled engines and is designated by the subscript 0. The position of the sloshing mass in the body frame is given by

$$\mathbf{r}_{sj} = \mathbf{b}_{sj} + \boldsymbol{\delta}_{sj} \tag{2.1.1}$$

The subscript j (for tank j) is attached to these vectors, even though at this stage in the analysis there is only one tank. Here $\mathbf{b}_{sj}$ is a fixed vector and $\boldsymbol{\delta}_{sj}$ varies dynamically. $\mathbf{b}_{sj}$ aligns with the x_b axis if the tank is on the centerline. Under equilibrium conditions $\boldsymbol{\delta}_{sj} = 0$, and let $\mathbf{v}_{sj} \equiv \dot{\boldsymbol{\delta}}_{sj}$, the velocity of the sloshing mass relative to the body.

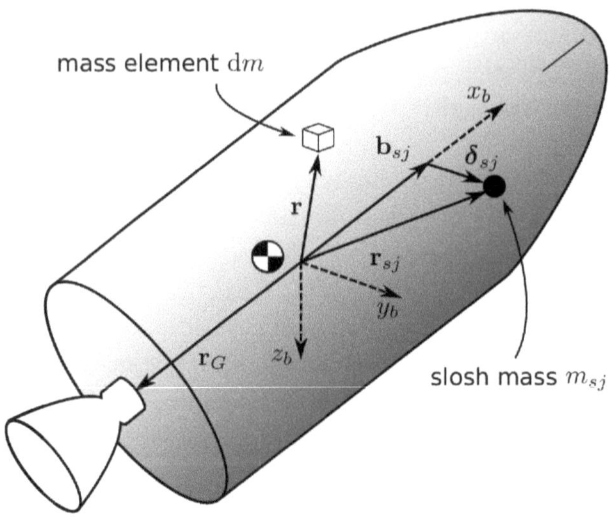

Figure 2.1 Rocket with sloshing fuel mass.

Let an inertial frame be temporarily assumed with an initial velocity that matches that of the vehicle. If there is no thrust, no gravitational acceleration, and no external force, the overall center of mass will remain stationary in inertial space. Rigid-body motion of the engine about the gimbal or displacement of the slosh mass from its equilibrium position will cause the rocket body to move in response to these motions, but the overall center of mass will remain fixed. The origin, however, remains attached to the rocket body, and thus moves relative to the inertial frame in response to these motions.

Retaining the assumption of no thrust or external force, we now add the additional assumptions that an integrated body FEM is being used, the engine gimbals are locked, and slosh masses are locked to the body. If there are elastic vibrations, the position of the center of mass will still remain stationary. The undeformed centerline (the body x_b axis) also remains stationary while the elastic vibrations take place. This means that the entire body axis frame and its origin remain stationary relative to an inertial frame. As the name implies, the undeformed centerline is always straight and represents the centerline of the rocket when all the elastic vibrations have decayed to zero. The origin is not fixed to any physical part of the rocket body but stays on the undeformed centerline. In the dynamics literature, this is sometimes referred to as a mean axis formulation.

The assumption that the rocket is stationary in the inertial frame can be removed. Let $\mathbf{v}_I$ be the velocity of the origin expressed in inertial coordinates. This velocity is defined by taking the time derivative of the location of the body origin with respect to the inertial origin. Let $\mathbf{v}$ correspond to the same vector expressed in the body frame. For a linearized analysis, the kinematic relationship of the inertial and body frames can be approximated using the expression

$$\mathbf{v} = \left(\mathbf{1} - \boldsymbol{\phi}^{\times}\right)\mathbf{v}_I \tag{2.1.2}$$

where

$$\boldsymbol{\phi} \equiv \begin{bmatrix} \phi_x & \phi_y & \phi_z \end{bmatrix}^T \tag{2.1.3}$$

is a column matrix containing the roll, pitch, and yaw of the body frame relative to the inertial frame, all of which are assumed to be small such that the dependency of that relationship on the order of rotations is negligible. If this assumption is not valid, one has instead

$$\mathbf{v} = \mathbf{C}_{bI}\mathbf{v}_I \tag{2.1.4}$$

where $\mathbf{C}_{bI}$ is the transformation from the inertial frame to the body frame.

The column matrix $\mathbf{v}$ has hybrid characteristics: it is defined by taking the time derivative relative to an inertial frame, but it is expressed in the body frame. The acceleration in the body frame is

$$\mathbf{a}_b = \mathbf{C}_{bI}\frac{\mathrm{d}\mathbf{v}_I}{\mathrm{d}t} = \mathbf{C}_{bI}\left(\mathbf{C}_{Ib}\dot{\mathbf{v}} + \dot{\mathbf{C}}_{Ib}\mathbf{v}\right) = \dot{\mathbf{v}} + \boldsymbol{\omega}^{\times}\mathbf{v} \tag{2.1.5}$$

where $\boldsymbol{\omega}$ is the angular rotation of the body frame with respect to the inertial frame, expressed in the body frame. Here, the kinematic differential equation

$$\dot{\mathbf{C}}_{Ib} = \mathbf{C}_{Ib}\boldsymbol{\omega}^{\times} \tag{2.1.6}$$

has been used to determine the time derivative of $\mathbf{v}_I$ in the body frame.

Lagrange's equation can be employed to derive general expressions for mechanical systems undergoing vibrations; it is given by

$$\frac{d}{dt}\left(\frac{\partial T}{\partial \dot{q}_i}\right) - \frac{\partial T}{\partial q_i} + \frac{\partial V}{\partial q_i} + \frac{\partial D}{\partial \dot{q}_i} = Q_i \tag{2.1.7}$$

where T is the kinetic energy, V is the potential energy, and D is the dissipation function. The generalized coordinates and generalized external forces are given by q_i and Q_i, respectively. It can be shown that when applied to problem under consideration, the second term in (2.1.7) is always zero. The third and fourth terms in (2.1.7) can be moved to the right hand side (RHS) of the equation. Thus Lagrange's equation can be expressed as

$$\frac{d}{dt}\left(\frac{\partial T}{\partial \dot{q}_i}\right) = Q_i - \frac{\partial V}{\partial q_i} - \frac{\partial D}{\partial \dot{q}_i} \tag{2.1.8}$$

This is the ultimate form of Lagrange's equation that is used in the development of a typical simulation. The left hand side (LHS) is used to generate a mass matrix. The solution procedure is to compute the RHS from the system state, solve the matrix equations to generate accelerations, and integrate the states forward in time. The remainder of this chapter is devoted to explaining in detail how the mass matrix is derived. Computation of the RHS is postponed to Chapter 5. Thus, in the present chapter, the generalized coordinates q_i do not appear, only their derivatives $\dot{q}_i$. A detailed description of how these equations are integrated is provided in Chapter 10.

For bookkeeping purposes, the analysis is simplified if the potential energy and dissipation terms are used for the sole purpose of representing inter-body forces. Thus, the gravitational potential is not included as part of V. Instead, the gravity force is included as a part of Q_i. The term "system" is used to denote the entire rocket, i.e., the system consisting of the rocket body, the sloshing masses, and the engine. The three translation and three rotation equations of the body frame do not contain any excitations on the RHS from inter-body forces, since these forces have an equal and opposite effect on the overall motion.

Strictly speaking, Lagrange's equation is not valid in a rotating body frame. Hughes [6] notes this fact, and provides a set of "quasi-Lagrangian" equations, valid for a rigid body, for the case in which the second, third, and fourth terms in Eq. (2.1.7) are zero.

$$\frac{d}{dt}\left(\frac{\partial T}{\partial \mathbf{v}}\right)+\boldsymbol{\omega}^{\times}\left(\frac{\partial T}{\partial \mathbf{v}}\right)=\mathbf{f} \tag{2.1.9}$$

$$\frac{d}{dt}\left(\frac{\partial T}{\partial \boldsymbol{\omega}}\right)+\boldsymbol{\omega}^{\times}\left(\frac{\partial T}{\partial \boldsymbol{\omega}}\right)+\mathbf{v}^{\times}\left(\frac{\partial T}{\partial \mathbf{v}}\right)=\mathbf{g} \tag{2.1.10}$$

Here $\mathbf{f}$ is the external force vector, and $\mathbf{g}$ is the external torque vector about the origin. For a linearized analysis, the equilibrium trajectory (e.g., solution of the motion equations) can be subtracted, thus converting the dynamic variables such as $\mathbf{v}$ and $\boldsymbol{\omega}$ to small quantities (perturbation variables) so that terms like $\mathbf{v}^{\times}\,(\partial T/\partial \mathbf{v})$ become second order and the equations revert to the Lagrangian form given by (2.1.7). What is called the "translation equation" in the following development is obtained by taking the time derivative of the linear momentum, $\partial T/\partial \mathbf{v}$. Later chapters provide a multibody Newton-Euler analysis in which rotating body effects are fully taken into account. It can be verified that when the rotation rates are sufficiently small, a linearized version of the Newton-Euler approach gives the same result as the present Lagrangian approach.

Eqs. (2.1.9) and (2.1.10) are also known to dynamicists as the Boltzmann-Hamel equations.

Mass properties

The total mass is divided into separate components for the rocket body (subscript 0), the sloshing fuel, and the engine.

$$m_T = m_0 + m_{sj} + m_E \tag{2.1.11}$$

Let ρ_0 be the density (mass per unit volume) of the rocket body, and ρ_E be the density of the engine. The slosh mass density is defined using δ, the Dirac delta, located at the slosh mass position $\mathbf{r}_{sj}$. This function has the property that its value is zero for every value of $\mathbf{r}$ except in an infinitesimal region around $\mathbf{r}=\mathbf{r}_{sj}$, and the value in this region is such that

$$\int \delta\left(\mathbf{r}-\mathbf{r}_{sj}\right)\mathrm{d}V = 1 \tag{2.1.12}$$

where $\mathrm{d}V$ is an element of volume, and the integration takes place over the entire volume of the rocket. The sloshing mass density is written as

$$\rho_{sj}(\mathbf{r}) = m_{sj}\delta\left(\mathbf{r}-\mathbf{r}_{sj}\right) \tag{2.1.13}$$

so that

$$m_{sj} = \int \rho_{sj}\, \mathrm{d}V \tag{2.1.14}$$

The Dirac delta is introduced in order to enable the entire rocket mass to be expressed in one integral. Whenever it is encountered inside an integral, it represents an opportunity for simplification by taking advantage of (2.1.12). The other masses are given by

$$\begin{aligned} m_0 &= \int \rho_0 dV \\ m_E &= \int_E \rho_E\, \mathrm{d}V \end{aligned} \tag{2.1.15}$$

The E on this last integral indicates that the integration takes place over the volume of the engine. If ρ_0 and ρ_E are defined to be zero in the region outside the boundaries of their respective bodies, then a mass density expression can be defined that is valid over the total rocket:

$$\rho_T(\mathbf{r}) = \rho_0(\mathbf{r}) + \rho_{sj}(\mathbf{r}) + \rho_E(\mathbf{r}) \tag{2.1.16}$$

Using this, the total mass is

$$m_T = \int \rho_T\, \mathrm{d}V \tag{2.1.17}$$

It is also convenient to define the mass element

$$\mathrm{d}m \equiv \rho_T\, \mathrm{d}V \tag{2.1.18}$$

Thus

$$m_T = \int \mathrm{d}m \tag{2.1.19}$$

The first moment of inertia of the system is defined as

$$\mathbf{s}_{TD} \equiv \int \mathbf{r}\rho_T\, \mathrm{d}V = \int \mathbf{r}\, \mathrm{d}m \tag{2.1.20}$$

The first moment of inertia is simply the total mass times the vector from the origin to the center of mass. The first subscript, T, indicates that this applies to the total body. The second subscript, D, indicates that this varies dynamically as the engine and slosh masses move around. The second moment of inertia about the origin can be written in either of the following forms;

$$\mathbf{I}_{TD} = \int \left(\mathbf{r}^T \mathbf{r} \mathbf{1} - \mathbf{r}\mathbf{r}^T \right) \mathrm{d}m = -\int \mathbf{r}^\times \mathbf{r}^\times \, \mathrm{d}m \tag{2.1.21}$$

where **1** is the identity matrix.

When the second moment of inertia is computed, if a propellant tank has circular symmetry about the x_b axis, then the roll component of the fluid inertia (sloshing and non-sloshing) is not included. That is, it is assumed that the rocket can roll about the x_b axis without the fluid mass rolling along with it. In actuality, there is some viscous coupling between the tank wall and the fluid that depends on the wall geometry and the fluid properties. In most cases, this coupling can be determined via secondary analyses, and accounted for by adjusting the rigid body roll inertia. If the tank is compartmented or radially segmented, this assumption must be modified.

Bauer [7] presented a model for calculating the effective moments of inertia of a cylindrical tank filled with liquid for the transverse (y and z) axes. For a smooth-walled tank, these inertias are only a small portion of what they would be for a solid mass of the same shape. If baffles are present or the wall has an orthogrid/isogrid structure, the liquid mass tends to move along with the walls of the tank, and hence increases the effective inertia. Bauer's work was extended by Dodge and Kana [8], who found that Bauer's model could be simplified in many practical circumstances. Theoretical and experimental results for a few baffle arrangements are also provided. Bauer [9] went on to provide more general formulas for different baffle geometries.

For small offset cylindrical tanks, the contribution to the roll inertia can be computed by treating the fluid mass as a point mass. There will be a contribution to the x_b axis inertia proportional to this mass times the square of the radial offset, but no significant contribution due to the fluid inertia about the tank centerline.

2.2 Structural dynamics

The dynamics and control engineer will require a working knowledge of how to use data from a Finite Element Model (FEM), even though he or she may not have the expertise to create such a model. It may also be necessary to interact with a specialist in structural dynamics in order to specify what idealizations should be employed during the creation of the FEM.

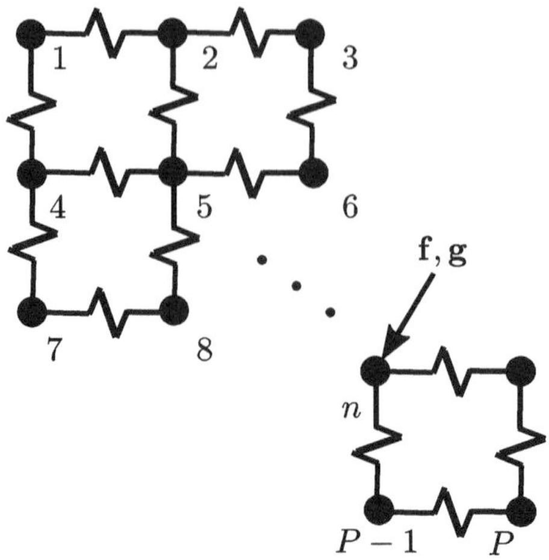

Figure 2.2 Finite element model.

At a conceptual level, a structural dynamic model is essentially an assemblage of masses and springs (Fig. 2.2). These mass elements usually correspond to a computational mesh of a large three dimensional model of the structure, and are called *nodes*. A point on the structure associated with a node may also be referred to as a *grid point*.

Each node may have a maximum of six degrees of freedom (DoFs), although the model may be set up to have fewer DoFs per node. For example, in the diagram in Fig. 2.2, there are P nodes. If each of these nodes has six DoFs, the system will have a total of $M = 6P$ DoFs. The diagram also shows an external load, consisting of a force $\mathbf{f}$ and a torque $\mathbf{g}$, acting on one node. In the more general case, there may be external forces acting on a number of nodes.

For a FEM of a real aerospace structure with a relatively fine mesh, it is not unusual for the number of DoFs to exceed one million. It is the job of the finite element analyst to reduce the problem to a numerically tractable size, using a technique such as Guyan reduction [10]. Each DoF has its own equation of motion, and these equations can be assembled into a matrix equation of the following form:

$$\mathbf{M}_B\ddot{\mathbf{x}} + \mathbf{K}_B\mathbf{x} = \mathcal{F} \tag{2.2.1}$$

In this equation, $\mathbf{M}_B$ is the mass matrix, $\mathbf{K}_B$ is the stiffness matrix, $\mathbf{x}$ is a column vector of physical displacement coordinates, and $\mathcal{F}$ is a column vector of physical loads. The term "physical" is added here for reasons that will be explained. Following tradition, the subscript B (for bending) is used, even though the FEM contains all types of flexible motion. The displacement vector $\mathbf{x}$ contains translations and rotations for nodes 1 through P;

$$\mathbf{x} = \left[x_1, y_1, z_1, \theta_{x1}, \theta_{y1}, \theta_{z1}, \ldots, x_P, y_P, z_P, \theta_{xP}, \theta_{yP}, \theta_{zP}\right]^T \tag{2.2.2}$$

All these elements are functions of time, based upon the solution of Eq. (2.2.1). It is assumed that the elastic displacements are sufficiently small that linear vibration theory can represent the structural dynamic response as an M-DoF linear system. If this is the case, the response can be decomposed into a linear combination of orthogonal solutions (vibration modes). Note that in this form, the system has no damping; that is, there is no coefficient of $\dot{\mathbf{x}}$ in Eq. (2.2.1). While it is possible to include a physical damping term, structural damping is very difficult to model and estimate in practice, and it significantly complicates the linear analysis. As such, it is usually assumed for the purposes of finding the initial modal response that either the damping is proportional to the mass and stiffness, or zero. A damping term is later added to the modal equations. This damping can be based on experience with similar structures, or correlated to test data.

Diagonalization

The solution of Eq. (2.2.1) proceeds first by finding the homogeneous solution corresponding to $\mathcal{F} = \mathbf{0}$, where $\mathbf{0}$ is the M-element null vector. There will also be $M = 6P$ modes, the same as the number of DoFs. For each mode i, a solution that varies sinusoidally with time is assumed:

$$\mathbf{x}_i = \boldsymbol{\phi}_i A_i \sin \Omega_{Bi} t \tag{2.2.3}$$

A vector $\boldsymbol{\phi}_i$ and a scalar A_i are included in this expression. The scalar makes resizing $\boldsymbol{\phi}_i$ more convenient. Substitution of the assumed solution converts (2.2.1) into the *generalized symmetric eigenvalue problem*

$$\mathbf{M}_B \boldsymbol{\Omega}_{Bi}^2 \boldsymbol{\phi}_i - \mathbf{K}_B \boldsymbol{\phi}_i = \mathbf{0} \tag{2.2.4}$$

This is solved for the M eigenvalues Ω_{Bi}^2 and associated eigenvectors $\boldsymbol{\phi}_i$. Note that $A_i \sin \Omega_{Bi} t$ has been canceled out of both terms of this expression. The eigenvectors are also called *mode shapes* and their spatial derivatives, taken with respect to the undeformed body axes, are called *mode slopes*. All the eigenvectors can be assembled into a single square matrix, of the same size as $\mathbf{K}_B$ and $\mathbf{M}_B$, such that

$$\boldsymbol{\Phi} = \begin{bmatrix} | & \cdots & | \\ \boldsymbol{\phi}_1 & \cdots & \boldsymbol{\phi}_M \\ \downarrow & \cdots & \downarrow \end{bmatrix} \tag{2.2.5}$$

The eigenvectors $\boldsymbol{\phi}_i$, with a subscript, are not to be confused with the $\boldsymbol{\phi}$ in (2.1.2). According to linear algebra (see, for example, Strang [11]), this is a congruence transformation that can be used to simultaneously diagonalize the matrices $\mathbf{K}_B$ and $\mathbf{M}_B$. A solution of the generalized symmetric eigenproblem will yield real eigenvalues and real eigenvectors if the associated matrices $\mathbf{K}_B$ and $\mathbf{M}_B$ are symmetric, positive definite. For physical structural dynamic systems, this is almost always the case. Thus

$$\boldsymbol{\Phi}^T\mathbf{M}_B\boldsymbol{\Phi} = \mathbf{m}_B \tag{2.2.6}$$

$$\boldsymbol{\Phi}^T\mathbf{K}_B\boldsymbol{\Phi} = \mathbf{m}_B\boldsymbol{\Omega}_B^2 \tag{2.2.7}$$

where

$$\mathbf{m}_B \equiv \text{diag}(m_{B1} \cdots m_{BM}) \tag{2.2.8}$$

and the *generalized mass* of each individual mode is defined as

$$m_{Bi} \equiv \boldsymbol{\phi}_i^T\mathbf{M}_B\boldsymbol{\phi}_i \tag{2.2.9}$$

The last matrix on the right hand side of (2.2.7) contains the eigenvalues of each mode:

$$\boldsymbol{\Omega}_B^2 = \text{diag}(\Omega_{B1}^2, \Omega_{B2}^2, \cdots \Omega_{BM}^2) \tag{2.2.10}$$

An eigenvector can be scaled (multiplied by a constant) and it will still satisfy Eq. (2.2.4). This allows the eigenvectors $\boldsymbol{\phi}_i$ to be chosen such that all the generalized masses are equal to one, i.e.,

$$\boldsymbol{\Phi}^T\mathbf{M}_B\boldsymbol{\Phi} = \mathbf{1} \tag{2.2.11}$$

where $\mathbf{1}$ is the $M \times M$ identity matrix. This is what is meant by "mass normalization." If $\boldsymbol{\Phi}$ is mass normalized, it follows that

$$\boldsymbol{\Phi}^T\mathbf{K}_B\boldsymbol{\Phi} = \boldsymbol{\Omega}_B^2 \tag{2.2.12}$$

It is worth emphasizing that the only reason to solve the homogeneous problem is to produce the crucial matrix $\boldsymbol{\Phi}$. Once this is obtained, the original problem can be diagonalized and the equations for each mode can be decoupled.

The units of the constants A_i in (2.2.3) must be chosen so as to produce the correct units for the elements of the vector x_i, i.e., they must undo the strange units that emerge in the eigenvectors from mass normalization. For example, it is typical for finite element analyses in the US aerospace industry to use inches for displacement and *slinches* (sometimes called *snails*, in contrast with slugs) for mass. A force of one pound acting on a mass of one slinch will produce an acceleration of one inch per second squared. Thus one slinch equals twelve slugs. For this system of units, the A_i all have units of inch $\cdot \sqrt{\text{slinch}}$.

It is important to note that the FEM may utilize a different system of units, and a different coordinate system (often the x axis points rearward) from that used in the dynamic model (in which the x axis points forward). It may be the responsibility of the dynamicist to convert the FEM data into compatible units and axes. More detail regarding these transformations is discussed in Appendix C.

The physical displacements of the structure can be expressed as a linear combination of the generalized coordinates;

$$\boldsymbol{\eta}(t) = [\eta_1 \cdots \eta_M]^T \tag{2.2.13}$$

Each element of this vector is a function of time. When mass normalization is used, they have the same units as A_i, that is, length $\cdot \sqrt{\text{mass}}$. The generalized coordinates are related to the physical displacements using the transformation (2.2.5);

$$\mathbf{x} = \boldsymbol{\Phi}\boldsymbol{\eta} \tag{2.2.14}$$

This amounts to separation of variables. The matrix $\boldsymbol{\Phi}$ gives the deflection as a function of the location in the structure, i.e., node number, and $\boldsymbol{\eta}$ gives the variation with time. In the parlance of finite element analysis, $\mathbf{x}$ represents the solution in physical space, $\boldsymbol{\eta}$ is the solution in eigenspace, and $\boldsymbol{\Phi}$ is the transformation between these solutions. In terms of Lagrange's equation, the elements of $\boldsymbol{\eta}$ represent the generalized degrees of freedom. Thus it is necessary to distinguish between the physical DoFs and the generalized DoFs. That is the reason why $\mathbf{x}$ in (2.2.1) is referred to as a vector of "physical" DoF's.

Substitution of (2.2.14) into (2.2.1) gives

$$\mathbf{M}_B\boldsymbol{\Phi}\ddot{\boldsymbol{\eta}} + \mathbf{K}_B\boldsymbol{\Phi}\boldsymbol{\eta} = \mathcal{F}. \tag{2.2.15}$$

Pre-multiplying by $\mathbf{\Phi}^T$ and using (2.2.11) plus (2.2.12), this yields the *modal equations*

$$\ddot{\boldsymbol{\eta}} + \mathbf{\Omega}_B^2 \boldsymbol{\eta} = \mathbf{\Phi}^T \mathcal{F} \tag{2.2.16}$$

Since the modal equations have been decoupled or *diagonalized*, each row of this matrix equation can be solved individually.

$$\ddot{\eta}_i + \Omega_{Bi}^2 \eta = \boldsymbol{\phi}_i^T \mathcal{F} \tag{2.2.17}$$

External loads

For a discrete load, such as that from the thrust of one engine, the external force vector $\mathcal{F}$ will typically be nonzero for only one node, in this case the node at the engine gimbal point. In the example of Fig. 2.2, the only nonzero load is at node n:

$$\mathcal{F} = \left[\mathbf{0}^T, \mathbf{0}^T, \ldots, \mathbf{f}_n^T, \mathbf{g}_n^T, \ldots, \mathbf{0}^T, \mathbf{0}^T\right]^T \tag{2.2.18}$$

Here, $\mathbf{0}$ represents the 3×1 null vector. Thus it is not necessary to include the entire eigenvector on the RHS of (2.2.16). Only the portion associated with the node under consideration is required. The entire eigenvector for mode i can be decomposed into P nodal components;

$$\boldsymbol{\phi}_i = \left[\boldsymbol{\phi}_{1i}^T \cdots \boldsymbol{\phi}_{Pi}^T\right]^T \tag{2.2.19}$$

where each $\boldsymbol{\phi}_{ni}$ is a 6×1 column vector that maps generalized displacements η_i into the associated 6 physical degrees of freedom of each node n. If applying forces and torques only to node n, Eq. (2.2.16) becomes

$$\ddot{\eta}_i + \Omega_{Bi}^2 \eta = \boldsymbol{\phi}_{ni}^T \begin{bmatrix} \mathbf{f}_n \\ \mathbf{g}_n \end{bmatrix} \tag{2.2.20}$$

It is useful to further decompose the node vector into translation and rotation vectors (each 3×1) such that

$$\boldsymbol{\phi}_{ni} = \begin{bmatrix} \boldsymbol{\psi}_{ni} \\ \boldsymbol{\sigma}_{ni} \end{bmatrix} \tag{2.2.21}$$

This allows one to write

$$\ddot{\eta}_i + \Omega_{Bi}^2 \eta_i = \boldsymbol{\psi}_{ni}^T \mathbf{f} + \boldsymbol{\sigma}_{ni}^T \mathbf{g} \tag{2.2.22}$$

In general, one has

$$\ddot{\eta}_i + \Omega_{Bi}^2 \eta_i = \sum_n \left(\boldsymbol{\psi}_{ni}^T \mathbf{f}_n + \boldsymbol{\sigma}_{ni}^T \mathbf{g}_n \right) \tag{2.2.23}$$

where the summation takes place over all the nodes that have applied loads.

As discussed earlier, a common practice is to simply add an equivalent viscous damping term of the following form:

$$\ddot{\eta}_i + 2\zeta_{Bi}\Omega_{Bi}\dot{\eta}_i + \Omega_{Bi}^2 \eta_i = \sum_n \left(\boldsymbol{\psi}_{ni}^T \mathbf{f}_n + \boldsymbol{\sigma}_{ni}^T \mathbf{g}_n \right) \tag{2.2.24}$$

The damping ratio ζ_{Bi} is chosen to match experimental data, if available, or experience with similar structures. A typical value is $\zeta_{Bi} = .005$. The virtue of this approach is that it is completely linear and fits very nicely into a linearized control system analysis. Actual damping is a combination of nonlinear (amplitude-dependent) structural damping and coulomb friction damping due to joints in the structure. Damping is discussed in many texts on structural dynamics, for example Hurty and Rubenstein [12].

Eq. (2.2.24) is valid for the problem consisting of the number of DoF's in the finite element model. If additional DoF's are added that are external to the FEM, such as slosh motion or engine motion, additional terms are added, as described in subsequent sections. In practice, the numerical solution of a FEM involves many intermediate reductions and truncation operations that remove the contributions of elements that are insignificant to the motions of interest. As such, the eigenvectors and eigenvalues delivered for dynamic analysis will typically only represent a few hundred modes of the lowest frequency response.

Continuous form of elastic displacement

In the solution of Eq. (2.2.4), there will appear six degrees of freedom that involve no relative displacement of structural nodes. These six degrees of freedom result from the orthogonalization implicit in solving the generalized eigenvalue problem; that is, separation of the elastic (relative) motion from the rigid-body motion of the entire structure. Thus it is straightforward to truncate the resultant model to include only elastic motion, or only a subset of elastic motion in a frequency range of interest. A much higher fidelity model can be constructed by replacing the six linearized rigid-body degrees of freedom with a linear or nonlinear, perhaps time-varying, representation of the rigid-body dynamics. This methodology, known as modal superposition, is the approach discussed in this book.

A common practice for modeling rockets is to create a number of centerline nodes along the x axis. The displacement of each centerline node is equal to the average displacement of the nodes that surround it at the same x position.[1] For an axisymmetric rocket, these nodes will be distributed in a circle around the centerline. If the configuration includes components such strap-on solid rocket boosters, these items may have centerlines of their own.

Mode shapes for the centerline nodes can be used in Eq. (2.2.24). The slosh force is applied at the centerline node nearest to the slosh mass, even though there is not actually any structure there. This is physically equivalent to distributing the force to the associated nodes at a given x position. For distributed loads, such as aerodynamic loads, the right side of (2.2.24) typically includes all the centerline nodes. An axisymmetric rocket can be sliced into circular sections, one per centerline node, and the aerodynamic force associated with each section can be computed and inserted into the vector $\mathcal{F}$.

Let the vector $\mathbf{x}$ in Eq. (2.2.2) be decomposed into translational and rotational components

$$\mathbf{x} = [\mathbf{w}_1, \boldsymbol{\theta}_1 \cdots \mathbf{w}_P, \boldsymbol{\theta}_P]^T \tag{2.2.25}$$

This allows the translational deflection at node j to be written using (2.2.14) and (2.2.21) as

$$\mathbf{w}_n = \sum \boldsymbol{\psi}_{ni}\eta_i(t) \tag{2.2.26}$$

It becomes convenient to write the continuous form of this equation

$$\mathbf{w}(\mathbf{r}) = \sum \boldsymbol{\psi}_i(\mathbf{r})\eta_i(t) \tag{2.2.27}$$

This transforms the equation into one that uses the location $\mathbf{r}$ (see Fig. 2.1) rather than the node number n. This would require a node map, giving the location of each node in the structure. However, the following analysis does not actually require the use of a complete map – it is sufficient to know that such a mapping can be done. The mapping is only required for points at which forces are applied. The continuous form allows the use of integrals, rather than summations. The continuous form is equivalent in

[1] In finite element analysis software, this can be accomplished by inserting a special massless rigid body element (RBE) that connects multiple nearby points on the structure to a single point on the centerline, or by the creation of *virtual grids* in postprocessing.

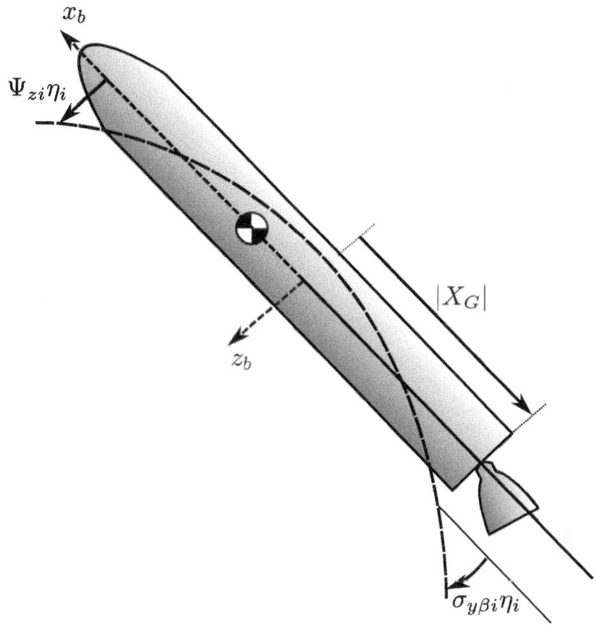

Figure 2.3 Deformed centerline of vehicle for the pitch plane.

the limit as the units of the *spatial discretization* approach zero; that is, as a truly continuous model is contemplated, the number of nodes becomes infinite. The historical development of structural dynamics occurred in the reverse order of the present discussion. The earliest analyses were done on continuous models of beams, and only later was the notion introduced of representing structures as a mesh of nodes and finite elements.

The vector quantities in (2.2.27) are each column matrices. Thus

$$\boldsymbol{\psi}_i(\mathbf{r}) = \begin{bmatrix} \psi_{xi}(\mathbf{r}) \\ \psi_{yi}(\mathbf{r}) \\ \psi_{zi}(\mathbf{r}) \end{bmatrix} \tag{2.2.28}$$

gives three coordinates of each mode shape. Fig. 2.3 shows a typical mode shape along the centerline.

It is a property of any body undergoing free-free vibrations that in the absence of external forces no linear momentum is generated. This statement can be expressed mathematically as

$$\dot{\eta}_i \int \boldsymbol{\psi}_i(\mathbf{r})\, \mathrm{d}m = 0 \,\forall i \tag{2.2.29}$$

Since $\dot{\eta}_i$ may be non-zero, the integral by itself must be zero. It is also true that no net angular momentum about the center of mass is generated;

$$\dot{\eta}_i \int \mathbf{r}^{\times} \boldsymbol{\psi}_i(\mathbf{r})\, \mathrm{d}m = 0\, \forall i. \tag{2.2.30}$$

Again, the integral by itself must be zero. Appendix E shows that it is not necessary for $\mathbf{r}$ to be defined using a coordinate system with the origin at the center of mass. That is, this equation is valid for any origin.

The continuous form of the orthogonality condition (2.2.6) is

$$\int \boldsymbol{\psi}_i^T \boldsymbol{\psi}_k \mathrm{d}m = 0, \; i \neq k \tag{2.2.31}$$

$$\int \boldsymbol{\psi}_i^T \boldsymbol{\psi}_k \mathrm{d}m = m_{Bi}, \; i = k \tag{2.2.32}$$

where m_{Bi} is the generalized mass of the ith mode. $m_{Bi} = 1$ for all i if the modes are mass normalized. Here, the shorthand that $\boldsymbol{\psi}_i = \boldsymbol{\psi}_i(\mathbf{r})$ has been introduced.

From each eigenvector, it is necessary to pick out nodes representing particular physical points on the structure. For example, the subscript j is used to represent the point of application of the force from sloshing propellant mass j, and the subscript β is used to represent the location of the engine gimbal. The modal parameters for the slosh mass and the engine gimbal are thus extracted from the eigenvector

$$\boldsymbol{\phi}_i^T = \begin{bmatrix} \cdots & \boldsymbol{\phi}_{ji}^T & \cdots & \boldsymbol{\phi}_{\beta i}^T & \cdots \end{bmatrix} \tag{2.2.33}$$

where

$$\boldsymbol{\phi}_{ji}^T = \begin{pmatrix} \psi_{xji} & \psi_{yji} & \psi_{zji} & \sigma_{xji} & \sigma_{yji} & \sigma_{zji} \end{pmatrix} \tag{2.2.34}$$

for the sloshing mass, and

$$\boldsymbol{\phi}_{\beta i}^T = \begin{pmatrix} \psi_{x\beta i} & \psi_{y\beta i} & \psi_{z\beta i} & \sigma_{x\beta i} & \sigma_{y\beta i} & \sigma_{z\beta i} \end{pmatrix} \tag{2.2.35}$$

for the engine. The lateral components of the sloshing mass and engine gimbal location degrees of freedom are labeled as ψ_{yji}, ψ_{zji}, $\psi_{y\beta i}$, $\psi_{z\beta i}$. Likewise, the modal rotation components of the eigenvector about the body y and z axes are labeled σ_{yji}, σ_{zji}, $\sigma_{y\beta i}$, $\sigma_{z\beta i}$, and so on. It is worth noting that for a single-engine vehicle, the only time two index subscripts, j and i, are required on $\boldsymbol{\psi}$ is when dealing with the sloshing propellants. The first index, j, refers to the tank number, and the second index, i, refers to the mode number. An additional subscript s is not necessary and is not included in the derivations that follow.

2.3 Kinetic energy

As with the mass, the kinetic energy is divided into separate components for the rocket body, the sloshing fuel, and the engine.

$$T = T_0 + T_s + T_E \tag{2.3.1}$$

$$T_0 = \frac{1}{2}\int \rho_0 \left[\mathbf{v} + \boldsymbol{\omega}^\times \mathbf{r} + \sum \boldsymbol{\psi}_{;i}\dot{\eta}_i\right]^T \left[\mathbf{v} + \boldsymbol{\omega}^\times \mathbf{r} + \sum \boldsymbol{\psi}_{;i}\dot{\eta}_i\right] \mathrm{d}V \tag{2.3.2}$$

$$T_s = \frac{1}{2}\int m_{sj}\left[(\mathbf{v} + \boldsymbol{\omega}^\times \mathbf{r} + \sum \boldsymbol{\psi}_{;i}\dot{\eta}_i) + \mathbf{v}_{sj}\right]^T \left[(\mathbf{v} + \boldsymbol{\omega}^\times \mathbf{r} + \sum \boldsymbol{\psi}_{;i}\dot{\eta}_i) + \mathbf{v}_{sj}\right] \delta(\mathbf{r} - \mathbf{r}_{sj})\,\mathrm{d}V \tag{2.3.3}$$

$$T_E = \frac{1}{2}\int_E \rho_E \left[(\mathbf{v} + \boldsymbol{\omega}^\times \mathbf{r} + \sum \boldsymbol{\psi}_{;i}\dot{\eta}_i) + \boldsymbol{\omega}_{Eb}^\times(\mathbf{r} - \mathbf{r}_G)\right]^T \left[(\mathbf{v} + \boldsymbol{\omega}^\times \mathbf{r} + \sum \boldsymbol{\psi}_{;i}\dot{\eta}_i) + \boldsymbol{\omega}_{Eb}^\times(\mathbf{r} - \mathbf{r}_G)\right] \mathrm{d}V \tag{2.3.4}$$

where $\mathbf{r}_G$ is the location of the gimbal point (see Fig. 2.1), and $\boldsymbol{\omega}_{Eb}$ is the angular rate of the engine relative to the body. The subscript b is added to emphasize that this quantity must be expressed in the body frame. At this point in the analysis, this subscript is unnecessary, since everything is in the body frame. However, it often turns out to be convenient to express each engine angular rate in its own engine frame. Chapter 6 discusses this issue in more detail.

The slosh energy is shown as an integral, even though it can be readily evaluated using the δ function defined in (2.1.12).

The integrated body energy

There are some unnecessary parentheses within the brackets of (2.3.3) and (2.3.4), placed there to emphasize that the same grouping appears in all three energy components. In order to take advantage of this fact, it is desired to consolidate these groupings into one integral. To do this, (2.3.1) is rearranged as follows;

$$T = T_{IB} + \Delta T_s + \Delta T_E \tag{2.3.5}$$

where T_{IB} is the *integrated body* energy, ΔT_s is the additional energy due to $\mathbf{v}_{sj}$, and ΔT_E is the additional energy due to $\boldsymbol{\omega}_{Eb}$. T_{IB} includes the kinetic energy of all the masses due to rotation, translation, and elastic motion. It is the energy that would be present if the slosh mass and engine mass were locked to the rocket body. T_{IB} has contributions from (2.3.2), (2.3.3), and (2.3.4). The aforementioned groupings in these three integrals are consolidated by using (2.1.16).

$$T_{IB} = \frac{1}{2}\int \rho_T\left[\mathbf{v}+\boldsymbol{\omega}^\times\mathbf{r}+\sum\boldsymbol{\psi}_i\dot{\eta}_i\right]^T\left[\mathbf{v}+\boldsymbol{\omega}^\times\mathbf{r}+\sum\boldsymbol{\psi}_k\dot{\eta}_k\right]\mathrm{d}V \quad (2.3.6)$$

$$\Delta T_s = \frac{1}{2}\int m_{sj}\left[\left(2\mathbf{v}+2\boldsymbol{\omega}^\times\mathbf{r}+2\sum\boldsymbol{\psi}_i\dot{\eta}_i\right)+\mathbf{v}_{sj}\right]^T\mathbf{v}_{sj}\delta\left(\mathbf{r}-\mathbf{r}_{sj}\right)\mathrm{d}V \quad (2.3.7)$$

$$\Delta T_E = \frac{1}{2}\int \rho_E\left[\left(2\mathbf{v}+2\boldsymbol{\omega}^\times\mathbf{r}+2\sum\boldsymbol{\psi}_i\dot{\eta}_i\right)+\boldsymbol{\omega}_{Eb}^\times\left(\mathbf{r}-\mathbf{r}_G\right)\right]^T\boldsymbol{\omega}_{Eb}^\times\left(\mathbf{r}-\mathbf{r}_G\right)\mathrm{d}V \quad (2.3.8)$$

These equations were obtained by expanding (2.3.2) through (2.3.4), keeping the grouping within parentheses intact in order to avoid an insufferable proliferation of terms.

Expanding (2.3.6) and using (2.1.14), it follows that

$$T_{IB} = \frac{1}{2}\int\Big[\mathbf{v}^T\mathbf{v}+2\mathbf{v}^T\boldsymbol{\omega}^\times\mathbf{r}+2\mathbf{v}^T\sum\boldsymbol{\psi}_i\dot{\eta}_i + 2\left(\boldsymbol{\omega}^\times\mathbf{r}\right)^T\sum\boldsymbol{\psi}_i\dot{\eta}_i+\left(\boldsymbol{\omega}^\times\mathbf{r}\right)^T\boldsymbol{\omega}^\times\mathbf{r} + \left(\sum\boldsymbol{\psi}_i\dot{\eta}_i\right)^T\left(\sum\boldsymbol{\psi}_k\dot{\eta}_k\right)\Big]\mathrm{d}m \quad (2.3.9)$$

$$= \frac{1}{2}m_T\mathbf{v}^T\mathbf{v}+\mathbf{v}^T\boldsymbol{\omega}^\times\int\mathbf{r}\,\mathrm{d}m+\sum\dot{\eta}_i\mathbf{v}^T\int\boldsymbol{\psi}_i\mathrm{d}m + \sum\dot{\eta}_i\boldsymbol{\omega}^T\int\mathbf{r}^\times\boldsymbol{\psi}_i\,\mathrm{d}m-\frac{1}{2}\int\boldsymbol{\omega}^T\mathbf{r}^\times\mathbf{r}^\times\boldsymbol{\omega}\,\mathrm{d}m + \frac{1}{2}\sum\dot{\eta}_i\sum\dot{\eta}_k\int\boldsymbol{\psi}_i^T\boldsymbol{\psi}_k\,\mathrm{d}m \quad (2.3.10)$$

By taking advantage of (2.1.20), (2.1.21), and (2.2.29) through (2.2.32) this reduces to

$$T_{IB} = \frac{1}{2}m_T\mathbf{v}^T\mathbf{v}+\mathbf{v}^T\boldsymbol{\omega}^\times\mathbf{s}_{TD}+\frac{1}{2}\boldsymbol{\omega}^T\mathbf{I}_{TD}\boldsymbol{\omega}+\frac{1}{2}\sum\dot{\eta}_i^2 m_{Bi} \quad (2.3.11)$$

This illustrates the significant advantage of the integrated body approach over the reduced body approach. The expression for T_{IB} consolidates into just four terms. Furthermore, the rotational and translational motions for this portion of the energy are decoupled from the slosh, elastic, and engine motion.

Slosh energy increment

Applying the Dirac delta function in (2.3.7), the sloshing energy increment is given by

$$\Delta T_s = m_{sj}\left(\mathbf{v}^T\mathbf{v}_{sj} + \left(\boldsymbol{\omega}^\times\mathbf{r}_{sj}\right)^T\mathbf{v}_{sj} + \mathbf{v}_{sj}^T\sum\boldsymbol{\psi}_{ji}\dot{\eta}_i + \frac{1}{2}\mathbf{v}_{sj}^T\mathbf{v}_{sj}\right) \tag{2.3.12}$$

where the modal amplitude of mode i at the jth slosh mass location is given by

$$\boldsymbol{\psi}_{ji} \equiv \boldsymbol{\psi}_i\left(\mathbf{r}_{sj}\right). \tag{2.3.13}$$

In practical terms, this is the nearest finite element centerline node to the equilibrium position of the slosh mass at a given flight condition. As the sloshing propellant equivalent mechanical model parameters change with propellant liquid level, it is sometimes necessary to select, interpolate, or otherwise assign different centerline nodes as a function of time. For long, structurally integral tanks, it may also be prudent to select the finite element model nodes according to the characteristics of the liquid force distribution on the wall. This is discussed further in Chapter 3.

It is relatively straightforward to take the derivative of (2.3.12) with respect to $\mathbf{v}$, $\mathbf{v}_{sj}$, or $\dot{\eta}_i$. Derivatives with respect to $\boldsymbol{\omega}$ can be obtained more easily by first operating on the second term as follows, using (1.8)

$$\left(\boldsymbol{\omega}^\times\mathbf{r}_{sj}\right)^T\mathbf{v}_{sj} = \boldsymbol{\omega}^T\mathbf{r}_{sj}^\times\mathbf{v}_{sj} \tag{2.3.14}$$

Engine energy increment

The engine energy can be simplified by defining

$$\mathbf{r}_1 \equiv \mathbf{r} - \mathbf{r}_G, \tag{2.3.15}$$

that is, the displacement relative to the undeformed location of the engine gimbal $\mathbf{r}_G$. Eq. (2.3.8) has four terms, which can be expressed as

$$\Delta T_E = A_1 + A_2 + A_3 + A_4 \tag{2.3.16}$$

where

$$A_1 \equiv \int_E \mathbf{v}^T \boldsymbol{\omega}_{Eb}^{\times} \mathbf{r}_1 dm = \mathbf{v}^T \boldsymbol{\omega}_{Eb}^{\times} \int_E \mathbf{r}_1 dm \tag{2.3.17}$$

$$\begin{aligned} A_2 &\equiv \int_E \left(\boldsymbol{\omega}^{\times} (\mathbf{r}_1 + \mathbf{r}_G)\right)^T \boldsymbol{\omega}_{Eb}^{\times} \mathbf{r}_1 dm \\ &= \int_E \left(\boldsymbol{\omega}^{\times} \mathbf{r}_1\right)^T \boldsymbol{\omega}_{Eb}^{\times} \mathbf{r}_1 dm + \int_E \left(\boldsymbol{\omega}^{\times} \mathbf{r}_G\right)^T \boldsymbol{\omega}_{Eb}^{\times} \mathbf{r}_1 dm \end{aligned} \tag{2.3.18}$$

$$A_3 \equiv \int_E \left(\sum \dot{\eta}_i \boldsymbol{\psi}_i\right)^T \boldsymbol{\omega}_{Eb}^{\times} \mathbf{r}_1 dm \tag{2.3.19}$$

$$A_4 \equiv \frac{1}{2} \int_E \left(\boldsymbol{\omega}_{Eb}^{\times} \mathbf{r}_1\right)^T \boldsymbol{\omega}_{Eb}^{\times} \mathbf{r}_1 dm \tag{2.3.20}$$

All these integrals take place over the engine volume and these terms can be integrated by defining the engine first moment of inertia about the gimbal as

$$\mathbf{s}_{Eb} \equiv \int_E \mathbf{r}_1 dm \tag{2.3.21}$$

and its second moment as

$$\mathbf{I}_{Eb} \equiv -\int_E \mathbf{r}_1^{\times} \mathbf{r}_1^{\times} dm \tag{2.3.22}$$

The subscript b indicates these mass properties are expressed in the body frame (see the discussion following Eq. (2.3.4)). It is also useful to take advantage of the fact that the engine is approximately a rigid body. For each mode, the displacement field can be represented by the combined effect of modal translation and modal rotation

$$\boldsymbol{\psi}_i = \boldsymbol{\psi}_{\beta i} + \boldsymbol{\sigma}_{\beta i}^{\times} \mathbf{r}_1 \tag{2.3.23}$$

where $\boldsymbol{\psi}_{\beta i}$ is the modal deflection at the gimbal point, and $\boldsymbol{\sigma}_{\beta i}$ is the modal rotation at the gimbal point. Since both of these terms are constants, they can be taken outside the integrals. Using (2.1.16), (2.1.17), and (2.3.23), the quantities

$$A_1 = \mathbf{v}^T \boldsymbol{\omega}_{Eb}^{\times} \mathbf{s}_{Eb} \tag{2.3.24}$$

$$A_2 = \boldsymbol{\omega}^T \mathbf{I}_{Eb} \boldsymbol{\omega}_{Eb} + (\boldsymbol{\omega}^{\times} \mathbf{r}_G)^T \boldsymbol{\omega}_{Eb}^{\times} \mathbf{s}_{Eb} \tag{2.3.25}$$

$$A_3 = \left(\sum \dot{\eta}_i \boldsymbol{\psi}_{\beta i}\right)^T \boldsymbol{\omega}_{Eb}^{\times} \mathbf{s}_{Eb} + \left(\sum \dot{\eta}_i \boldsymbol{\sigma}_{\beta i}\right)^T \mathbf{I}_{Eb} \boldsymbol{\omega}_{Eb} \tag{2.3.26}$$

$$A_4 = \frac{1}{2} \boldsymbol{\omega}_{Eb}^T \boldsymbol{I}_{Eb} \boldsymbol{\omega}_{Eb} \tag{2.3.27}$$

are obtained. Inserting all this into (2.3.16) gives

$$\Delta T_E = \mathbf{v}^T \boldsymbol{\omega}_{Eb}^{\times} \mathbf{s}_{Eb} + \boldsymbol{\omega}^T \mathbf{I}_{Eb} \boldsymbol{\omega}_{Eb} + \left(\boldsymbol{\omega}^{\times} \mathbf{r}_G\right)^T \boldsymbol{\omega}_{Eb}^{\times} \mathbf{s}_{Eb} + \left(\sum \dot{\eta}_i \boldsymbol{\psi}_{\beta i}\right)^T \boldsymbol{\omega}_{Eb}^{\times} \mathbf{s}_{Eb} + \left(\sum \dot{\eta}_i \boldsymbol{\sigma}_{\beta i}\right)^T \mathbf{I}_{Eb} \boldsymbol{\omega}_{Eb} + \frac{1}{2} \boldsymbol{\omega}_{Eb}^T \mathbf{I}_{Eb} \boldsymbol{\omega}_{Eb} \quad (2.3.28)$$

2.4 Lagrangian accelerations

The *Lagrangian accelerations* are defined herein as those given by the LHS of (2.1.8), and are used to construct the mass matrix.

For translation, the derivative with respect to $\mathbf{v}$ is needed. From (2.3.11), (2.3.12), and (2.3.28), the components related to translation are

$$\frac{d}{dt}\left(\frac{\partial T_{IB}}{\partial \mathbf{v}}\right) = m_T \dot{\mathbf{v}} + \dot{\boldsymbol{\omega}}^{\times} \mathbf{s}_{TD} \quad (2.4.1)$$

$$\frac{d}{dt}\left(\frac{\partial \Delta T_s}{\partial \mathbf{v}}\right) = m_{sj} \dot{\mathbf{v}}_{sj} \quad (2.4.2)$$

$$\frac{d}{dt}\left(\frac{\partial \Delta T_E}{\partial \mathbf{v}}\right) = \dot{\boldsymbol{\omega}}_{Eb}^{\times} \mathbf{s}_{Eb} \quad (2.4.3)$$

Using Eq. (2.3.5), this becomes

$$\frac{d}{dt}\left(\frac{\partial T}{\partial \mathbf{v}}\right) = m_T \dot{\mathbf{v}} + m_{sj} \dot{\mathbf{v}}_{sj} - \mathbf{s}_{TD}^{\times} \dot{\boldsymbol{\omega}} - \mathbf{s}_{Eb}^{\times} \dot{\boldsymbol{\omega}}_{Eb} \quad (2.4.4)$$

For rotation, the derivative with respect to $\boldsymbol{\omega}$ is computed. From (2.3.11), (2.3.12), and (2.3.28), the expressions

$$\frac{d}{dt}\left(\frac{\partial T_{IB}}{\partial \boldsymbol{\omega}}\right) = I_{TD} \dot{\boldsymbol{\omega}} + \mathbf{s}_{TD}^{\times} \dot{\mathbf{v}} \quad (2.4.5)$$

$$\frac{d}{dt}\left(\frac{\partial \Delta T_s}{\partial \boldsymbol{\omega}}\right) = m_{sj} \mathbf{r}_{sj}^{\times} \dot{\mathbf{v}}_{sj} \quad (2.4.6)$$

$$\frac{d}{dt}\left(\frac{\partial \Delta T_E}{\partial \boldsymbol{\omega}}\right) = I_{Eb} \dot{\boldsymbol{\omega}}_{Eb} + r_G^{\times} \dot{\boldsymbol{\omega}}_{Eb}^{\times} \mathbf{s}_{Eb} \quad (2.4.7)$$

are obtained. Rearranging the last term and summing the result,

$$\frac{d}{dt}\left(\frac{\partial T}{\partial \omega}\right) = I_{TD} \dot{\boldsymbol{\omega}} + s_{TD}^{\times} \dot{\mathbf{v}} + m_{sj} \mathbf{r}_{sj}^{\times} \dot{\mathbf{v}}_{sj} + \mathbf{I}_{Eb} \dot{\boldsymbol{\omega}}_{Eb} - \mathbf{r}_G^{\times} \mathbf{s}_{Eb}^{\times} \dot{\boldsymbol{\omega}}_{Eb} \quad (2.4.8)$$

For slosh, the derivatives of (2.3.11), (2.3.12), and (2.3.28) are computed with respect to $\mathbf{v}_{sj}$.

$$\frac{d}{dt}\left(\frac{\partial T_{IB}}{\partial \mathbf{v}_{sj}}\right)=0 \tag{2.4.9}$$

$$\frac{\partial \Delta T_s}{\partial \mathbf{v}_{sj}}=m_{sj}\left(\mathbf{v}+\boldsymbol{\omega}^{\times}\mathbf{r}_{sj}+\sum \boldsymbol{\psi}_{ji}\dot{\eta}_i+\mathbf{v}_{sj}\right) \tag{2.4.10}$$

$$\frac{d}{dt}\left(\frac{\partial \Delta T_s}{\partial \mathbf{v}_{sj}}\right)=m_{sj}\left(\dot{\mathbf{v}}+\dot{\boldsymbol{\omega}}^{\times}\mathbf{r}_{sj}\ +\boldsymbol{\omega}^{\times}\dot{\mathbf{r}}_{sj}+\sum \boldsymbol{\psi}_{si}\ddot{\eta}_i+\dot{\mathbf{v}}_{sj}\right) \tag{2.4.11}$$

$$\frac{d}{dt}\left(\frac{\partial \Delta T_E}{\partial \mathbf{v}_{sj}}\right)=0 \tag{2.4.12}$$

The $\boldsymbol{\omega}^{\times}\dot{\mathbf{r}}_{sj}$ term on the right hand side of (2.4.11) is the product of two small velocities and can be omitted in a linearized or quasi-linear analysis. Rearranging and adding,

$$\frac{d}{dt}\left(\frac{\partial T}{\partial \mathbf{v}_{sj}}\right)=m_{sj}\left(\dot{\mathbf{v}}-\mathbf{r}_{sj}^{\times}\dot{\boldsymbol{\omega}}+\sum \psi_{ji}\ddot{\eta}_i+\dot{\mathbf{v}}_{sj}\right). \tag{2.4.13}$$

For bending, the generalized coordinates are the modal amplitudes η_i. Each mode has a separate equation. From (2.3.11), (2.3.12), and (2.3.28),

$$\frac{d}{dt}\left(\frac{\partial T_{IB}}{\partial \dot{\eta}_i}\right)=\ddot{\eta}_i m_{Bi} \tag{2.4.14}$$

$$\frac{d}{dt}\left(\frac{\partial \Delta T_s}{\partial \dot{\eta}_i}\right)=m_{sj}\dot{\mathbf{v}}_{sj}^{T}\boldsymbol{\psi}_{ji} \tag{2.4.15}$$

$$\frac{d}{dt}\left(\frac{\partial \Delta T_E}{\partial \dot{\eta}_i}\right)=\boldsymbol{\psi}_{\beta i}^{T}\dot{\boldsymbol{\omega}}_{Eb}^{\times}\mathbf{s}_{Eb}+\boldsymbol{\sigma}_{\beta i}^{T}I_{Eb}\dot{\boldsymbol{\omega}}_{Eb} \tag{2.4.16}$$

Again rearranging and adding,

$$\frac{d}{dt}\left(\frac{\partial T}{\partial \dot{\eta}_i}\right)=m_{Bi}\ddot{\eta}_i+m_{sj}\boldsymbol{\psi}_{ji}^{T}\dot{\mathbf{v}}_{sj}+\left(\boldsymbol{\sigma}_{\beta i}^{T}\mathbf{I}_{EG}-\boldsymbol{\psi}_{\beta i}^{T}\mathbf{s}_{E}^{\times}\right)\dot{\boldsymbol{\omega}}_{Eb}. \tag{2.4.17}$$

For the engine, one could define two new generalized coordinates β_{Ey} and β_{Ez} for the engine local pitch and yaw rotations, under the assumption that the engines are nominally aligned to the vehicle symmetry axis. However, it turns out to be more convenient to define the vector $\boldsymbol{\beta}_{Eb}=\begin{bmatrix}\beta_{Ex} & \beta_{Ey} & \beta_{Ez}\end{bmatrix}^{T}$ and then set $\beta_{Ex}=0$. Eq. (2.1.8) becomes

$$\frac{d}{dt}\left(\frac{\partial T}{\partial \dot{\boldsymbol{\beta}}_{Eb}}\right)=\frac{d}{dt}\left(\frac{\partial T}{\partial \boldsymbol{\omega}_{Eb}}\right)=\mathbf{g}_{Eb} \tag{2.4.18}$$

where $\boldsymbol{g}_{Eb}$ is the total moment on the engine about the gimbal point, including moments from the thrust vector control (TVC) system. The derivative of $\boldsymbol{\beta}_{Eb}$ is approximately equal to the engine angular velocity relative to the body frame as long as the engine gimbal rotations are sufficiently small, which is usually the case. Thus the order of rotation does not matter. Using the vector relations appearing in Chapter 1, (2.3.28) can be manipulated into the form

$$\Delta T_E = \boldsymbol{\omega}_{Eb}^T \Big(-\mathbf{v}^\times \mathbf{s}_{Eb} + \mathbf{I}_{Eb}\boldsymbol{\omega} + \mathbf{s}_{Eb}^\times \boldsymbol{\omega}^\times \mathbf{r}_G + \mathbf{s}_{Eb}^\times \sum \dot{\eta}_i \boldsymbol{\psi}_{\beta i} + \mathbf{I}_{Eb} \sum \dot{\eta}_i \boldsymbol{\sigma}_{\beta i} \Big) + \frac{1}{2} \boldsymbol{\omega}_{Eb}^T \mathbf{I}_{Eb} \boldsymbol{\omega}_{Eb} \quad (2.4.19)$$

Since $\boldsymbol{\beta}_{Eb}$ and $\boldsymbol{\omega}_{Eb}$ do not appear in the expressions for T_{IB} or ΔT_s, it follows that

$$\frac{\partial \Delta T_E}{\partial \omega_{Eb}} = \frac{\partial T}{\partial \omega_{Eb}} \quad (2.4.20)$$

Thus

$$\frac{d}{dt}\left(\frac{\partial T}{\partial \boldsymbol{\omega}_{Eb}}\right) = \mathbf{s}_{Eb}^\times \dot{\mathbf{v}} + \mathbf{I}_{Eb}\dot{\boldsymbol{\omega}} - \mathbf{s}_{Eb}^\times \mathbf{r}_G^\times \dot{\boldsymbol{\omega}} + \mathbf{s}_{Eb}^\times \sum \boldsymbol{\psi}_{\beta i} \ddot{\eta}_i + \mathbf{I}_{Eb} \sum \boldsymbol{\sigma}_{\beta i} \ddot{\eta}_i + \mathbf{I}_{Eb} \dot{\boldsymbol{\omega}}_{Eb} \quad (2.4.21)$$

The engine/elastic coupling vector for mode i is defined as

$$\mathbf{c}_{EFi} \equiv \mathbf{s}_{Eb}^\times \boldsymbol{\psi}_{\beta i} + \mathbf{I}_{Eb} \boldsymbol{\sigma}_{\beta i} \quad (2.4.22)$$

along with the *tail wags dog* (TWD) inertia

$$\mathbf{I}_{TWD} \equiv \mathbf{I}_{Eb} - \mathbf{r}_G^\times \mathbf{s}_{Eb}^\times \quad (2.4.23)$$

Since $\mathbf{I}_{Eb}$ is symmetric, the transpose of this quantity is

$$\mathbf{I}_{TWD}^T \equiv \mathbf{I}_{Eb} - \mathbf{s}_{Eb}^\times \mathbf{r}_G^\times \quad (2.4.24)$$

Using these quantities along with (2.4.18), (2.4.21) becomes

$$\frac{d}{dt}\left(\frac{\partial T}{\partial \boldsymbol{\omega}_{Eb}}\right) = \mathbf{s}_{Eb}^\times \dot{\mathbf{v}} + \mathbf{I}_{TWD}^T \dot{\boldsymbol{\omega}} + \mathbf{I}_{Eb} \dot{\boldsymbol{\omega}}_{Eb} + \sum \mathbf{c}_{EFi} \ddot{\eta}_i = \mathbf{g}_{Eb} \quad (2.4.25)$$

2.5 Assembled equations of motion

Lagrange's equation (2.1.7) is written with generalized forces on the right hand side. In the first four of the following equations, the linear acceleration term $\dot{\mathbf{v}}$ has been replaced by its nonlinear counterpart $\mathbf{a}_b$ defined in Eq. (2.1.5). The two differ by the nonlinear term $\boldsymbol{\omega}^\times\mathbf{v}$. Nonlinear effects are more fully discussed in Chapters 3 and 4. At this stage, it is sufficient to know that these issues can be handled by adding nonlinear terms (subscript *NL*) to the right hand side. Nonlinear terms are usually too small to be of significance for the engine and bending equations.

The assembled description of the vehicle motion dynamics consists of five principal groups of equations. These equations model the translation and rotation of the rigid body, the motion of the engines and sloshing propellants relative to the body, and the bending of the airframe.

Translation (2.4.4)

$$m_T\mathbf{a}_b - \mathbf{s}_{TD}^\times\dot{\boldsymbol{\omega}} - \mathbf{s}_{Eb}^\times\dot{\boldsymbol{\omega}}_{Eb} + \sum_j m_{sj}\ddot{\boldsymbol{\delta}}_{sj} = \mathbf{f} + \mathbf{f}_{NL} \tag{2.5.1}$$

Rotation (2.4.8)

$$\mathbf{s}_{TD}^\times\mathbf{a}_b + \mathbf{I}_{TD}\dot{\boldsymbol{\omega}} + \mathbf{I}_{TWD}\dot{\boldsymbol{\omega}}_{Eb} + \sum_j m_{sj}\mathbf{r}_{sj}^\times\ddot{\boldsymbol{\delta}}_{sj} = \mathbf{g} + \mathbf{g}_{NL} \tag{2.5.2}$$

Engine (2.4.25)

$$\mathbf{s}_{Eb}^\times\mathbf{a}_b + \mathbf{I}_{TWD}^T\dot{\boldsymbol{\omega}} + \mathbf{I}_{Eb}\dot{\boldsymbol{\omega}}_{Eb} + \sum \mathbf{c}_{EFi}\ddot{\eta}_i = \mathbf{g}_{Eb} \tag{2.5.3}$$

Slosh (2.4.13)

$$m_{sj}\left(\mathbf{a}_b - \mathbf{r}_{sj}^\times\dot{\omega} + \ddot{\boldsymbol{\delta}}_{sj} + \sum \psi_{ji}\ddot{\eta}_i\right) = \mathbf{f}_{sjNL} \tag{2.5.4}$$

Bending (2.4.17)

$$\mathbf{c}_{EFi}^T\dot{\boldsymbol{\omega}}_{Eb} + \sum_j m_{sj}\boldsymbol{\psi}_{ji}^T\ddot{\boldsymbol{\delta}}_{sj} + m_{Bi}\ddot{\eta}_i = f_{Bi} \tag{2.5.5}$$

Note that f_{Bi}, the elastic *generalized* force, is a scalar rather than a vector. It is typically computed as follows, using Eq. (2.2.24).

$$f_{Bi} = -m_{Bi}\left(\Omega_{Bi}^2\eta_i + 2\zeta_{Bi}\Omega_{Bi}\dot{\eta}_i\right) + \sum_n \left(\boldsymbol{\psi}_{ni}^T\mathbf{f}_n + \boldsymbol{\sigma}_{ni}^T\mathbf{g}_n\right) \tag{2.5.6}$$

The RHS terms are defined as follows. The quantities $\mathbf{f}$ and $\mathbf{g}$ are the sum of all forces and torques, respectively, applied to the rocket, including thrust and aerodynamic forces. With the exception of nonlinear forces and torques $\mathbf{f}_{NL}$ and $\mathbf{g}_{NL}$, internal, e.g., slosh, forces are excluded as they arise implicitly via the slosh acceleration terms on the LHS of (2.5.5).

The quantity $\mathbf{g}_{Eb}$ is the sum of all external torques on the engine about the gimbal point, including actuator torques, aerodynamic load torques, thrust misalignment torque, propellant feedline effects, gyroscopic torques due to rotating turbomachinery, and so on. The quantity $\mathbf{f}_{sj}$ is the sum of all forces on slosh mass j.

If the slosh spring model is being used, the system translation equation and the engine equation are the only equations for which it is necessary to consider a coordinate frame other than the body frame. The translation equation is sometimes expressed in an inertial frame. The engine equation is sometimes expressed in the engine frame. These are the reasons for the subscript b on $\mathbf{a}_b$, $\boldsymbol{\omega}_{Eb}$, $\mathbf{g}_{Eb}$, etc. For the latter two variables, when this subscript is dropped, the implication is that these variables are in the engine frame, i.e., the subscript E, by itself, means two things: this is an engine variable, and it is expressed in the engine frame. As stated earlier, it must be understood that everything else ($\boldsymbol{\omega}$, $\boldsymbol{\delta}_{sj}$, etc.) is in the body frame.

Mass matrix for prescribed engine motion

The engine motion may be considered either prescribed or not prescribed. The phrase "prescribed engine motion" means $\boldsymbol{\omega}_{Eb}$ and its time derivative are external variables that are supplied to the dynamic equations. For a simulation, this would require that the TVC dynamics be computed in a separate module with a separate set of state variables. The alternative is to include the engine motion $\boldsymbol{\beta}_{Eb}$ as part of the state vector.

In order to be able to adapt the analysis to various situations, additional notation is introduced. Variables with a tilde may be redefined as needed for different model assumptions. For an analysis in which all nonlinear forces and torques can be neglected and the engine motion is not prescribed, it follows that $\tilde{\mathbf{f}}=\mathbf{f}$, $\tilde{\mathbf{g}}=\mathbf{g}$, $\tilde{\mathbf{g}}_{Eb}=\mathbf{g}_{Eb}$, etc. On the other hand, if the vehicle rotation rate about any axis is large, it may be necessary to include nonlinear terms. These quantities would then be expressed as $\tilde{\mathbf{f}}=\mathbf{f}+\mathbf{f}_{NL}$, $\tilde{\mathbf{g}}=\mathbf{g}+\mathbf{g}_{NL}$, $\tilde{\mathbf{g}}_{Eb}=\mathbf{g}_{Eb}+\mathbf{g}_{Eb,NL}$, etc.

For prescribed engine motion, the engine equation is deleted from the above set. All but the slosh equations have terms with the variable $\boldsymbol{\omega}_{Eb}$, which are moved to the right hand side. The system forces and torques are therefore

$$\tilde{\mathbf{f}} = \mathbf{f} + \mathbf{f}_{\text{presc}} + \mathbf{f}_{NL} \tag{2.5.7}$$

$$\tilde{\mathbf{g}} = \mathbf{g} + \mathbf{g}_{\text{presc}} + \mathbf{g}_{NL} \tag{2.5.8}$$

$$\tilde{\mathbf{f}}_{sj} = \mathbf{f}_{sj} + \mathbf{f}_{sj,NL} \tag{2.5.9}$$

$$\tilde{f}_{Bi} = f_{Bi} + f_{Bi,\text{presc}} \tag{2.5.10}$$

The prescribed engine forces and torques, here with subscript presc, can be obtained from (2.5.1) through (2.5.5) by simply negating any term in which $\dot{\boldsymbol{\omega}}_{Eb}$ appears and including it on the RHS of the dynamic equation. Thus,

$$\mathbf{f}_{\text{presc}} = \mathbf{s}_{Eb}^{\times} \dot{\boldsymbol{\omega}}_{Eb} \tag{2.5.11}$$

$$\mathbf{g}_{\text{presc}} = -\mathbf{I}_{TWD} \dot{\boldsymbol{\omega}}_{Eb} \tag{2.5.12}$$

$$f_{Bi,\text{presc}} = -\mathbf{c}_{EFi}^{T} \dot{\boldsymbol{\omega}}_{Eb} \tag{2.5.13}$$

Note that there is no prescribed engine motion term that contributes to the slosh generalized force expression. Eqs. (2.5.1) through (2.5.5) can be assembled into a large matrix equation. For prescribed engine motion, Eq. (2.5.3) is excluded. It is useful to define **1** as the 3 × 3 identity matrix. It is also useful to define the 3 × 3 null matrix

$$\mathbf{O} \equiv \begin{bmatrix} \mathbf{0} & \mathbf{0} & \mathbf{0} \end{bmatrix} \tag{2.5.14}$$

where

$$\mathbf{0} = \begin{bmatrix} 0 & 0 & 0 \end{bmatrix}^T, \tag{2.5.15}$$

the null vector.

The assembled system equations may be written in matrix form as

$$\mathcal{M}\ddot{\mathbf{x}} = \mathcal{F}. \tag{2.5.16}$$

For a vehicle with N fuel tanks and M flexible modes, the acceleration vector is given by

$$\ddot{\mathbf{x}} = \begin{bmatrix} \mathbf{a}_b^T & \dot{\boldsymbol{\omega}}^T & \ddot{\boldsymbol{\delta}}_{s1}^T & \ddot{\boldsymbol{\delta}}_{s2}^T & \dots & \ddot{\boldsymbol{\delta}}_{sN}^T & \ddot{\eta}_1 & \ddot{\eta}_2 & \dots & \ddot{\eta}_M \end{bmatrix}^T \tag{2.5.17}$$

and the vector $\mathcal{F}$ is formed from the right hand sides of Eqs. (2.5.1), (2.5.2), (2.5.4), and (2.5.5);

$$\mathcal{F} = \begin{bmatrix} \tilde{\mathbf{f}}^T & \tilde{\mathbf{g}}^T & \tilde{\mathbf{f}}_{s1}^T & \tilde{\mathbf{f}}_{s2}^T & \ldots & \tilde{\mathbf{f}}_{sN}^T & \tilde{f}_{B1} & \tilde{f}_{B2} & \ldots & \tilde{f}_{BM} \end{bmatrix}^T. \quad (2.5.18)$$

The mass matrix is given by Eq. (2.5.19). The state vector (2.5.17) can be used as a guide to help understand how many rows and columns are represented in this matrix. In particular, each mode has only one row and one column. For example, if the rocket in question has no sloshing propellant and only a single flexible mode, the mass matrix would be 7 by 7.

Mass matrix with engine motion included

The assumption that the engine motion is prescribed is used in many control system analyses, and is often a good starting point for designing a control system. The initial analysis is based on the presumption that the TVC system is very stiff; that is, the resonances of the engines, with their positioning loops closed, are at higher natural frequencies than the control frequencies and significant global structural modes. This assumption may implicitly drive the design of the TVC system to increase its bandwidth and stiffen the engine support and actuator attach ("backup") structure.

For final verification of the design, it may be necessary to recognize the fact that both the rigid and elastic motion of the gimbal point will affect the resulting engine gimbal angle. Usually, as structural mass is allocated and traded during the vehicle design, a vehicle program concurrently matures to include the necessary analysis and testing to support detailed modeling of the engine dynamics. The details of such modeling and testing are a subject of Chapter 6.

Incorporation of the engine equation is accomplished by first setting the prescribed motion forces in Eqs. (2.5.7) through (2.5.10) to zero. The nonlinear torques on the engine are usually very small and these too can be set to zero. In the engine Eq. (2.5.3), $\tilde{\mathbf{g}}_E = \mathbf{g}_E$.

There are no direct engine effects due to slosh (Eq. (2.5.4)), and no slosh effects on the engine. While the engines and slosh couple indirectly through the rigid body and elasticity, this coupling is usually small. The primary interest is in the coupling of the engine and flexibility. In order to highlight this, consider a rocket with no sloshing propellant. For a rocket with M modes, Eqs. (2.5.1), (2.5.2), (2.5.3), and (2.5.5) can be combined into a single matrix equation, given by (2.5.20).

For a rocket with sloshing fuel masses, it is a straightforward matter to insert the slosh rows and columns appearing in (2.5.19) into this system of equations. Also, it is usually desirable, if the engines are aligned to the

$$\mathcal{M} = \begin{bmatrix} m_T\mathbf{1} & -\mathbf{s}_{TD}^{\times} & m_{s1}\mathbf{1} & m_{s2}\mathbf{1} & \dots & m_{sN}\mathbf{1} & \mathbf{0} & \mathbf{0} & \dots & \mathbf{0} \\ \mathbf{s}_{TD}^{\times} & \mathbf{I}_{TD} & m_{s1}\mathbf{r}_{s1}^{\times} & m_{s2}\mathbf{r}_{s2}^{\times} & \dots & m_{sN}\mathbf{r}_{sN}^{\times} & \mathbf{0} & \mathbf{0} & \dots & \mathbf{0} \\ m_{s1}\mathbf{1} & -m_{s1}\mathbf{r}_{s1}^{\times} & m_{s1}\mathbf{1} & \mathbf{O} & \dots & \mathbf{O} & m_{s1}\psi_{11} & m_{s1}\psi_{12} & \dots & m_{s1}\psi_{1M} \\ m_{s2}\mathbf{1} & -m_{s2}\mathbf{r}_{s2}^{\times} & \mathbf{O} & m_{s2}\mathbf{1} & \dots & \mathbf{O} & m_{s2}\psi_{21} & m_{s2}\psi_{22} & \dots & m_{s2}\psi_{2M} \\ \vdots & \vdots & \vdots & \vdots & \ddots & \vdots & \vdots & \vdots & \ddots & \vdots \\ m_{sN}\mathbf{1} & -m_{sN}\mathbf{r}_{sN}^{\times} & \mathbf{O} & \mathbf{O} & \dots & m_{sN}\mathbf{1} & m_{sN}\psi_{N1} & m_{sN}\psi_{N2} & \dots & m_{sN}\psi_{NM} \\ \mathbf{0}^T & \mathbf{0}^T & m_{s1}\psi_{11}^T & m_{s2}\psi_{21}^T & \dots & m_{sN}\psi_{N1}^T & m_{B1} & 0 & \dots & 0 \\ \mathbf{0}^T & \mathbf{0}^T & m_{s1}\psi_{12}^T & m_{s2}\psi_{22}^T & \dots & m_{sN}\psi_{N2}^T & 0 & m_{B2} & \dots & 0 \\ \vdots & \vdots & \vdots & \vdots & \ddots & \vdots & \vdots & \vdots & \ddots & \vdots \\ \mathbf{0}^T & \mathbf{0}^T & m_{s1}\psi_{1M}^T & m_{s2}\psi_{2M}^T & \dots & m_{sN}\psi_{NM}^T & 0 & 0 & \dots & m_{BM} \end{bmatrix} \quad (2.5.19)$$

Eq. (2.5.19): Mass matrix for prescribed engine motion

$$
\begin{bmatrix}
m_T\mathbf{1} & -\mathbf{s}_{TD}^{\times} & -\mathbf{s}_{Eb}^{\times} & \mathbf{0} & \mathbf{0} & \cdots & \mathbf{0} \\
\mathbf{s}_{TD}^{\times} & \mathbf{I}_{TD} & \mathbf{I}_{TWD} & \mathbf{0} & \mathbf{0} & \cdots & \mathbf{0} \\
\mathbf{s}_{Eb}^{\times} & \mathbf{I}_{TWD}^{T} & \mathbf{I}_{Eb} & \mathbf{c}_{EF1} & \mathbf{c}_{EF2} & \cdots & \mathbf{c}_{EFM} \\
\mathbf{0}^T & \mathbf{0}^T & \mathbf{c}_{EF1}^{T} & m_{B1} & 0 & \cdots & 0 \\
\mathbf{0}^T & \mathbf{0}^T & \mathbf{c}_{EF2}^{T} & 0 & m_{B2} & \cdots & 0 \\
\vdots & \vdots & \vdots & \vdots & \vdots & \ddots & \vdots \\
\mathbf{0}^T & \mathbf{0}^T & \mathbf{c}_{EFM}^{T} & 0 & 0 & \cdots & m_{BM}
\end{bmatrix}
\begin{bmatrix}
\mathbf{a}_b \\ \dot{\boldsymbol{\omega}} \\ \dot{\boldsymbol{\omega}}_{Eb} \\ \ddot{\eta}_1 \\ \ddot{\eta}_2 \\ \vdots \\ \ddot{\eta}_M
\end{bmatrix}
=
\begin{bmatrix}
\tilde{\mathbf{f}} \\ \tilde{\mathbf{g}} \\ \mathbf{g}_{Eb} \\ \tilde{f}_{B1} \\ \tilde{f}_{B2} \\ \vdots \\ \tilde{f}_{BM}
\end{bmatrix}
\tag{2.5.20}
$$

Eq. (2.5.20): Mass matrix for non-prescribed engine motion (no slosh) body x axis, to apply the constraint that $\beta_{Ex} = 0$ (no rotation of the engine about the x axis). This can be accomplished approximately by deleting the corresponding row and column for engine x rotation. Section 2.7 below provides a more formal mathematical approach that is applied to a similar problem of truncating the slosh motion.

If the engine motion is prescribed, the TVC model includes the engine inertia and provides the engine gimbal angles, angular rates, and angular accelerations as a function of the commanded inputs. If the engine motion is not prescribed, the TVC model must be modified so as to provide the torques g_{Ey} and g_{Ez} about the gimbal point for given values of the gimbal angles and gimbal angle commands. The engine inertia is included in the system equations, and is excluded from the TVC model.

As a general rule, the most accurate way to solve a set of dynamic equations is to solve all of the equations simultaneously. The prescribed engine motion approach is a departure from this ideal, since it means the equations for the engine motion are solved separately from those for the rest of the dynamics. From the standpoint of these remaining dynamics, the engine motion is one step late. This piecemeal approach to a solution may require a smaller time step in order to achieve numerical stability.

An approach that is intermediate between prescribed and not prescribed is to treat the engine as prescribed in the system equations, but to add external torques to a separate TVC model that represent the effects of motion of the gimbal point. Eq. (2.5.3) can be rearranged as follows:

$$\dot{\boldsymbol{\omega}}_{Eb} = \mathbf{I}_{Eb}^{-1}\left[\mathbf{g}_{Eb} - \left(\mathbf{s}_{Eb}^{\times}\mathbf{a}_b + \mathbf{I}_{TWD}^{T}\dot{\boldsymbol{\omega}} + \sum \mathbf{c}_{EFi}\ddot{\eta}_i\right)\right] \tag{2.5.21}$$

This expression is inserted into the TVC model to compute the engine acceleration. The terms in parentheses represent the torques on the engine due to gimbal point motion. Including these torques improves the accuracy of the solution, but it is still necessary for the dynamic motion to be integrated separately from the TVC motion, so the numerical issues described above will remain.[2]

[2] Thrust vector control engine dynamics models for launch vehicles typically include an angular stiffness term that represents the combined effect of linearized gravitational restoring torque on the engine at 1-g conditions along with torques due to propellant feedlines or flexible bearing stiffness (for liquid and solid rocket motors, respectively). The use of a 1-g value is chosen to correlate with TVC test data at ambient conditions on the surface. In order to incorporate the engine dynamics from an existing TVC model using Eqs. (2.5.20) or (2.5.21), the gimbal angular stiffness should be corrected to zero acceleration since these

2.6 Reduced body modes

There is no guarantee that the mass matrices defined above are positive-definite. Indeed, it can be shown that if the mode shapes $\boldsymbol{\psi}_{ji}$ are large enough, or if a very large number of modes are included, the determinant of $\mathcal{M}$ may become negative. The solution to this difficulty is to use modes for the reduced body consisting of the total rocket minus the sloshing fuel mass, and in some cases it may also be necessary to subtract the engine mass. This requires three changes to the analysis. Eq. (2.2.29) was previously used for the integrated body to establish that in the absence of external forces, no linear momentum is generated from flexible motion. The first change is that when reduced-body modes are used, (2.2.29) becomes

$$\int \boldsymbol{\psi}_i \mathrm{d}m = \int_R \boldsymbol{\psi}_i(\mathbf{r})\,\mathrm{d}m + \sum_j \int m_{sj}\boldsymbol{\psi}_i(\mathbf{r})\,\delta\left(\mathbf{r}-\mathbf{r}_{sj}\right)\mathrm{d}V + \int_E \boldsymbol{\psi}_i(\mathbf{r})\,\mathrm{d}m, \ \forall i \quad (2.6.1)$$

The subscript R on the first integral of the RHS represents an integration over the reduced body. Since free-free modes are being used, this integral is equal to zero. The previous convention is retained that a volume integral with no subscript represents an integration over the complete body. The second integral on the RHS employs the Dirac delta function, so it is an integral over just the sloshing masses and is readily integrated. The third integral, for the engine, is evaluated using (2.4.22) through (2.4.24). Defining

$$\boldsymbol{\chi}_{Ei} = \boldsymbol{\psi}_{\beta_i} + \frac{\boldsymbol{\sigma}_{\beta i}\mathbf{s}_E}{m_E}, \quad (2.6.2)$$

the overall integral becomes

$$\int \boldsymbol{\psi}_i \mathrm{d}m = \sum_j m_{sj}\boldsymbol{\psi}_{ji} + m_E\boldsymbol{\chi}_{Ei}, \ \forall i. \quad (2.6.3)$$

Previously, this integral was zero for the complete body, meaning that there could be an oscillation of any of the modes and the undeformed centerline (the body x axis) remains fixed in inertial space as long as the external forces are zero (see Fig. 2.4). In the reduced body approach, a modal

effects are accounted for in the body inertial acceleration $\mathbf{a}_b$. Care should be exercised in implementation to ensure proper bookkeeping of gravitational acceleration.

oscillation means that the center of mass of the reduced body oscillates in one direction and the net center of mass of the slosh masses and engine oscillates in the opposite direction.

The second required change is that the same kind of modification is necessary for Eq. (2.6.1) dealing with the angular momentum. Through a similar argument, it is shown that

$$\int \mathbf{r}^{\times}\boldsymbol{\psi}_i \mathrm{d}m = \sum_j \mathbf{r}_{sj}^{\times} m_{sj}\boldsymbol{\psi}_{ji} + \int_E (\mathbf{r}_G + \mathbf{r}_1)^{\times}\left(\boldsymbol{\psi}_{\beta i} + \boldsymbol{\sigma}_{\beta i}^{\times}\mathbf{r}_1\right)\mathrm{d}m,\ \forall i \tag{2.6.4}$$

$$\int \mathbf{r}^{\times}\boldsymbol{\psi}_i \mathrm{d}m = \sum_j \mathbf{r}_{sj}^{\times} m_{sj}\boldsymbol{\psi}_{ji} + \mathbf{r}_G^{\times}\left(m_E\boldsymbol{\psi}_{\beta i} + \boldsymbol{\sigma}_{\beta i}^{\times}\mathbf{s}_E\right) + \mathbf{s}_E^{\times}\boldsymbol{\psi}_{\beta i} + \mathbf{I}_{Eb}\boldsymbol{\sigma}_{\beta i},\ \forall i \tag{2.6.5}$$

Using (2.4.22) and (2.6.2) this becomes

$$\int \mathbf{r}^{\times}\boldsymbol{\psi}_i \mathrm{d}m = \sum_j \mathbf{r}_{sj}^{\times} m_{sj}\boldsymbol{\psi}_{ji} + m_E\mathbf{r}_G^{\times}\boldsymbol{\chi}_{Ei} + \mathbf{c}_{EFi},\ \forall i. \tag{2.6.6}$$

The third change is that the orthogonality conditions (2.2.31) and (2.2.32) must be modified in the same way and become

$$\int \boldsymbol{\psi}_i^T\boldsymbol{\psi}_k \mathrm{d}m = \sum_j \boldsymbol{\psi}_{ji}^T m_{sj}\boldsymbol{\psi}_{jk} + \int_E \left(\boldsymbol{\psi}_{\beta i} + \boldsymbol{\sigma}_{\beta i}^{\times}\mathbf{r}_1\right)^T\left(\boldsymbol{\psi}_{\beta k} + \boldsymbol{\sigma}_{\beta k}^{\times}\mathbf{r}_1\right)\mathrm{d}m \quad i \neq k \tag{2.6.7}$$

$$\int \boldsymbol{\psi}_i^T\boldsymbol{\psi}_k \mathrm{d}m = m_{Bi} + \sum_j \boldsymbol{\psi}_{ji}^T m_{sj}\boldsymbol{\psi}_{jk} + \int_E \left(\boldsymbol{\psi}_{\beta i} + \boldsymbol{\sigma}_{\beta i}^{\times}\mathbf{r}_1\right)^T\left(\boldsymbol{\psi}_{\beta k} + \boldsymbol{\sigma}_{\beta k}^{\times}\mathbf{r}_1\right)\mathrm{d}m \quad i = k \tag{2.6.8}$$

These expressions are very similar so only the first, for $i \neq k$, will be developed to the final form;

$$\int \boldsymbol{\psi}_i^T\boldsymbol{\psi}_k \mathrm{d}m = \sum_j \boldsymbol{\psi}_{ji}^T m_{sj}\boldsymbol{\psi}_{jk} + \boldsymbol{\psi}_{\beta i}^T\boldsymbol{\psi}_{\beta k} m_E + \boldsymbol{\psi}_{\beta i}^T\boldsymbol{\sigma}_{\beta k}^{\times}\mathbf{s}_E + \int_E \left(\boldsymbol{\sigma}_{\beta i}^{\times}\mathbf{r}_1\right)^T\boldsymbol{\psi}_{\beta k}\mathrm{d}m + \int_E \left(\boldsymbol{\sigma}_{\beta i}^{\times}\mathbf{r}_1\right)^T\boldsymbol{\sigma}_{\beta k}^{\times}\mathbf{r}_1 \mathrm{d}m \tag{2.6.9}$$

From the matrix identity $(\mathbf{u}^{\times}\mathbf{v})^T\mathbf{w} = \mathbf{u}^T\mathbf{v}^{\times}\mathbf{w}$ it follows that

$$\left(\boldsymbol{\sigma}_{\beta i}^{\times}\mathbf{r}_1\right)^T\boldsymbol{\sigma}_{\beta k}^{\times}\mathbf{r}_1 = \boldsymbol{\sigma}_{\beta i}^T\mathbf{r}_1^{\times}\boldsymbol{\sigma}_{\beta k}^{\times}\mathbf{r}_1 = -\boldsymbol{\sigma}_{\beta i}^T\mathbf{r}_1^{\times}\mathbf{r}_1^{\times}\boldsymbol{\sigma}_{\beta k}. \tag{2.6.10}$$

Thus for $i \neq k$,

$$\int \boldsymbol{\psi}_i^T\boldsymbol{\psi}_k \mathrm{d}m = \sum_j \boldsymbol{\psi}_{ji}^T m_{sj}\boldsymbol{\psi}_{jk} + \boldsymbol{\psi}_{\beta i}^T\boldsymbol{\psi}_{\beta k}m_E + \left(\boldsymbol{\psi}_{\beta i}^T\boldsymbol{\sigma}_{\beta k}^{\times} + \boldsymbol{\psi}_{\beta k}^T\boldsymbol{\sigma}_{\beta i}^{\times}\right)\mathbf{s}_E + \boldsymbol{\sigma}_{\beta i}^T\mathbf{I}_{Eb}\boldsymbol{\sigma}_{\beta k}. \tag{2.6.11}$$

The case of $i = k$ is obtained simply by adding m_{Bi} to the RHS, as in Eq. (2.4.17).

The previous development of Lagrange's equation is unchanged up to Eq. (2.3.9). This is repeated here for convenience;

$$T_{IB} = \frac{1}{2}m_T\mathbf{v}^T\mathbf{v} + \mathbf{v}^T\boldsymbol{\omega}^{\times}\int \mathbf{r}\,\mathrm{d}m + \sum \dot{\eta}_i\mathbf{v}^T\int \boldsymbol{\psi}_i\mathrm{d}m + \sum \dot{\eta}_i\boldsymbol{\omega}^T\int \mathbf{r}^{\times}\boldsymbol{\psi}_i\,\mathrm{d}m - \frac{1}{2}\int \boldsymbol{\omega}^T\mathbf{r}^{\times}\mathbf{r}^{\times}\boldsymbol{\omega}\,\mathrm{d}m + \frac{1}{2}\sum \dot{\eta}_i\sum \dot{\eta}_k\int \boldsymbol{\psi}_i^T\boldsymbol{\psi}_k\,\mathrm{d}m \tag{2.6.12}$$

For the reduced-body modes, Eqs. (2.6.6) to (2.6.11), instead of Eqs. (2.2.31) and (2.2.32), must be used to simplify Eq. (2.6.12). The revised integrated body energy is given by

$$\begin{aligned} T'_{IB} = {} & \frac{1}{2}m_T\mathbf{v}^T\mathbf{v} + \mathbf{v}^T\boldsymbol{\omega}^{\times}\mathbf{s}_{TD} + \frac{1}{2}\boldsymbol{\omega}^T\mathbf{I}_{TD}\boldsymbol{\omega} + \frac{1}{2}\sum \dot{\eta}_i^2 m_{Bi} \\ & + \sum_i \dot{\eta}_i\mathbf{v}^T\left(\sum_j m_{sj}\boldsymbol{\psi}_{ji} + m_E\boldsymbol{\chi}_{Ei}\right) \\ & + \sum_i \dot{\eta}_i\boldsymbol{\omega}^T\left(\sum_j m_{sj}\mathbf{r}_{sj}^{\times}\boldsymbol{\psi}_{ji} + m_E\mathbf{r}_G^{\times}\boldsymbol{\chi}_{Ei} + \mathbf{c}_{EFi}\right) \\ & + \frac{1}{2}\sum_i \dot{\eta}_i\sum_k \dot{\eta}_k\left(\sum_j m_{sj}\boldsymbol{\psi}_{ji}^T\boldsymbol{\psi}_{jk} + m_E\boldsymbol{\psi}_{\beta i}^T\boldsymbol{\psi}_{\beta k}\right. \\ & \left. + \left(\boldsymbol{\psi}_{\beta i}^T\boldsymbol{\sigma}_{\beta k}^{\times} + \boldsymbol{\psi}_{\beta k}^T\boldsymbol{\sigma}_{\beta i}^{\times}\right)\mathbf{s}_E + \boldsymbol{\sigma}_{\beta i}^T\mathbf{I}_{Eb}\boldsymbol{\sigma}_{\beta k}\right) \end{aligned} \tag{2.6.13}$$

where $T_{IB} = T'_{IB}$; that is, the energies are equivalent, although the finite element model corresponding to T'_{IB} has sloshing and engine mass omitted from its modes. The prime is dropped in the development below. The first line of this expression is the same as Eq. (2.3.11), and the newly added terms appear in the remaining lines. Using this result, the Lagrangian acceleration for the translation equation becomes

$$\begin{aligned}\frac{d}{dt}\left(\frac{\partial T}{\partial \mathbf{v}}\right) &= m_T\dot{\mathbf{v}} + \sum_j m_{sj}\dot{\mathbf{v}}_{sj} - \mathbf{s}_{TD}^{\times}\dot{\boldsymbol{\omega}} - \mathbf{s}_{Eb}^{\times}\dot{\boldsymbol{\omega}}_{Eb} \\ &\quad + \sum_i \ddot{\eta}_i\left(\sum_j m_{sj}\boldsymbol{\psi}_{ji} + m_E\boldsymbol{\chi}_{Ei}\right)\end{aligned} \tag{2.6.14}$$

This is a modified version of Eq. (2.4.4), extended to multiple tanks, with the newly added terms appearing on the second line. The Lagrangian term for the rotation equation becomes

$$\begin{aligned}\frac{d}{dt}\left(\frac{\partial T}{\partial \omega}\right) &= I_{TD}\dot{\boldsymbol{\omega}} + s_{TD}^{\times}\dot{\mathbf{v}} + \sum_j m_{sj}\mathbf{r}_{sj}^{\times}\dot{\mathbf{v}}_{sj} \\ &\quad + \mathbf{I}_{Eb}\dot{\boldsymbol{\omega}}_{Eb} - \mathbf{r}_G^{\times}\mathbf{s}_{Eb}^{\times}\dot{\boldsymbol{\omega}}_{Eb} \\ &\quad + \sum_i \ddot{\eta}_i\left(\sum_j m_{sj}\mathbf{r}_{sj}^{\times}\boldsymbol{\psi}_{ji} + m_E\mathbf{r}_G^{\times}\boldsymbol{\chi}_{Ei} + \mathbf{c}_{EFi}\right)\end{aligned} \tag{2.6.15}$$

There is no change to the slosh equation. Finally, the bending equation becomes

$$\begin{aligned}\frac{d}{dt}\left(\frac{\partial T}{\partial \dot{\eta}_i}\right) &= m_{Bi}\ddot{\eta}_i + \sum_j m_{sj}\boldsymbol{\psi}_{ji}^T\dot{\mathbf{v}}_{sj} + \left(\boldsymbol{\sigma}_{\beta i}^T\mathbf{I}_{Eb} - \boldsymbol{\psi}_{\beta i}^T\mathbf{s}_{Eb}^{\times}\right)\dot{\boldsymbol{\omega}}_{Eb} \\ &\quad + \dot{\mathbf{v}}^T\left(\sum_j m_{sj}\boldsymbol{\psi}_{ji} + m_E\boldsymbol{\chi}_{Ei}\right) \\ &\quad + \dot{\boldsymbol{\omega}}^T\left(\sum_j m_{sj}\mathbf{r}_{sj}^{\times}\boldsymbol{\psi}_{ji} + m_E\mathbf{r}_G^{\times}\boldsymbol{\chi}_{Ei} + \mathbf{c}_{EFi}\right) \\ &\quad + \sum_k \ddot{\eta}_k\left(\sum_j m_{sj}\boldsymbol{\psi}_{ji}^T\boldsymbol{\psi}_{jk} + m_E\boldsymbol{\psi}_{\beta i}^T\boldsymbol{\psi}_{\beta k}\right. \\ &\quad \left. + \left(\boldsymbol{\psi}_{\beta i}^T\boldsymbol{\sigma}_{\beta k}^{\times} + \boldsymbol{\psi}_{\beta k}^T\boldsymbol{\sigma}_{\beta i}^{\times}\right)\mathbf{s}_{Eb} + \boldsymbol{\sigma}_{\beta i}^T\mathbf{I}_{Eb}\boldsymbol{\sigma}_{\beta k}\right)\end{aligned} \tag{2.6.16}$$

It is apparent that the same groupings of terms are making a repeated appearance. It is helpful to define

$$\mathbf{p}_{\psi i} = \sum_j m_{sj}\boldsymbol{\psi}_{ji} + m_E \boldsymbol{\chi}_{Ei}$$

$$\mathbf{h}_{\psi i} = \sum_j m_{sj}\mathbf{r}_{sj}^{\times}\boldsymbol{\psi}_{ji} + m_E \mathbf{r}_G^{\times}\boldsymbol{\chi}_{Ei} + \mathbf{c}_{EFi}$$

$$d_{\psi ik} = \sum_j m_{sj}\boldsymbol{\psi}_{ji}^T\boldsymbol{\psi}_{jk} + m_E\boldsymbol{\psi}_{\beta i}^T\boldsymbol{\psi}_{\beta k} + \left(\boldsymbol{\psi}_{\beta i}^T\boldsymbol{\sigma}_{\beta k}^{\times} + \boldsymbol{\psi}_{\beta k}^T\boldsymbol{\sigma}_{\beta i}^{\times}\right)\mathbf{s}_E + \boldsymbol{\sigma}_{\beta i}^T\mathbf{I}_{Eb}\boldsymbol{\sigma}_{\beta k} \quad (2.6.17)$$

The last parameter is a scalar. Note that for any values of i, k, $d_{\psi ik} = d_{\psi ki}$. In order to cover the case of more than one engine, let r be the engine index number. The above parameters then become

$$\mathbf{p}_{\psi i} = \sum_j m_{sj}\boldsymbol{\psi}_{ji} + \sum_r m_{Er}\boldsymbol{\chi}_{Eri} \quad (2.6.18)$$

$$\mathbf{h}_{\psi i} = \sum_j m_{sj}\mathbf{r}_{sj}^{\times}\boldsymbol{\psi}_{ji} + \sum_r m_{Er}\mathbf{r}_{Gr}^{\times}\boldsymbol{\chi}_{Eri} + \mathbf{c}_{EFri}$$

$$d_{\psi ik} = \sum_j m_{sj}\boldsymbol{\psi}_{ji}^T\boldsymbol{\psi}_{jk} + \sum_r \left(m_{Er}\boldsymbol{\psi}_{\beta ri}^T\boldsymbol{\psi}_{\beta rk} + \left(\boldsymbol{\psi}_{\beta ri}^T\boldsymbol{\sigma}_{\beta rk}^{\times} + \boldsymbol{\psi}_{\beta rk}^T\boldsymbol{\sigma}_{\beta i}^{\times}\right)\mathbf{s}_{Ebr} + \boldsymbol{\sigma}_{\beta ri}^T\mathbf{I}_{Ebr}\boldsymbol{\sigma}_{\beta rk}\right) \quad (2.6.19)$$

In practice, the phrase "reduced body model" may have more than one meaning, and caution is advised when communicating model needs to finite element model developers, integrating models, and evaluating results. A common configuration is that in the finite element model, the mass of the rocket has been reduced by the engine masses, but the rocket body includes everything else (structure and liquid propellants). In that case, all terms involving the sloshing mass (subscript j) should be deleted from Eqs. (2.6.18) through (2.6.19). That is, only the summations over the engine masses (the summations over r) will remain. A second possibility is that the finite element model of the rocket body includes neither the slosh masses nor the engine. In that case, both the slosh and engine terms should be included in the above definitions. In summary, every mass should be included only once, in either the finite element model or in Eqs. (2.6.18) through (2.6.19).

$$\mathcal{M}'=\begin{bmatrix} m_T\mathbf{1} & -\mathbf{s}_{TD}^{\times} & m_{s1}\mathbf{1} & m_{s2}\mathbf{1} & \cdots & m_{sN}\mathbf{1} & \mathbf{p}_{\psi 1} & \mathbf{p}_{\psi 2} & \cdots & \mathbf{p}_{\psi M} \\ \mathbf{s}_{TD}^{\times} & \mathbf{I}_{TD} & m_{s1}\mathbf{r}_{s1}^{\times} & m_{s2}\mathbf{r}_{s2}^{\times} & \cdots & m_{sN}\mathbf{r}_{sN}^{\times} & \mathbf{h}_{\psi 1} & \mathbf{h}_{\psi 2} & \cdots & \mathbf{h}_{\psi M} \\ m_{s1}\mathbf{1} & -m_{s1}\mathbf{r}_{s1}^{\times} & m_{s1}\mathbf{1} & \mathbf{O} & \cdots & \mathbf{O} & m_{s1}\psi_{11} & m_{s1}\psi_{12} & \cdots & m_{s1}\psi_{1M} \\ m_{s2}\mathbf{1} & -m_{s2}\mathbf{r}_{s2}^{\times} & \mathbf{O} & m_{s2}\mathbf{1} & \cdots & \mathbf{O} & m_{s2}\psi_{21} & m_{s2}\psi_{22} & \cdots & m_{s2}\psi_{2M} \\ \vdots & \vdots & \vdots & \vdots & \ddots & \vdots & \vdots & \vdots & \ddots & \vdots \\ m_{sN}\mathbf{1} & -m_{sN}\mathbf{r}_{sN}^{\times} & \mathbf{O} & \mathbf{O} & \cdots & m_{sN}\mathbf{1} & m_{sN}\psi_{N1} & m_{sN}\psi_{N2} & \cdots & m_{sN}\psi_{NM} \\ \mathbf{p}_{\psi 1}^{T} & \mathbf{h}_{\psi 1}^{T} & m_{s1}\psi_{11}^{T} & m_{s2}\psi_{21}^{T} & \cdots & m_{sN}\psi_{N1}^{T} & m_{B1}+d_{\psi 11} & d_{\psi 12} & \cdots & d_{\psi 1M} \\ \mathbf{p}_{\psi 2}^{T} & \mathbf{h}_{\psi 2}^{T} & m_{s1}\psi_{12}^{T} & m_{s2}\psi_{22}^{T} & \cdots & m_{sN}\psi_{N2}^{T} & d_{\psi 21} & m_{B2}+d_{\psi 22} & \cdots & d_{\psi 2M} \\ \vdots & \vdots & \vdots & \vdots & \ddots & \vdots & \vdots & \vdots & \ddots & \vdots \\ \mathbf{p}_{\psi M}^{T} & \mathbf{h}_{\psi M}^{T} & m_{s1}\psi_{1M}^{T} & m_{s2}\psi_{2M}^{T} & \cdots & m_{sN}\psi_{NM}^{T} & d_{\psi M1} & d_{\psi M2} & \cdots & m_{BM}+d_{\psi MM} \end{bmatrix} \quad (2.6.20)$$

Eq. (2.6.20): Mass matrix for reduced-body prescribed motion model

$$\begin{bmatrix} m_T\mathbf{1} & -\mathbf{s}_{TD}^{\times} & -\mathbf{s}_{Eb}^{\times} & \mathbf{p}_{\psi 1} & \mathbf{p}_{\psi 2} & \cdots & \mathbf{p}_{\psi M} \\ \mathbf{s}_{TD}^{\times} & \mathbf{I}_{TD} & \mathbf{I}_{TWD} & \mathbf{h}_{\psi 1} & \mathbf{h}_{\psi 2} & \cdots & \mathbf{h}_{\psi M} \\ \mathbf{s}_{Eb}^{\times} & \mathbf{I}_{TWD}^{T} & \mathbf{I}_{Eb} & \mathbf{c}_{EF1} & \mathbf{c}_{EF2} & \cdots & \mathbf{c}_{EFM} \\ \mathbf{p}_{\psi 1}^{T} & \mathbf{h}_{\psi 1}^{T} & \mathbf{c}_{EF1}^{T} & m_{B1}+d_{\psi 11} & d_{\psi 12} & \cdots & d_{\psi 1M} \\ \mathbf{p}_{\psi 2}^{T} & \mathbf{h}_{\psi 2}^{T} & \mathbf{c}_{EF2}^{T} & d_{\psi 21} & m_{B2}+d_{\psi 22} & \cdots & d_{\psi 2M} \\ \vdots & \vdots & \vdots & \vdots & \vdots & \ddots & \vdots \\ \mathbf{p}_{\psi M}^{T} & \mathbf{h}_{\psi M}^{T} & \mathbf{c}_{EFM}^{T} & d_{\psi M1} & d_{\psi M2} & \cdots & m_{BM}+d_{\psi MM} \end{bmatrix} \begin{bmatrix} \mathbf{a}_b \\ \dot{\boldsymbol{\omega}} \\ \dot{\boldsymbol{\omega}}_{Eb} \\ \ddot{\eta}_1 \\ \ddot{\eta}_2 \\ \vdots \\ \ddot{\eta}_M \end{bmatrix} = \begin{bmatrix} \tilde{\mathbf{f}} \\ \tilde{\mathbf{g}} \\ \mathbf{g}_{Eb} \\ \tilde{f}_{B1} \\ \tilde{f}_{B2} \\ \vdots \\ \tilde{f}_{BM} \end{bmatrix} \quad (2.6.21)$$

Eq. (2.6.21): System equation for reduced-body model (no slosh, non prescribed engine motion)

For the case of prescribed engine motion, the integrated-body mass matrix (2.5.19) is shown in its reduced body form in Eq. (2.6.20).

Note that neither the slosh rows nor the slosh columns have been affected by the switch from integrated body modes to reduced body modes – all of the changes appear in the translation, rotation, and bending equations. If the engine motion is not prescribed and it is necessary to insert an engine equation, it is found that the engine rows and columns are similarly unaffected. Thus the matrix equation (2.5.20) corresponding to the previously analyzed case of just engine motion and no slosh motion is given by (2.6.21).

Integrated versus reduced body model

A simple example can be used to obtain greater insight into the difference between using an integrated body FEM and a reduced body FEM. Assume a single slosh mass located at the center of a uniform beam. First consider the case in which the slosh mass is locked to the beam, as shown in the figure below. The undeformed centerline defines the x axis of the coordinate frame. With an integrated body FEM, the elastic equation without external excitation is

$$m_{B1}\ddot{\eta}_1 + m_{B1}\Omega_{B1}^2\eta_1 = 0 \qquad (2.6.22)$$

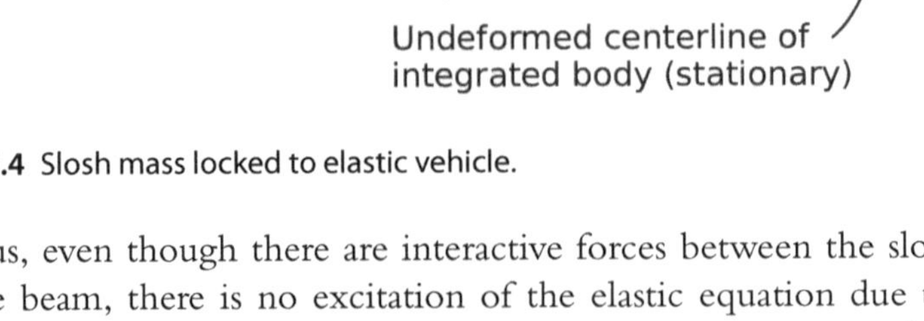

Figure 2.4 Slosh mass locked to elastic vehicle.

Thus, even though there are interactive forces between the slosh mass and the beam, there is no excitation of the elastic equation due to these slosh forces, since these interactive forces are already built into the FEM. Another important point is that as the beam oscillates, the coordinate frame of the integrated body remains stationary.

On the other hand, with a reduced body model, the slosh mass is not part of the FEM and the forces on the beam due to slosh motion will give rise to terms that must be added to the elastic equation. As shown in Fig. 2.4, the undeformed centerline of the reduced body oscillates. The coordinate frame is defined by this centerline. Thus there will be an acceleration $\mathbf{a}_b$ of this frame. The elastic equation is, from Eq. (2.6.16),

$$\bar{m}_{B1}\ddot{\eta}_1 + \bar{m}_{B1}\bar{\Omega}_{B1}^2\eta_1 = -\bar{\boldsymbol{\psi}}_{11}^T f_{s1} \tag{2.6.23}$$

where overbars have been introduced to represent the reduced-body FEM parameters. The slosh force f_{s1} is the slosh mass multiplied by the combined acceleration of the coordinate frame and the flexible motion;

$$f_{s1} = m_{s1}\left(\mathbf{a}_b + \bar{\boldsymbol{\psi}}_{11}\ddot{\eta}_1\right). \tag{2.6.24}$$

Substituting this into Eq. (2.6.23) yields

$$\bar{m}_{B1}\ddot{\eta}_1 + \bar{m}_{B1}\bar{\Omega}_{B1}^2\eta_1 = -\bar{\boldsymbol{\psi}}_{11}^T m_{s1}\left(\mathbf{a}_b + \bar{\boldsymbol{\psi}}_{11}\ddot{\eta}_1\right). \tag{2.6.25}$$

For this locked case, there are two terms present in the reduced body expression (2.6.25) that are not present in the integrated body expression (2.6.22).

Retaining all of the foregoing assumptions, consider the case in which the slosh mass is free to oscillate relative to the beam. With the integrated body model, the elastic equation reads

$$m_{B1}\ddot{\eta}_1 + m_{B1}\Omega_{B1}^2\eta_1 = -\boldsymbol{\psi}_{11}^T m_{s1}\ddot{\delta}_{s1}. \tag{2.6.26}$$

The reduced-body approach requires three parameters defined by Eqs. (2.6.18) through (2.6.19), which become

$$\mathbf{p}_{\psi 1} = m_{s1}\bar{\boldsymbol{\psi}}_{11} \tag{2.6.27}$$

$$\mathbf{h}_{\psi 1} = \mathbf{0} \tag{2.6.28}$$

$$d_{\psi 11} = m_{s1}\bar{\boldsymbol{\psi}}_{11}^T\bar{\boldsymbol{\psi}}_{11} \tag{2.6.29}$$

The bending equation of the first mode can be inferred from the first bending row of the matrix in Eq. (2.6.20);

$$\mathbf{p}_{\psi 1}^T\mathbf{a}_b + \mathbf{h}_{\psi 1}^T\dot{\boldsymbol{\omega}} + m_{s1}\bar{\boldsymbol{\psi}}_{11}^T\ddot{\delta}_{s1} + \left(\bar{m}_{B1} + d_{\psi 11}\right)\ddot{\eta}_1 = -\bar{m}_{B1}\bar{\Omega}_{B1}^2\eta_1. \tag{2.6.30}$$

Substituting from above and rearranging,

$$\bar{m}_{B1}\ddot{\eta}_1 + \bar{m}_{B1}\bar{\Omega}_{B1}^2\eta_1 = -\bar{\boldsymbol{\psi}}_{11}^T m_{s1}\left(\mathbf{a}_b + \ddot{\delta}_{s1} + \bar{\boldsymbol{\psi}}_{11}\ddot{\eta}_1\right). \tag{2.6.31}$$

This equation could have also been obtained by taking Eq. (2.6.25) and adding the $\ddot{\delta}_{s1}$ term, where $\ddot{\delta}_{s1}$ is the slosh acceleration relative to the point of attachment on the beam, which itself is accelerating due to flexible motion and the acceleration of the frame. Thus, $\mathbf{a}_b + \ddot{\delta}_{s1} + \bar{\boldsymbol{\psi}}_{11}\ddot{\eta}_1$ represents the total inertial acceleration of the slosh mass. This acceleration multiplied by the slosh mass m_{s1} is exactly the same as the force on the slosh mass (i.e., the spring force, plus a damping force if that is included). Thus, if a reduced body model is used, the beam excitation is equal and opposite to the slosh force. Note that the equation can be expressed in the same form as Eq. (2.2.23). If a mass matrix formulation is used, the terms must be arranged as in (2.6.30), with all the accelerations on the LHS.

This simple problem does not include gravity. A detailed discussion of the effect of gravity on the bending equation is given in Chapter 5. The net result is that gravity affects the left and right sides of Eq. (2.6.30) in equal measure, with no net change.

To summarize, one could say that the integrated body approach is more Lagrangian, and the reduced body approach is more Newtonian. Thus if the FEM does not contain the slosh mass and the rearranged reduced body equations are used, then it becomes possible to apply Newton's third law such that the excitation of the bending equation is equal and opposite to the force on the slosh mass. However, if the FEM does contain the slosh mass and the integrated body approach is used, then the excitation of the bending equation due to slosh comes from the relative acceleration of the slosh mass. Newton's third law still applies, even though it may be hidden in the Lagrangian formulation.

2.7 Truncating the slosh motion

Up to this point, the slosh mass m_{sj} of Fig. 2.1 has been allowed three degrees of freedom. However, it has been recognized for almost two centuries that longitudinal slosh motion (in a direction parallel to the equilibrium acceleration) must be treated as a special case. NASA SP-106 [13] provides an excellent summary of the work prior to 1966 on this subject, and goes on to provide its own contributions. The important point is that longitudinal slosh motion only occurs after sustained excitation and is not normally a factor in the dynamics.

One form of longitudinal resonance, known as "pogo," involves a coupled longitudinal oscillation of the liquid propellant, the axial elasticity of the rocket and tank structure, and the engine thrust dynamics that change as a function of propellant inlet pressure. This issue has all but been eliminated since the mid-1960's through the application of standard design methods and suppression hardware devices such as accumulators.

Another way to view this issue is that if there are sufficient longitudinal oscillations to excite slosh motion in the x direction, then the rocket is unlikely to be acceptable regardless of the ensuing slosh activity. Ref. [14], a recent update of Ref. [13], provides the following illuminating quote:

> *This update does not cover all the subjects included in NASA SP-106. Some topics were omitted because they discuss marginally-related material accessible elsewhere (e.g., the principles of similitude as applied to scale models) or the material is quite specialized or does not now appear to be as important as it was in 1966 (e.g., vertical excitation of tanks).*

"Vertical excitation" in this context is synonymous with motion in the x direction. Thus it becomes desirable to eliminate the x degree of freedom of the slosh motion. To do this, define a truncation matrix

$$\bar{\mathbf{U}} = \begin{bmatrix} 0 & 0 \\ 1 & 0 \\ 0 & 1 \end{bmatrix} \tag{2.7.1}$$

Every slosh column is post-multiplied by $\bar{\mathbf{U}}$, and every slosh row is pre-multiplied by $\bar{\mathbf{U}}^T$. Using Eq. (2.5.19) as an example, the mass matrix is given as shown in Eq. (2.7.2), noting now that $\mathbf{0} = \begin{bmatrix} 0 & 0 \end{bmatrix}^T$ and $\mathbf{O} = \begin{bmatrix} \mathbf{0} & \mathbf{0} \end{bmatrix}$, where appropriate. This form of the mass matrix is suggested as a starting point for analyzing coupled slosh/flexible motion. Note that $\bar{\mathbf{U}}^T\bar{\mathbf{U}}$ is the 2×2 identity matrix. This process can be applied to any of the mass matrices given above, and can also be applied to truncate engine motion as suggested earlier. The slosh perturbation $\boldsymbol{\delta}_{sj}$ is redefined as $\begin{bmatrix} \delta_{sjy} & \delta_{sjz} \end{bmatrix}^T$, thus $\bar{\mathbf{U}}\boldsymbol{\delta}_{sj} = \begin{bmatrix} 0 & \delta_{sjy} & \delta_{sjz} \end{bmatrix}^T$.

A functional starting point for numerical integration of these equations in a simulation is presented in Chapter 10, along with some detail regarding typical analysis processes.

In the following chapter, a detailed discussion of the use of mechanical analogs for liquid sloshing is presented, along with details supporting

$$
\mathcal{M} \equiv \begin{bmatrix}
m_T \mathbf{1} & -\mathbf{s}_{TD}^{\times} & m_{s1}\bar{\mathbf{U}} & m_{s2}\bar{\mathbf{U}} & \dots & m_{sN} & \mathbf{0} & \mathbf{0} & \dots & \mathbf{0} \\
\mathbf{s}_{TD}^{\times} & \mathbf{I}_{TD} & m_{s1}\mathbf{r}_{s1}^{\times}\bar{\mathbf{U}} & m_{s2}\mathbf{r}_{s2}^{\times}\bar{\mathbf{U}} & \dots & m_{sN}\mathbf{r}_{sN}^{\times}\bar{\mathbf{U}} & \mathbf{0} & \mathbf{0} & \dots & \mathbf{0} \\
m_{s1}\bar{\mathbf{U}}^T & -m_{s1}\bar{\mathbf{U}}^T\mathbf{r}_{s1}^{\times} & m_{s1}\bar{\mathbf{U}}^T\bar{\mathbf{U}} & \mathbf{O} & \dots & \mathbf{O} & m_{s1}\bar{\mathbf{U}}^T\psi_{11} & m_{s1}\bar{\mathbf{U}}^T\psi_{12} & \dots & m_{s1}\bar{\mathbf{U}}^T\psi_{1M} \\
m_{s2}\mathbf{1} & -m_{s2}\mathbf{r}_{s2}^{\times} & \mathbf{0} & m_{s2}\bar{\mathbf{U}}^T\bar{\mathbf{U}} & \dots & \mathbf{O} & m_{s2}\bar{\mathbf{U}}^T\psi_{21} & m_{s2}\bar{\mathbf{U}}^T\psi_{22} & \dots & m_{s2}\bar{\mathbf{U}}^T\psi_{2M} \\
\vdots & \vdots & \vdots & \vdots & \ddots & \vdots & \vdots & \vdots & \ddots & \vdots \\
m_{sN}\mathbf{1} & -m_{sN}\mathbf{r}_{sN}^{\times} & \mathbf{0} & \mathbf{0} & \dots & m_{sN}\bar{\mathbf{U}}^T\bar{\mathbf{U}} & m_{sN}\bar{\mathbf{U}}^T\psi_{N1} & m_{sN}\bar{\mathbf{U}}^T\psi_{N2} & \dots & m_{sN}\bar{\mathbf{U}}^T\psi_{NM} \\
\mathbf{0}^T & \mathbf{0}^T & m_{s1}\psi_{11}^T\bar{\mathbf{U}} & m_{s2}\psi_{21}^T\bar{\mathbf{U}} & \dots & m_{sN}\psi_{N1}^T\bar{\mathbf{U}} & m_{B1} & 0 & \dots & 0 \\
\mathbf{0}^T & \mathbf{0}^T & m_{s1}\psi_{12}^T\bar{\mathbf{U}} & m_{s2}\psi_{22}^T\bar{\mathbf{U}} & \dots & m_{sN}\psi_{N2}^T\bar{\mathbf{U}} & 0 & m_{B2} & \dots & 0 \\
\vdots & \vdots & \vdots & \vdots & \ddots & \vdots & \vdots & \vdots & \ddots & \vdots \\
\mathbf{0}^T & \mathbf{0}^T & m_{s1}\psi_{1M}^T\bar{\mathbf{U}} & m_{s2}\psi_{2M}^T\bar{\mathbf{U}} & \dots & m_{sN}\psi_{NM}^T\bar{\mathbf{U}} & 0 & 0 & \dots & m_{BM}
\end{bmatrix} \tag{2.7.2}
$$

Eq. (2.7.2): Mass matrix with prescribed engine motion and slosh truncation the incorporation of nonlinear terms associated with the spring-mass slosh model. This provides insight into the nonlinear dynamics effects that lead to the use of the nonlinear spherical pendulum model discussed in Chapter 4.

CHAPTER 3

Slosh modeling

There are two models that are commonly used to represent slosh dynamics – the spring model and the pendulum model, as shown in Fig. 3.1. For the small motions that are typical of conventional liquid propellant launch vehicles, a linearized analysis can be used. In this case, these two models give essentially the same results. In some cases, however, one can expect large changes in the rocket attitude, or a trim condition (i.e., average acceleration) that is not collinear with the axis of symmetry (counter to the assumptions of Section 2.7). A good example would be an air-launched rocket, which may require large pitch changes after release from the carrier aircraft. This will result in large motions of the slosh mass, requiring a nonlinear analysis to properly capture the dynamic effects. The spring model

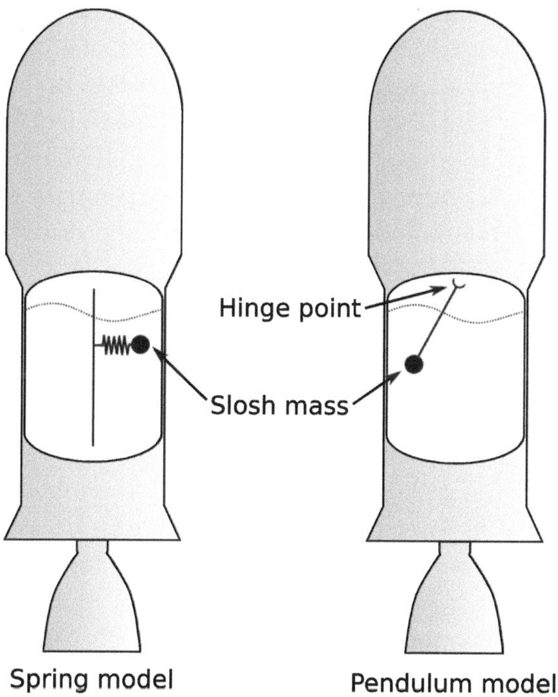

Figure 3.1 Slosh models.

Dynamics and Simulation of Flexible Rockets
https://doi.org/10.1016/B978-0-12-819994-7.00008-X

may give relatively inaccurate results during such maneuvers, since there is nothing in the model to prevent the slosh mass from moving far outside the boundaries of the tank walls. In contrast, with the pendulum model the slosh mass is always constrained to be within a fixed distance of the hinge point.

If the rocket has a finite roll rate, there may be significant nonlinear effects even if the slosh motion is small. The spring model can be used with the addition of nonlinear terms. The present chapter focuses on this situation. Chapter 4 addresses nonlinear effects using the pendulum model.

Analytical solutions for surface wave motion in tanks can be obtained from basic inviscid, incompressible fluid mechanics. Mode shapes and frequencies for a rectangular tank were derived by Lamb [15], and similar approaches have since been used to derive solutions for a variety of container geometries and fill levels. An excellent summary of early work on slosh modeling is contained in NASA special publication SP-106, Ref. [13], which was previously mentioned in Chapter 2. This comprehensive study discusses all aspects of slosh modeling and testing, including slosh damping, and summarizes many test results. Analyses of surface wave motion in partially filled rectangular, cylindrical, spheroidal, and conical tanks are presented. This publication has come to be known in the launch vehicle community as the "Old Testament." The "New Testament" is a more recent publication of Southwest Research Institute [14]. These monographs both discuss the very instructive (albeit not very common) case of a rectangular tank, which is a good starting point for the reader interested in understanding the analytical fluid mechanics approach.

The analytical solution for the mechanical model parameters for a cylindrical tank is valid only when the fluid is in the barrel section and the liquid depth is sufficient that the fluid does not interact with the bottom dome. The analytical solutions for a bare flat-bottomed cylindrical tank and an infinite-depth cylindrical tank converge when the fluid depth is about twice the tank radius, or $h > 2r$, as shown in Fig. 3.2. If the tank is integral to the structure and relatively long, flexible motion of the tank may be an important consideration. There are two types of motion: fundamental bending modes of the rocket, which can occur close to the sloshing frequencies, and higher frequency modes of the tank structure. More advanced finite element techniques can be used to analyze hydroelastic modes, which include both fluid and structural motion. The higher frequency modes are seldom of importance to the overall response of the rocket.

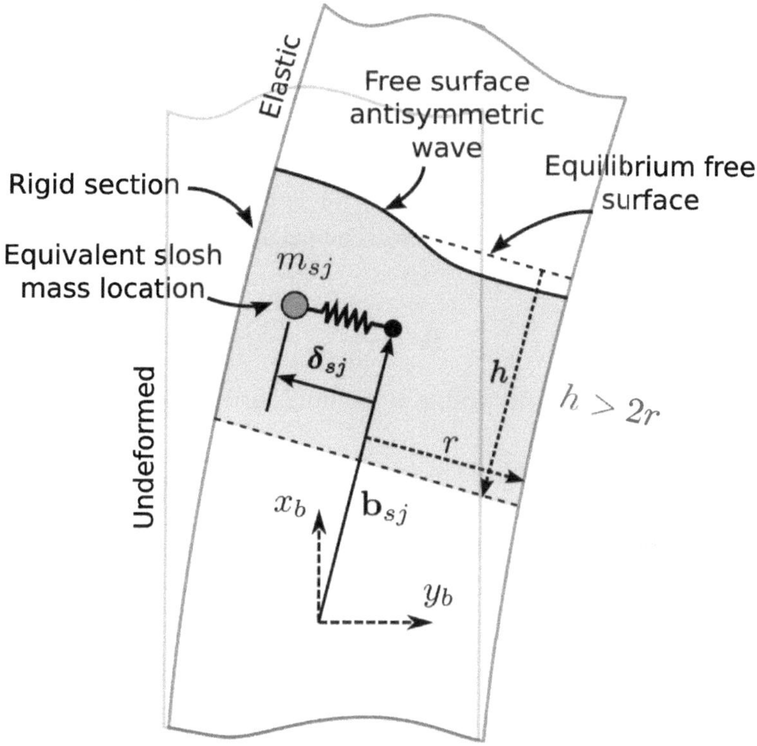

Figure 3.2 Slosh spring-mass model for an elastic tank.

Long, integral rocket structural tanks can be treated as an equivalent rigid cylindrical section attached to a flexible structure, as depicted in Fig. 3.2. Since the condition $h > 2r$ is satisfied, the effects of the bottom geometry are negligible. For lower liquid levels or for liquid levels in an upper dome, corrections must be made to the parameters. In practice, the parameters can be generated for arbitrarily-shaped axially symmetric tanks using the numerical method outlined by Lomen [16,17], which also gives more accurate results when the liquid is not in the barrel section. The Lomen method is a numerical solution under the same incompressible, inviscid flow assumptions used in the simplified container geometries.

If the vehicle motions are large and coupled in three axes, there is a risk of exciting rotary sloshing motion. Some rotary behavior is always expected but is usually insignificant unless the liquid damping is very small. In addition to modeling large amplitudes, the pendulum model is better suited to predicting rotary effects, although more advanced models such as the Kana pendulum [18] and the Bauer parabolic model [19] should be considered if the tank configuration or vehicle maneuvers strongly suggest a tendency toward rotary phenomena.

There is a rich literature regarding the dynamics of liquids in reduced gravity conditions. The mechanical models discussed herein can in fact be used, with appropriate parameters, for conditions where the average acceleration is very small but not zero. The parameter of importance in this case is the Bond number,

$$Bo = \frac{\rho \bar{g} r^2}{\sigma} \tag{3.0.1}$$

where ρ is the fluid density and σ is the fluid surface tension, and $\bar{g}$ is the quasi-steady acceleration.

There are valid mechanical models for Bond numbers near $Bo = 30$. The principal feature in such conditions is a very low liquid natural frequency. As such, the pendulum model of Chapter 4 is usually more appropriate, and the analyst must be wary of capillary and interface stability effects. For additional details regarding mechanical models in near zero-g conditions, the interested reader is referred to Ref. [14].

3.1 Fluid mechanics model

The fluid analysis is used to determine the equivalent slosh mass, spring stiffness, and equilibrium location of the slosh mass with respect to the undisturbed free surface, such that the mechanical analog generates approximately the same forces and moments as the fluid (except at resonance). Given the mechanical model parameters, the dynamic interactions of slosh with the other motions of a rocket can be computed without further need for surface wave analysis. It is important to avoid taking the mass-spring analog too literally. For example, when using a spring-mass model of a flat-bottomed cylindrical tank for liquid levels of $0 < h < r$, the appropriate location of the slosh mass is actually above the liquid free surface [20]. This apparent inconsistency is necessary to properly account for the *moment* of the sloshing fluid caused by its pressure distribution on the tank bottom, which for low liquid levels, is more important than the forces on the tank walls. This does have some implications for the coupling of slosh into flex, which will be discussed in Chapter 5.

In the wave model, the distance the wave moves back and forth, herein called the displacement amplitude, is fixed by the width of the tank. If the tank is accelerated laterally in a way that causes the wave motion to increase, this is manifested by the height of the wave increasing, but no change takes place in the displacement amplitude. One might say that the "mass" of the wave increases, and the displacement amplitude remains constant. The reverse is true in the spring model – the slosh mass remains constant, and the displacement amplitude increases as the result of continued excitation. Thus if a dynamic analysis reveals that the slosh mass has traveled outside the boundaries of the tank, this does not necessarily mean that there is something wrong with the model, it simply means that the wave height in the corresponding fluid model is large. However, standard practice in the industry is to limit the spring mass model to situations in which the displacement is 20% of the tank radius. With larger displacements, use of a more accurate model, such as the pendulum model, could be considered. However, the pendulum model also has some limitations that require consideration; particularly, the effective slosh damping in high-g conditions increases rapidly above a certain threshold value of the pendulum angle when slosh wave crash-over occurs, since energy is more rapidly dissipated in the resultant turbulent breakup of the free surface [21].

Despite these caveats, the success of the spring model is based on the premise that if the analogy has been properly established, the product of the mass times the displacement amplitude matches that for the wave model. It is always this product, and not the individual factors, that appears in the equations of motion.

Two simplified examples will now be presented to illustrate this principle. For the first example, suppose the slosh mass m_s is displaced in the y direction by a distance δ_{sy} and there is a net force $F - D$ through the rocket centerline, as shown in Fig. 3.3. It is convenient to define the net quasi-steady acceleration of the rocket as

$$\bar{g} = \frac{F - D}{m_T}. \tag{3.1.1}$$

The yaw moment created from this offset is

$$g_z = \bar{g} m_s \delta_{sy} \tag{3.1.2}$$

For the second example, suppose the slosh mass is oscillating at its natural frequency in the y direction. The resultant lateral force is given by

$$f_y = m_s \ddot{\delta}_{sy} = -\Omega_s^2 m_s \delta_{sy} \tag{3.1.3}$$

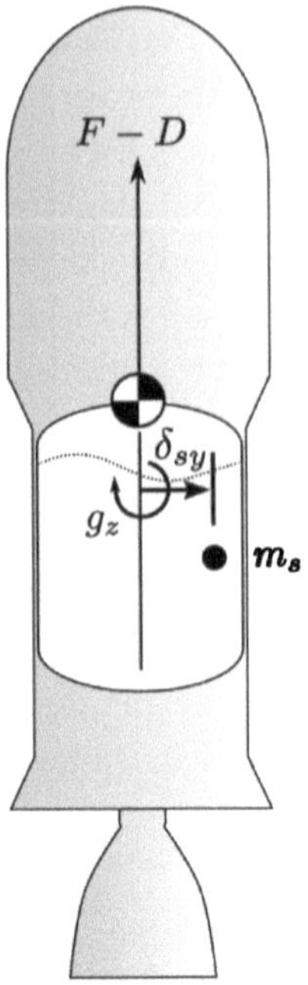

Figure 3.3 Moment generated from offset slosh mass.

In both examples, we only see the product of the slosh mass times the slosh displacement. This holds true throughout the dynamic equations.

In a typical development program, a slosh natural frequency is supplied that corresponds to test conditions on the ground, with a gravitational acceleration of one earth gravity, denoted g_0 or "1 g". This must be scaled to the actual acceleration on the rocket. Thus

$$\Omega_s^2 = \Omega_{s,1g}^2 \frac{\bar{g}}{g_0} \tag{3.1.4}$$

with $\bar{g}$ defined as in Eq. (3.1.1).

The fact that the natural frequency is proportional to the square root of the acceleration is analogous to the situation of a pendulum with a fixed length l on the earth's surface, whose natural frequency near equilibrium is proportional to $\sqrt{g_0/l}$. In this pendulum relation, g_0 plays the role of $\bar{g}$ in (3.1.4).

These and similar equivalent relationships hold for the spring mass and pendulum models when the pendulum is linearized about zero relative angular displacement.

3.2 Spring slosh model with nonlinear terms

This section extends the slosh analysis of Chapter 2 to the case in which the rotation rate of the rocket body is finite, i.e., it is not just an infinitesimal perturbation from an equilibrium state with zero rotation about all axes. This extension results in nonlinear terms being added to the RHS of the equations. Flexibility is not included. However, it is valid to superimpose the flexibility effects of the previous chapter with the nonlinear effects of the present chapter.

A possible advantage of the present approach over various other approaches is that the velocity $\mathbf{v}$ is eliminated from the final equations. Inclusion of this velocity may be problematic for computations while simulating the later stages of flight when the velocity becomes very large. Computations involving a velocity commensurate with orbital maneuvers, for example, several km/s, may introduce numerical difficulties.

Fig. 2.1 defines three vectors (which are treated as column matrices) that are related by

$$\mathbf{r}_{sj} = \mathbf{b}_{sj} + \boldsymbol{\delta}_{sj}. \tag{3.2.1}$$

Here, $\mathbf{b}_{sj}$ is a fixed vector at the equilibrium position of the slosh mass, and $\boldsymbol{\delta}_{sj}$ is the dynamic displacement. The subscript j is attached to the slosh parameters in anticipation of extending the results to N masses, and summing j from 1 to N. There is one vector equation for the slosh mass itself, a vector equation for the system translational momentum, and a vector equation for the system angular momentum.

The "rigid body" in this section means the rocket body plus the nozzle(s) firmly fixed in place. The analysis starts with the assumption that the slosh mass is free to move with three relative translational degrees of freedom. The final constraint condition is that the slosh mass is only allowed to move in the body yz plane. This is imposed by eliminating the slosh x displacement from the final system of equations, as described in Section 2.7.

The subscript 0 represents the rigid body. The definitions used in the following analysis are given in Table 3.1.

Since $\mathbf{b}_{sj}$ is constant, it can be incorporated into $\mathbf{s}_T$;

$$\mathbf{s}_{TD} = \mathbf{s}_0 + m_{sj}\left(\mathbf{b}_{sj} + \boldsymbol{\delta}_{sj}\right) \tag{3.2.2}$$

$$\mathbf{s}_T = \mathbf{s}_0 + m_{sj}\mathbf{b}_{sj} \tag{3.2.3}$$

$$\mathbf{s}_{TD} = \mathbf{s}_T + m_{sj}\boldsymbol{\delta}_{sj} \tag{3.2.4}$$

Table 3.1 Slosh analysis notation.

$\mathbf{s}_0$	first moment of inertia of body 0 about the origin
$\mathbf{s}_T$	total first moment of inertia, with slosh mass in equilibrium position
$\mathbf{s}_{TD}$	dynamic first moment of inertia, including slosh displacement
$\mathbf{I}_0$	second moment of inertia of body 0 about the origin
$\mathbf{I}_T$	total second moment of inertia, with slosh mass in equilibrium position
$\mathbf{I}_{TD}$	dynamic second moment of inertia, including slosh displacement

The linear momentum of body 0 and the slosh mass are, respectively,

$$\mathbf{p}_0 = \int \left(\mathbf{v} - \mathbf{r}^{\times}\boldsymbol{\omega}\right) dm = m_0\mathbf{v} - \mathbf{s}_0^{\times}\boldsymbol{\omega} \tag{3.2.5}$$

$$\mathbf{p}_{sj} = m_{sj}\left(\mathbf{v} - \mathbf{r}_{sj}^{\times}\boldsymbol{\omega} + \dot{\boldsymbol{\delta}}_{sj}\right) \tag{3.2.6}$$

where the variable $\mathbf{r}$ is defined in Fig. 2.1. The total linear momentum becomes

$$\mathbf{p} = \mathbf{p}_0 + \mathbf{p}_{sj} = m_T\mathbf{v} - \mathbf{s}_{TD}^{\times}\boldsymbol{\omega} + m_{sj}\dot{\boldsymbol{\delta}}_{sj} \tag{3.2.7}$$

where $m_T = m_0 + m_{sj}$. The angular momentum of body 0 is given by

$$\mathbf{h}_0 = \int \mathbf{r}^{\times}\left(\mathbf{v} + \boldsymbol{\omega}^{\times}\mathbf{r}\right) dm = \mathbf{s}_0^{\times}\mathbf{v} + \mathbf{I}_0\boldsymbol{\omega} \tag{3.2.8}$$

The total angular momentum about the origin defined in Fig. 2.1 becomes

$$\mathbf{h} = \mathbf{h}_0 + \mathbf{r}_{sj}^{\times}\mathbf{p}_{sj} = \mathbf{s}_{TD}^{\times}\mathbf{v} + \mathbf{I}_{TD}\boldsymbol{\omega} + m_{sj}\mathbf{r}_{sj}^{\times}\dot{\boldsymbol{\delta}}_{sj}. \tag{3.2.9}$$

Eqs. (3.2.6), (3.2.7), and (3.2.9) can be combined into a single matrix equation;

$$\begin{bmatrix} \mathbf{p} \\ \mathbf{h} \\ \mathbf{p}_{sj} \end{bmatrix} = \begin{bmatrix} m_T\mathbf{1} & -\mathbf{s}_{TD}^{\times} & m_{sj}\mathbf{1} \\ \mathbf{s}_{TD}^{\times} & \mathbf{I}_{TD} & m_{sj}\mathbf{r}_{sj}^{\times} \\ m_{sj}\mathbf{1} & -m_{sj}\mathbf{r}_{sj}^{\times} & m_{sj}\mathbf{1} \end{bmatrix} \begin{bmatrix} \mathbf{v} \\ \boldsymbol{\omega} \\ \dot{\boldsymbol{\delta}}_{sj} \end{bmatrix} \tag{3.2.10}$$

The matrix in this equation is the same as a mass matrix. Note that this identical to the upper left corner of the matrix in Eq. (2.5.19).

The rates of change of linear and angular momentum for a system such as this, consisting of a set of bodies, have been derived in many texts. See, for example, the derivation by Hughes [6], Section 3.5;

$$\dot{\mathbf{p}} = -\boldsymbol{\omega}^{\times}\mathbf{p} + \mathbf{f} \tag{3.2.11}$$

$$\dot{\mathbf{h}} = -\boldsymbol{\omega}^\times \mathbf{h} - \mathbf{v}^\times \mathbf{p} + \mathbf{g} \tag{3.2.12}$$

When $\mathbf{p}$ and $\mathbf{h}$ are properly defined, these two equations are valid for either a rigid body or a multiple-body system. In particular, Eq. (3.2.11) is valid for the point mass model that is being used to represent the slosh mass. Thus, we can substitute (3.2.6) into (3.2.11) and get

$$m_{sj}\left(\dot{\mathbf{v}} - \dot{\mathbf{r}}_{sj}^\times \boldsymbol{\omega} - \mathbf{r}_{sj}^\times \dot{\boldsymbol{\omega}} + \ddot{\boldsymbol{\delta}}_{sj}\right) = -m_{sj}\boldsymbol{\omega}^\times\left(\mathbf{v} - \mathbf{r}_{sj}^\times \boldsymbol{\omega} + \dot{\boldsymbol{\delta}}_{sj}\right) + \mathbf{f}_{sj} \tag{3.2.13}$$

where $\mathbf{f}_{sj}$ is the force on the slosh mass. From (3.2.1) we have $\dot{\mathbf{r}}_{sj} = \dot{\boldsymbol{\delta}}_{sj}$. From (2.1.5), we can eliminate $\mathbf{v}$ by using the acceleration of the rocket origin in body coordinates:

$$\mathbf{a}_b = \dot{\mathbf{v}} + \boldsymbol{\omega}^\times \mathbf{v} \tag{3.2.14}$$

The resulting slosh equation becomes

$$m_{sj}\left(\mathbf{a}_b - \mathbf{r}_{sj}^\times \dot{\boldsymbol{\omega}} + \ddot{\boldsymbol{\delta}}_{sj}\right) = m_{sj}\left(\boldsymbol{\omega}^\times \mathbf{r}_{sj}^\times \boldsymbol{\omega} - 2\boldsymbol{\omega}^\times \dot{\boldsymbol{\delta}}_{sj}\right) + \mathbf{f}_{sj} \tag{3.2.15}$$

The first and second terms on the right side of this equation can be identified as the centrifugal force and the Coriolis force. For stability analysis of a rocket without a steady rotation rate about any axis, these terms are quite small, but if there is a steady roll rate the Coriolis force may be an important source of coupling between pitch and yaw. The LHS of (3.2.15) is absorbed into the mass matrix. To reduce the computational burden, it may be desirable to create a "quasi-steady" mass matrix, i.e. one that changes slowly as the propellant is depleted. It follows from (3.2.1) that

$$m_{sj}\left(\mathbf{a}_b - \mathbf{b}_{sj}^\times \dot{\boldsymbol{\omega}} + \ddot{\boldsymbol{\delta}}_{sj}\right) = m_{sj}\left(\boldsymbol{\delta}_{sj}^\times \dot{\boldsymbol{\omega}} + \boldsymbol{\omega}^\times \mathbf{r}_{sj}^\times \boldsymbol{\omega} - 2\boldsymbol{\omega}^\times \dot{\boldsymbol{\delta}}_{sj}\right) + \mathbf{f}_{sj} \tag{3.2.16}$$

In this form, all the factors on the LHS multiplying the accelerations $\mathbf{a}_b$, $\dot{\boldsymbol{\omega}}$, and $\ddot{\boldsymbol{\delta}}_{sj}$ are constant, or at least quasi-steady. This arrangement suffers the disadvantage that the first term on the RHS requires the acceleration $\dot{\boldsymbol{\omega}}$, which in a numerical simulation would have to be obtained from a previous time step. Such an approach may create a risk of numerical instability. This term could be omitted with a small loss in accuracy.

The linear and angular momentum equations for the system consisting of the body plus the slosh mass are obtained by substituting (3.2.7) and (3.2.9) into (3.2.11) and (3.2.12) to yield

$$m\dot{\mathbf{v}} - \dot{\mathbf{s}}_{TD}^\times \boldsymbol{\omega} - \mathbf{s}_{TD}^\times \dot{\boldsymbol{\omega}} + m_{sj}\ddot{\boldsymbol{\delta}}_{sj} = -\boldsymbol{\omega}^\times m\mathbf{v} + \boldsymbol{\omega}^\times \mathbf{s}_{TD}^\times \boldsymbol{\omega} - \boldsymbol{\omega}^\times m_{sj}\dot{\boldsymbol{\delta}}_{sj} + \mathbf{f} \tag{3.2.17}$$

$$
\begin{aligned}
&\mathbf{s}_{TD}^{\times}\dot{\mathbf{v}} + \dot{\mathbf{s}}_{TD}^{\times}\mathbf{v} + \dot{\mathbf{I}}_{TD}\boldsymbol{\omega} + \mathbf{I}_{TD}\dot{\boldsymbol{\omega}} \\
&\qquad + m_{sj}\mathbf{r}_{sj}^{\times}\ddot{\boldsymbol{\delta}}_{sj} + m_{sj}\dot{\mathbf{r}}_{sj}^{\times}\dot{\boldsymbol{\delta}}_{sj} = \\
&\qquad - \boldsymbol{\omega}^{\times}\left(\mathbf{s}_{TD}^{\times}\mathbf{v} + \mathbf{I}_{TD}\boldsymbol{\omega} + m_{sj}\mathbf{r}_{sj}^{\times}\dot{\boldsymbol{\delta}}_{sj}\right) \\
&\qquad\qquad - \mathbf{v}^{\times}\left(m\mathbf{v} - \mathbf{s}_{TD}^{\times}\boldsymbol{\omega} + m_{sj}\dot{\boldsymbol{\delta}}_{sj}\right) + \mathbf{g} \qquad (3.2.18)
\end{aligned}
$$

If (3.2.17) is used to create the first row of the mass matrix, the presence of $\mathbf{s}_{TD}$ means this matrix must be updated at each time step. To eliminate this requirement, we use

$$
\dot{\mathbf{s}}_{TD} = m_{sj}\dot{\boldsymbol{\delta}}_{sj} \qquad (3.2.19)
$$

Inserting this plus (3.2.4) and (3.2.14) into (3.2.17) gives

$$
m\mathbf{a}_b - \mathbf{s}_T^{\times}\dot{\boldsymbol{\omega}} + m_{sj}\ddot{\boldsymbol{\delta}}_{sj} = \boldsymbol{\omega}^{\times}\mathbf{s}_T^{\times}\boldsymbol{\omega} + m_{sj}\left(\boldsymbol{\omega}^{\times}\boldsymbol{\delta}_{sj}^{\times}\boldsymbol{\omega} + \boldsymbol{\delta}_{sj}^{\times}\dot{\boldsymbol{\omega}} - 2\boldsymbol{\omega}^{\times}\dot{\boldsymbol{\delta}}_{sj}\right) + \mathbf{f} \qquad (3.2.20)
$$

Let us now examine whether the nonlinear forces acting on the slosh mass turn out to be the same as those in this equation. The Coriolis term $-2m_{sj}\omega^{\times}\dot{\boldsymbol{\delta}}_{sj}$ matches that from (3.2.15), but the centrifugal force term in (3.2.20) is $\boldsymbol{\omega}^{\times}\boldsymbol{\delta}_{sj}^{\times}\boldsymbol{\omega}$ as opposed to $\boldsymbol{\omega}^{\times}\mathbf{r}_{sj}^{\times}\boldsymbol{\omega}$ in (3.2.15). To see why, it is instructive to rewrite the first two terms on the RHS of (3.2.20) using $\mathbf{s}_0$ rather than $\mathbf{s}_T$

$$
m\mathbf{a}_b - \mathbf{s}_T^{\times}\dot{\boldsymbol{\omega}} + m_{sj}\ddot{\boldsymbol{\delta}}_{sj} = \boldsymbol{\omega}^{\times}\mathbf{s}_0^{\times}\boldsymbol{\omega} + m_{sj}\boldsymbol{\omega}^{\times}\mathbf{r}_{sj}^{\times}\boldsymbol{\omega} + m_{sj}\boldsymbol{\delta}_{sj}^{\times}\dot{\boldsymbol{\omega}} - 2m_{sj}\boldsymbol{\omega}^{\times}\dot{\boldsymbol{\delta}}_{sj} + \mathbf{f} \qquad (3.2.21)
$$

In this form, the centrifugal force term now matches that from (3.2.15). This highlights the fact that $\boldsymbol{\omega}^{\times}\mathbf{s}_T^{\times}\boldsymbol{\omega}$ contains part of the centrifugal force from the slosh mass. Although (3.2.21) offers this additional insight, for computational work the preferred equation is (3.2.20).

Turning now to the angular momentum, in (3.2.18) the second term on the LHS is $\dot{\mathbf{s}}_{TD}^{\times}\mathbf{v} = m_{sj}\dot{\boldsymbol{\delta}}_{sj}^{\times}\mathbf{v}$ which cancels a similar term on the RHS. The last term on the LHS is

$$
m_{sj}\dot{\mathbf{r}}_{sj}^{\times}\dot{\boldsymbol{\delta}}_{sj} = m_{sj}\dot{\boldsymbol{\delta}}_{sj}^{\times}\dot{\boldsymbol{\delta}}_{sj} = 0 \qquad (3.2.22)
$$

Thus (3.2.18) simplifies to

$$\mathbf{s}_{TD}^{\times}\dot{\mathbf{v}} + \dot{\mathbf{I}}_{TD}\boldsymbol{\omega} + \mathbf{I}_{TD}\dot{\boldsymbol{\omega}} + m_{sj}\mathbf{r}_{sj}^{\times}\ddot{\boldsymbol{\delta}}_{sj} = - \boldsymbol{\omega}^{\times}\left(\mathbf{s}_{TD}^{\times}\mathbf{v} + \mathbf{I}_{TD}\boldsymbol{\omega} + m_{sj}\mathbf{r}_{sj}^{\times}\dot{\boldsymbol{\delta}}_{sj}\right) - \mathbf{v}^{\times}\left(-\mathbf{s}_{TD}^{\times}\boldsymbol{\omega}\right) + \mathbf{g} \quad (3.2.23)$$

Using the triple vector product, the first and fourth terms on the RHS can be combined:

$$-\boldsymbol{\omega}^{\times}\mathbf{s}_{TD}^{\times}\mathbf{v} - \mathbf{v}^{\times}\boldsymbol{\omega}^{\times}\mathbf{s}_{TD} = \mathbf{s}_{TD}^{\times}\mathbf{v}^{\times}\boldsymbol{\omega} \quad (3.2.24)$$

This term can be put on the LHS, after which we use (3.2.14) to get

$$\mathbf{s}_{TD}^{\times}\mathbf{a}_b + \dot{\mathbf{I}}_{TD}\boldsymbol{\omega} + \mathbf{I}_{TD}\dot{\boldsymbol{\omega}} + m_{sj}\mathbf{r}_{sj}^{\times}\ddot{\boldsymbol{\delta}}_{sj} = -\boldsymbol{\omega}^{\times}\left(\mathbf{I}_{TD}\boldsymbol{\omega} + m_{sj}\mathbf{r}_{sj}^{\times}\dot{\boldsymbol{\delta}}_{sj}\right) + \mathbf{g} \quad (3.2.25)$$

The moment of inertia is

$$\mathbf{I}_{TD} = \mathbf{I}_0 - m_{sj}\mathbf{r}_{sj}^{\times}\mathbf{r}_{sj}^{\times} = \mathbf{I}_T - m_{sj}\left(\mathbf{b}_{sj}^{\times}\boldsymbol{\delta}_{sj}^{\times} + \boldsymbol{\delta}_{sj}^{\times}\mathbf{b}_{sj}^{\times} + \boldsymbol{\delta}_{sj}^{\times}\boldsymbol{\delta}_{sj}^{\times}\right) \quad (3.2.26)$$

where

$$\mathbf{I}_T = \mathbf{I}_0 - m_{sj}\mathbf{b}_{sj}^{\times}\mathbf{b}_{sj}^{\times} \quad (3.2.27)$$

Using $\dot{\mathbf{r}}_{sj} = \dot{\boldsymbol{\delta}}_{sj}$ in the first equation of (3.2.26), the time derivative is

$$\dot{\mathbf{I}}_{TD} = -m_{sj}\dot{\boldsymbol{\delta}}_{sj}^{\times}\mathbf{r}_{sj}^{\times} - m_{sj}\mathbf{r}_{sj}^{\times}\dot{\boldsymbol{\delta}}_{sj}^{\times} \quad (3.2.28)$$

The expression for $\dot{\mathbf{I}}_{TD}$ can be inserted on the RHS of (3.2.25):

$$\mathbf{s}_{TD}^{\times}\mathbf{a}_b + \mathbf{I}_{TD}\dot{\boldsymbol{\omega}} + m_{sj}\mathbf{r}_{sj}^{\times}\ddot{\boldsymbol{\delta}}_{sj} = -\boldsymbol{\omega}^{\times}\mathbf{I}_{TD}\boldsymbol{\omega} + m_{sj}\left(\boldsymbol{\omega}^{\times}\dot{\boldsymbol{\delta}}_{sj}^{\times}\mathbf{r}_{sj} + \dot{\boldsymbol{\delta}}_{sj}^{\times}\mathbf{r}_{sj}^{\times}\boldsymbol{\omega} + \mathbf{r}_{sj}^{\times}\dot{\boldsymbol{\delta}}_{sj}^{\times}\boldsymbol{\omega}\right) + \mathbf{g} \quad (3.2.29)$$

Again using the triple vector product, all three terms multiplying m_{sj} on the RHS can be combined. Thus

$$\mathbf{s}_{TD}^{\times}\mathbf{a}_b + \mathbf{I}_{TD}\dot{\boldsymbol{\omega}} + m_{sj}\mathbf{r}_{sj}^{\times}\ddot{\boldsymbol{\delta}}_{sj} = -\boldsymbol{\omega}^{\times}\mathbf{I}_{TD}\boldsymbol{\omega} - 2m_{sj}\mathbf{r}_{sj}^{\times}\boldsymbol{\omega}^{\times}\dot{\boldsymbol{\delta}}_{sj} + \mathbf{g} \quad (3.2.30)$$

This should be compared to (2.4.8). The second term on the RHS of this can be identified as the Coriolis torque. Note that all of the centrifugal torque is contained within the Euler coupling term $\boldsymbol{\omega}^{\times}\mathbf{I}_{TD}\boldsymbol{\omega}$. These two terms comprise part of the nonlinear torque $\mathbf{g}_{NL}$. If the dynamic equation set includes engine equations, there will be a nonlinear contribution from the engines. This is discussed in the next chapter.

Gravity will affect both sides of this equation in equal measure, with no net effect. That is, $\mathbf{a}_b$ contains the effect of gravity, but so does the external torque $\mathbf{g}$. The external torque due to gravity is discussed in Section 5.1.

Eq. (3.2.30) has been arranged so that all the accelerations are on the LHS. If (3.2.30) is used to create the second row of the mass matrix, it is found that all of the resulting entries in this row vary as the state vector varies, so the mass matrix changes at every time step. As with the slosh and translation equations, it may be desirable to create a version of this that can be used to create a quasi-steady mass matrix. To do this, one uses (3.2.1), (3.2.4) and (3.2.26) to decompose the terms on the LHS into larger steady parts and smaller unsteady parts. The latter are moved to the RHS. If this is done, we arrive at the following rotation equation:

$$\begin{aligned}\mathbf{s}_T^{\times}\mathbf{a}_b + \mathbf{I}_T\dot{\boldsymbol{\omega}} + m_{sj}\mathbf{b}_{sj}^{\times}\ddot{\boldsymbol{\delta}}_{sj} = &-\boldsymbol{\omega}^{\times}\mathbf{I}_T\boldsymbol{\omega} \\ &- 2m_{sj}\mathbf{r}_{sj}^{\times}\boldsymbol{\omega}^{\times}\dot{\boldsymbol{\delta}}_{sj} + \mathbf{g} - m_{sj}\boldsymbol{\delta}_{sj}^{\times}\ddot{\boldsymbol{\delta}}_{sj} \\ &+ m_{sj}\Big[-\boldsymbol{\delta}_{sj}^{\times}\mathbf{a}_b + \left(\mathbf{b}_{sj}^{\times}\boldsymbol{\delta}_{sj}^{\times} + \boldsymbol{\delta}_{sj}^{\times}\mathbf{b}_{sj}^{\times} + \boldsymbol{\delta}_{sj}^{\times}\boldsymbol{\delta}_{sj}^{\times}\right)\dot{\boldsymbol{\omega}} \\ &\quad + \boldsymbol{\omega}^{\times}\left(\mathbf{b}_{sj}^{\times}\boldsymbol{\delta}_{sj}^{\times} + \boldsymbol{\delta}_{sj}^{\times}\mathbf{b}_{sj}^{\times} + \boldsymbol{\delta}_{sj}^{\times}\boldsymbol{\delta}_{sj}^{\times}\right)\boldsymbol{\omega}\Big]\end{aligned} \quad (3.2.31)$$

In this expression, all the terms that have been created as the result of switching the subscripts from TD to T appear on the third and fourth lines. Of these, the first is by far the most significant, since $\mathbf{a}_b$ includes the effect of steady-state thrust in the x direction. This first term is the linear "slosh offset" that is illustrated in Fig. 3.3, and further discussed in Section 5.2. The next three terms (from the first set of parentheses) are the torques produced from the combination of a fuel offset and an angular acceleration. If both $\dot{\boldsymbol{\omega}}$ and $\boldsymbol{\delta}_{sj}$ are considered small quantities, these are either second or third-order effects. They are included here for the sole purpose of illustrating the magnitude of the terms that should be dropped if a quasi-steady mass matrix is to be used. Retaining these unsteady acceleration terms on the RHS may create a risk of numerical instability. The same comment applies to the $m_{sj}\boldsymbol{\delta}_{sj}^{\times}\ddot{\boldsymbol{\delta}}_{sj}$ term on the second line. If there is a genuine need to accurately capture the effect of such small terms, it is preferable to abandon a quasi-steady approach and recompute the mass matrix at each time step. The remaining terms, from the set of parentheses in the fourth line, are only included for the sake of completeness, since they are tiny – either third order or fourth order. They have some similarity in nature to the centrifugal force in the system translation equation. Based on the centrifugal force from (3.2.15), we might expect the centrifugal torque to be $m_{sj}\mathbf{r}_{sj}^{\times}\boldsymbol{\omega}^{\times}\mathbf{r}_{sj}^{\times}\boldsymbol{\omega}$, but since the bulk of this expression is contained within $\boldsymbol{\omega}^\times \mathbf{I}_T \boldsymbol{\omega}$, we end up with something different, and smaller in magnitude. See Appendix B for the resolution of an apparent contradiction between $m_{sj}\mathbf{r}_{sj}^\times \boldsymbol{\omega}^\times \mathbf{r}_{sj}^\times \boldsymbol{\omega}$ and the contribution of the slosh mass to the Euler coupling term.

It is useful to write

$$\mathbf{a}_b = \mathbf{a}_{b0} + \mathbf{a}_b' \tag{3.2.32}$$

where, using Eq. (3.1.1),

$$\mathbf{a}_{b0} = \begin{bmatrix} \bar{g} \\ 0 \\ 0 \end{bmatrix} \tag{3.2.33}$$

is the sensed acceleration. The remainder is given by $\mathbf{a}_b'$ (including gravity). If we drop the third- and fourth-order terms from (3.2.31), plus all of the unsteady acceleration terms on the RHS (i.e. $\ddot{\boldsymbol{\delta}}_{sj}$ and $\dot{\boldsymbol{\omega}}$ terms), we are left with

$$\mathbf{s}_T^\times \mathbf{a}_b + \mathbf{I}_T \dot{\boldsymbol{\omega}} + m_{sj}\mathbf{b}_{sj}^\times \ddot{\boldsymbol{\delta}}_{sj} = -\boldsymbol{\omega}^\times \mathbf{I}_T \boldsymbol{\omega} - 2m_{sj}\mathbf{r}_{sj}^\times \boldsymbol{\omega}^\times \dot{\boldsymbol{\delta}}_{sj} + \mathbf{g} - m_{sj}\boldsymbol{\delta}_{sj}^\times \mathbf{a}_{b0} \tag{3.2.34}$$

This is recommended form for the system angular momentum for use with a quasi-steady mass matrix (sloshing fuel, locked nozzle). Chapters 4 and 5 discuss the incorporation of nonlinear terms from the nozzle motion.

3.3 Hydrodynamic model in the FEM

A sophisticated finite element package can represent a combination of structural elements and hydrodynamic models. For the problem of simulating the dynamics of a rocket, such a package can include the slosh modes of liquid fuel tanks, modeled using special fluid elements in the finite element analysis code. Traditional rocket analysis has not had such tools available, so the procedure was to compute a FEM that did not include any slosh motion, and to incorporate this into a dynamic analysis in which the slosh motion is modeled explicitly, as described in Chapter 2. With the expanded capability of finite element modeling, the question naturally arises as to what are the advantages and disadvantages of including hydrodynamic motion in the FEM.

The initial equations for a structural dynamic model are written in physical coordinates, also called physical degrees of freedom (DOF's). At the time the FEM is created, selected physical DOF's can be constrained or "locked." It is at least theoretically possible to lock the DOF's associated with the hydrodynamic motion in such a way that the fluid elements do not add to the structural stiffness. In a modern finite element model, this would involve locking thousands of DOF's. Once the appropriate constraints for each DOF have been defined, it is standard practice to set up the equations as a generalized eigenvalue problem. The result is a set of decoupled modal equations written in terms of generalized coordinates. In this decoupled form, selected modes can be truncated out of (removed from) the model. Thus physical coordinates are locked at the time the FEM is created, and generalized coordinates are truncated after the FEM is created. In what follows, the phrase "not included" means the motion is either locked or truncated. If there is some motion that is considered important, such as the lowest-frequency slosh mode, and this motion is not included in the FEM, then it must be explicitly included in the dynamic model, as described in Chapter 2.

As a practical matter, there must be some truncation. The term "truncation" normally refers to removing high-frequency modes from the FEM. In the simplest case, a cutoff frequency is defined, and any modes having frequencies exceeding this threshold are removed. In the present discussion, however, truncation refers to removing the low-frequency modes that are associated with slosh motion.

The issue becomes deciding which slosh motions should be included in the FEM and which should not be included. Some insight can be gained by considering two possible extremes. Including all the slosh modes means the FEM handles all the slosh dynamics. This means that flex and slosh dynamics emerge in one diagonal set of equations that are decoupled from the rigid-body dynamics. The simplicity of this situation can be enhanced by assuming (1) the state vector does not include engine DOF's (the engine motion is prescribed), (2) the origin is at the center of mass, and (3) the inertia matrix is diagonal. Under these simplifications, the overall mass matrix becomes diagonal, i.e. there is no coupling whatsoever. However, there is a price for all this simplicity. It is not possible to identify any mode as purely a slosh mode. There may be some modes that exhibit more slosh motion than others, but the modes will represent combinations of slosh motion and structural motion. Thus it is not obvious how to turn slosh on or off for diagnostic purposes, or how to adjust the slosh mode frequencies for conditions other than at 1 g.[1]

[1] This is a fundamental limitation of any sloshing or *hydroelastic* modes included in a finite element model. Even though the modes are orthogonal, the overall similarity transforma-

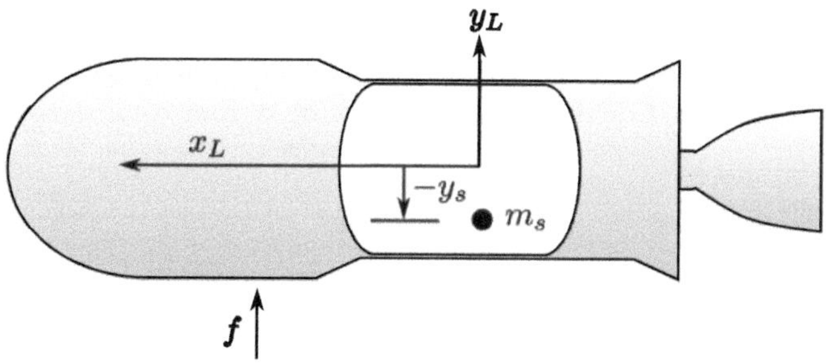

Figure 3.4 Slosh mass locked in the FEM and free in the dynamic equation.

At the opposite extreme, all the slosh modes can be explicitly modeled in the dynamics, and not included in the FEM. This gives the dynamicist more visibility into, and control over, the slosh model for each tank. For example, the sloshing frequency can be adjusted based upon the vehicle axial acceleration, slosh damping can be specified that is different from the structural damping, and individual slosh motions can be turned on or off.

Aside from these issues, it may be important to understand certain details about the mathematics. All of the analyses below employ what called an integrated body model. That is, the FEM includes the slosh mass, which may or may not be locked.

Case 1 - locked vs. unlocked

This case focuses on the decision to lock the slosh motion during creation of the FEM. As mentioned above, locking a hydrodynamic model may involve removing thousands of DOF's. For the following discussion, however, it may help to think in terms of a simple slosh model. A very simple rocket is analyzed using two methods. Fig. 3.4 shows the diagram of the rocket used for a locked-in-the-FEM analysis, i.e., the traditional approach. The rocket body is rigid. There is a coordinate frame whose origin is located at the equilibrium center of mass, on the centerline. There is a single slosh mass attached by a spring (not shown) to the rocket body. The equilibrium position of the slosh mass is at the center of mass, thus the position shown is

tion between the physical and generalized coordinates is not preserved if any eigenvalues (modal frequencies) are modified. Physically, this means that the coupled response of a fluid-structural dynamic system at conditions other than 1 g manifests as a change in the response of *all* of the modes, since the structural response depends on the fluid response, and vice-versa. This issue is discussed later in this Chapter.

non-equilibrium. There is an external force f that acts on the rocket body, due to environmental factors such as wind. The engine is rigidly attached and is considered part of the rocket body. The rocket total mass is defined as

$$m_T = m_0 + m_s \tag{3.3.1}$$

where m_0 is the rocket body mass and m_s is the slosh mass. In both methods, the FEM contains only two elements – the rocket body and the slosh mass. In the first method, the slosh mass is locked, so all that remains in the FEM is a rigid body containing all the mass. This eliminates the need for a bending equation. The slosh motion is modeled explicitly in a separate equation.

Let y_L be the inertial position of the origin, let y_s be the slosh position relative to the origin, and let k_s be the slosh spring constant. The subscript L stands for "locked" in the FEM. Assuming that the external force f acts at the center of mass, there is nothing to cause any yaw motion. Thus y_L and y_s are the only degrees of freedom to be included. The matrix equation for the present problem can be obtained by reducing the more complicated equations of Chapter 2 to these two remaining degrees of freedom. One obtains

$$\begin{bmatrix} m_T & m_s \\ m_s & m_s \end{bmatrix} \begin{bmatrix} \ddot{y}_L \\ \ddot{y}_s \end{bmatrix} = \begin{bmatrix} f \\ -k_s y_s \end{bmatrix} \tag{3.3.2}$$

Consider now the case in which f is a step function. At the first instant, $y_s = 0$. Solving the second equation of (3.3.2) gives

$$\ddot{y}_L(0) = -\ddot{y}_s(0) \tag{3.3.3}$$

Remember, $\ddot{y}_s$ is the relative acceleration. The inertial acceleration of the slosh mass at this instant is zero. Inserting this into the first equation of (3.3.2) and using (3.3.1) gives

$$(m_0 + m_s)\ddot{y}_L(0) - m_s \ddot{y}_L(0) = f \tag{3.3.4}$$

Thus at the first instant we find

$$\ddot{y}_L(0) = f/m_0 \tag{3.3.5}$$

For the second method, we unlock the slosh mass in the FEM, resulting in the mode shape shown in Fig. 3.5. The origin x_U, y_U of the coordinate frame of this diagram is at the instantaneous center of mass of the

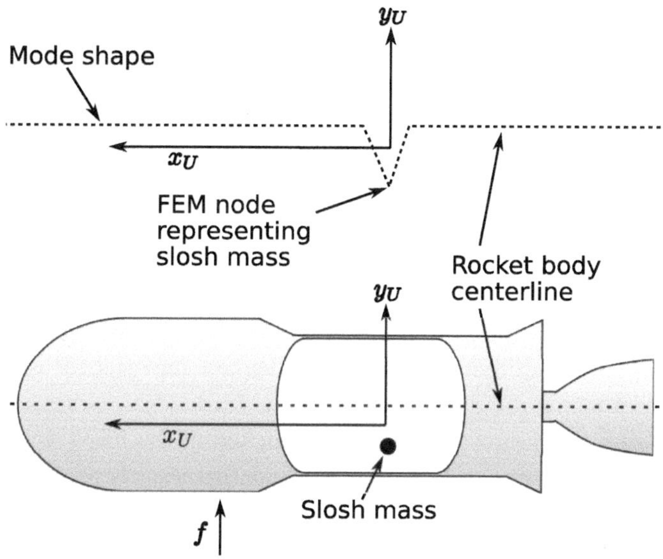

Figure 3.5 Rocket and mode shape when slosh mass is unlocked in the FEM.

rocket/slosh system. Fig. 3.5 also shows how this frame relates to the rocket body. The origin of this diagram is not the same as that in Fig. 3.4; it is displaced in the y direction.

The matrix equation using the unlocked method is

$$\begin{bmatrix} m_T & 0 \\ 0 & 1 \end{bmatrix}\begin{bmatrix} \ddot{y}_U \\ \ddot{\eta} \end{bmatrix} = \begin{bmatrix} f \\ -\Omega_B^2 \eta \end{bmatrix} \tag{3.3.6}$$

where η is the modal amplitude, and the mode frequency is given by

$$\Omega_B^2 = \frac{k_s}{m_s} \tag{3.3.7}$$

It can be seen from (3.3.6) that the lateral acceleration at all times is given by

$$\ddot{y}_U = \frac{f}{m_T} \tag{3.3.8}$$

Note that this is not the same as the acceleration given by (3.3.5). If the locked method is used, the measured acceleration at any sensor location on the rocket is given by $\ddot{y}_L$. If the unlocked method is used, $\ddot{y}_U$ gives a long-term average acceleration, assuming f remains constant. An additional term must be added to get the actual acceleration if there is modal motion. For this simple problem, in which there is no angular rotation, the sensed acceleration is the same everywhere on the rocket body and is given by

$$\ddot{y} = \ddot{y}_U + \psi_{11y}\ddot{\eta} \tag{3.3.9}$$

Here, following the notation of Chapter 2, ψ_{11y} is the mode shape in the y direction of the first mode at the location of the first slosh mass, as shown in Fig. 3.5.

Case 1 shows that the analysis using the locked approach uses a different coordinate frame than that used for the unlocked approach. It is important to understand that some terminology used in traditional rocket analysis becomes misleading when a slosh mass is unlocked in the FEM. In particular, the x_U axis herein corresponds to the "undeformed centerline" of traditional rocket terminology, even though it is not on the centerline. This phrase is perfectly appropriate when the slosh mass is locked. When the slosh mass is unlocked in the FEM, the x_U axis passes through the center of mass of the entire rocket including the liquid fuel in its displaced position. This axis does not coincide with the undeformed centerline of the structure, assuming the word structure does not include the liquid fuel. For this unlocked case, the x_U axis is stationary in the absence of external forces, even as the slosh mass oscillates.

Case 2 - truncated vs. not truncated

For this case, we alter the conditions from those of the previous case in two ways. First, a rocket body with one flexible mode is considered, rather than the rigid body of Case 1. Second, it is assumed that a FEM including a slosh mass has already been created, and the decision must be made as to whether to truncate the "slosh mode" from the FEM and explicitly model the slosh motion in the dynamic equations.

For simplicity, the rocket body in this case is considered as a uniform beam. The FEM analysis will yield two modes. In one mode, the slosh mass moves in the same direction as the center of the beam, and in the second mode, the slosh mass moves in the opposite direction (Fig. 3.6). The latter mode will be at a higher frequency. It is normally the case that the lowest-frequency slosh mode is at a lower frequency than the lowest-frequency bending mode. Thus the first mode in Fig. 3.6 would be identified as the slosh mode, and the second mode would be identified as the lowest-frequency bending mode.

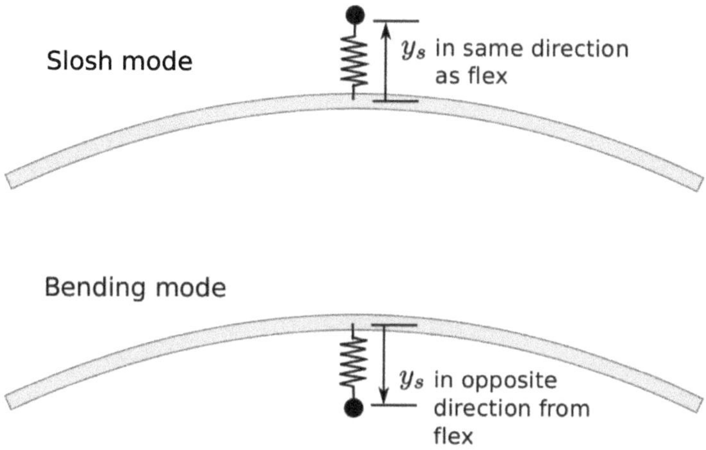

Figure 3.6 Simplified uniform beam with sloshing mass.

It is easiest to examine the un-truncated method first, even though this is the reverse of the order in which the methods were examined in Case 1. The matrix equation for the un-truncated method is

$$\begin{bmatrix} m_T & 0 & 0 \\ 0 & 1 & 0 \\ 0 & 0 & 1 \end{bmatrix} \begin{bmatrix} \ddot{y} \\ \ddot{\eta}_1 \\ \ddot{\eta}_2 \end{bmatrix} = \begin{bmatrix} f \\ -\Omega_{B1}^2 \eta_1 \\ -\Omega_{B2}^2 \eta_2 \end{bmatrix} \tag{3.3.10}$$

This is obtained by adding one more flex equation to (3.3.6). Assuming the modes are numbered in the order of ascending frequency, η_1 becomes assigned to the slosh mode, and η_2 is assigned to the bending mode.

Let us define the "isolated slosh frequency" as the frequency that is obtained when all the other degrees of freedom are locked. We can think of this as the frequency that would be obtained from ground experiments using a rigid rocket with the correct geometry, fill level, and so on. (One can see that a complete rocket would not actually be necessary – a rigid tank would suffice.) In the case of a flexible rocket, a finite element analysis will take into account the coupling to the structural modes and produce an Ω_{B1} that is shifted from the isolated slosh frequency. It is also true that Ω_{B2} is shifted from the isolated bending frequency, i.e. what would be obtained from a model in which the slosh motion is locked. The result is correct as long as the FEM is produced using an axial acceleration that matches the axial acceleration of interest. However, this is typically not the case; if the FEM is produced under some nominal acceleration g_0 (typically 1 g), the slosh frequency must be scaled to match the actual acceleration $\bar{g}$. Using an overbar to represent the scaled frequency, it follows that

$$\bar{\Omega}_{B1}^2 = \Omega_{B1}^2 \frac{\bar{g}}{g_0} \tag{3.3.11}$$

This is not perfectly accurate. The frequencies of both modes are derived from a slosh component, that should be scaled, and a structural component, that should not be scaled. This is most important for fundamental sloshing modes of large tanks that are strongly coupled with the structural dynamics. Thus, consideration might be given to the notion of weighted scaling. That is, weights are assigned based on the portion of the motion that is slosh. Any such scaling must be physically consistent with the solution of the finite element model, and is likely impractical in most cases.

In the truncated approach, one particular mode must be identified as the slosh mode and removed (truncated) from the FEM. The slosh dynamics are then explicitly added to the dynamic model, as described in Chapter 2. The matrix equation becomes

$$\begin{bmatrix} m_T & m_s & 0 \\ m_s & m_s & m_s\psi_{12y} \\ 0 & m_s\psi_{12y} & 1 \end{bmatrix} \begin{bmatrix} \ddot{y} \\ \ddot{y}_s \\ \ddot{\eta}_2 \end{bmatrix} = \begin{bmatrix} f \\ -m_s\Omega_{B1}^2 y_s \\ -\Omega_{B2}^2\eta_2 \end{bmatrix} \tag{3.3.12}$$

Here, following the setup in Chapter 2, we arrange the equations in the order translation, slosh, and bending. The subscript numbering has not been adjusted for truncation, i.e. the subscript 2 is still used to represent the bending mode, even though it would normally be thought of as the first bending mode. Thus η_2 is the same in (3.3.10) and (3.3.12). ψ_{12y} is the mode shape at the location of the first (and only) slosh mass of this "second" mode.

In the following discussion, it is assumed that we use the frequencies that are extracted from the FEM, which will have the same scaling issue just discussed. Since the slosh and bending equations are now coupled, the two frequencies that emerge from the coupled analysis will be shifted from the isolated slosh and bending frequencies. In other words, they will be shifted twice – once by the FEM, and again by the dynamic analysis. For the simplest case, in which the axial acceleration used while modeling the FEM matches the axial acceleration of interest, Ω_{B1} and Ω_{B2} would be correct as obtained from the FEM – they should not be shifted again. A similar situation arises in the engine interaction problem described in

Chapter 6. There, the engine "pendulum mode" is described. The notion is put forth that this could be removed from the FEM and introduced into the dynamics. The same frequency shift issue arises. Chapter 6 concludes that a more rigorous approach is to create a FEM in which the engine has been locked. Similarly, in the case of slosh motion, locking the slosh mass in the FEM or removing it entirely, avoids uncertainties in the analysis.

The truncation approach is the only one of the four examples in which an error is introduced that is present even for the idealized conditions described herein. It is tempting to eliminate the coupling terms ψ_{12y} in (3.3.12) to remove (or actually just reduce) the frequency shifting problem. Doing this with the lowest-frequency slosh mode is not recommended. It is better to use an independent slosh model, as described below, or else abandon the idea of truncation and use the original formulation (3.3.10).

The above analysis also suggests the notion of "reverse-shifting" the mode frequencies. That is, the frequency of the first slosh mode from the FEM is reverse-shifted before being inserted into the dynamic model, so that the dynamic model itself ends up shifting it back to the correct frequency. That is one way to avoid the problem of double shifting. The reverse-shifted frequency would in theory be very close to the isolated slosh frequency discussed above. This notion is relatively straightforward if there is only one tank but quickly becomes impractical with multiple modes and tanks. More importantly, it is difficult to verify the physical consistency of the coupled system when altering the finite element model data after they have been produced in a modal form.

There are some cases where the use of slosh modes directly from the FEM might be advantageous, and scaling can be performed without a significant loss of accuracy. Relatively small payload tanks or other liquid dynamics, such as propellant feedlines, can be included in the FEM, reducing the complexity of the simulation by requiring only large propellant tanks to be modeled in the dynamic equations. Such small tanks are usually not strongly coupled to the structural dynamics, especially if their frequencies can be shown to be separated from major bending modes.

Damping

The above discussion describes how the frequencies shift due to the coupling between slosh and bending. There is also some shifting in the amount of damping. When a slosh mode and a bending mode are dynamically coupled, the damping of one mode will increase, and that in the other will decrease relative to the uncoupled situation. This is problematic because both the slosh motion and the bending motion have very low damping, and the slosh damping depends on both the container geometry and the axial acceleration. Indeed, a primary challenge of the design of the control system is to guarantee that none of these lightly damped modes can create an instability.

Fuel tanks are typically fitted with ring-shape baffles that provide damping of the slosh motion. These baffles usually are designed iteratively with the dynamics and control analysis in order to minimize the required baffle mass. If the traditional approach is used (no slosh motion in the FEM), the FEM does not have to be changed during such analysis cycles. All the parameters are in the slosh equations, and all responsibility for inserting damping into the slosh equations falls on the dynamicist. If, on the other hand, the FEM contains a slosh model, the only practical approach for interactive design is truncation, which partially achieves the advantages just described for the traditional approach.

Independent slosh model

For the lowest-frequency slosh motion, it might be valid to assume that the tank is effectively rigid, in which case one can consider using an independent model to derive the lowest slosh frequency, which is then inserted into the dynamic equations as in Chapter 2. If one can make the assumption that the shifts in the bending mode frequencies are small for the higher-frequency slosh modes, these modes can be obtained from the FEM. This approach has the advantage that the scaling for the lowest frequency slosh mode becomes straightforward. At the same time, high-frequency slosh mode effects are captured. However, identifying and scaling these higher mode slosh frequencies is problematic. One way to minimize this uncertainty is to create the FEM using an axial acceleration that matches the nominal flight acceleration. Thus for the nominal case, scaling becomes unnecessary. Whether some kind of scaling would be necessary to cover Monte-Carlo dispersions is a matter that would have to be analyzed.

3.4 Summary of hydrodynamic models

In order to simplify the analysis, the example in Case 1 above uses a rigid rocket. If instead a rocket body with one flexible mode is used, then the unlocked choice of Case 1 becomes the same as the non-truncated choice of Case 2. Thus there are only three choices available: Unlocked (leave the hydrodynamic modes in the FEM), lock the hydrodynamic modes, or truncate the (unlocked) hydrodynamic modes. The traditional rocket analysis has the slosh masses locked in the FEM and explicitly modeled in the dynamics. Of the three choices, this is the one with the least theoretical difficulty and the greatest control over acceleration scaling and other issues. However, it essentially negates the notion of using a hydrodynamic model in the FEM. If the choice is made to unlock the slosh mass, then acceleration scaling either introduces errors or requires a complicated system of weighting. Truncating the model is even worse – the frequency shift due to coupling ends up being applied twice, and the results are not accurate even if the axial acceleration is the same as what is assumed during creation of the FEM. This creates a dilemma. From the standpoint of frequency shift, truncation is undesirable, but from the standpoint of damping, truncation may be unavoidable.

A good argument can be made that the ideal approach would have both kinds of FEM. A finite element model that includes hydrodynamics offers significant advantages for calculating loads, and might be used for spot checks of the dynamics. A traditional approach that has the slosh masses locked in the FEM has the least theoretical difficulty for dynamics and control purposes.

CHAPTER 4

Pendulum model

The spring slosh model of the previous chapter is the easiest to analyze and is the one that is usually the most convenient. However, it does suffer from the disadvantage that there is no restriction on the slosh displacement. Indeed, under conditions of low thrust, the slosh spring constant can be very small and it is entirely possible for the model to predict that the slosh mass travels far outside the boundaries of the rocket body. Such problems are less likely with a pendulum model. The full nonlinear equations of a system consisting of a central body with an appendage attached by a universal joint are presented. The appendage can either be a rocket nozzle or a sloshing fuel mass. The effects of a rotating body frame are included.

Notation

The notation in this chapter differs from that in the other chapters of this book in a few instances. In particular, the symbol $\mathbf{d}$ is used to denote the first moment of inertia, and the symbol $\boldsymbol{\rho}$ is used to represent what was previously $\mathbf{r}$ in Fig. 2.1.

This chapter is the first to employ the frame notation described in Chapter 1; the symbol $\mathscr{F}_n$ (with a subscript) denotes coordinate frame n. Thus $\mathscr{F}_1$ is the coordinate frame of body 1.

4.1 General pendulum model

The analysis begins with a single generalized appendage (body 1) connected to the rocket body (body 0). The problem thus reduces to that for a simple system consisting of two bodies. The analysis in this section is very similar to the one found in Section 3.6 of the textbook by Hughes [6]. However, the present derivation uses a different path to get to results that are in a simpler, but equivalent, form. Let $\mathbf{C}_{01}$ be the matrix that transforms a vector expressed in $\mathscr{F}_1$ to the same vector in $\mathscr{F}_0$. Hughes provides a matrix identity that is used repeatedly below:

$$(\mathbf{C}_{01}\mathbf{u})^{\times} = \mathbf{C}_{01}\mathbf{u}^{\times}\mathbf{C}_{10} \tag{4.1.1}$$

Dynamics and Simulation of Flexible Rockets
https://doi.org/10.1016/B978-0-12-819994-7.00009-1

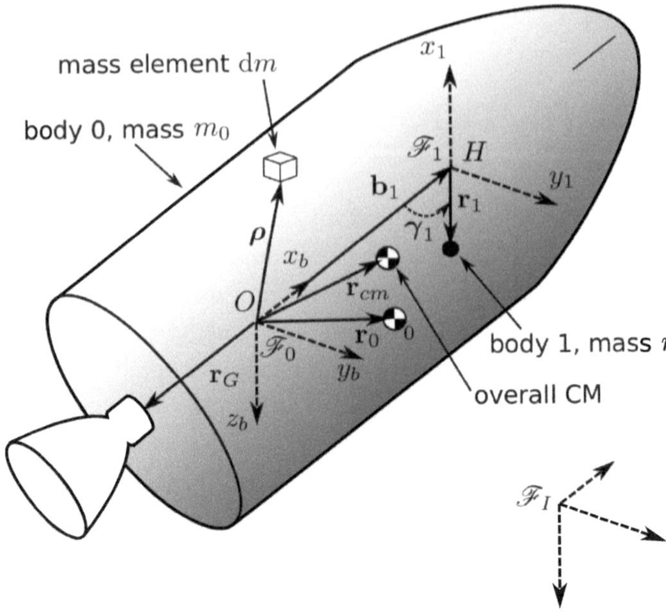

Figure 4.1 Rocket with a pendulum connected by a hinge H.

where $\mathbf{u}$ is any vector expressed in $\mathscr{F}_1$. In Fig. 4.1, $\mathscr{F}_0$ is fixed to body 0, with the origin at point O, and $\mathscr{F}_1$ is fixed to body 1, with its origin at the hinge point H. The definitions of the quantities used to describe the relative dynamic configuration of the two bodies are given in Table 4.1.

Table 4.1 Slosh pendulum analysis notation.

$\vec{\mathbf{v}}$	absolute velocity of point O
$\vec{\mathbf{v}}_H$	absolute velocity of point $H = \vec{\mathbf{v}} + \vec{\omega} \times \vec{\mathbf{b}}_1$
$\vec{\omega}$	absolute angular rate of body 0
$\vec{\omega}_1$	absolute angular rate of body 1
$\vec{\omega}_{r1} = \vec{\omega}_1 - \vec{\omega}$	angular rate of body 1 relative to body 0

Because of the fact that separate frames are used to describe body 0 and body 1, it is important to begin the analysis using vector notation. The reader may wish to review the brief discussion of the difference between a vector and a column matrix in Chapter 1.

Vector $\vec{\mathbf{b}}_1$ defines the location of H in Frame 0. Vectors $\vec{\mathbf{r}}_0$ and $\vec{\mathbf{r}}_1$ define the locations of the mass centers of bodies 0 and 1 relative to O and H

respectively. $\vec{\boldsymbol{\rho}}_0$ is a vector from O to an element of mass dm within body 0. $\vec{\boldsymbol{\rho}}_1$ is a similar vector from H to the mass elements within body 1.

The mass properties consist of the total system mass, $m_T = m_0 + m_1$ and the first moments of inertia of body 0, body 1, and the integrated system about O, H, and O, given by $\vec{\mathbf{d}}_0$, $\vec{\mathbf{d}}_1$, and $\vec{\mathbf{d}}$, respectively. These quantities are defined as

$$\vec{\mathbf{d}}_0 = \int_0 \vec{\boldsymbol{\rho}}_0 \, dm = m_0 \vec{\mathbf{r}}_0 \tag{4.1.2}$$

$$\vec{\mathbf{d}}_1 = \int_1 \vec{\boldsymbol{\rho}}_1 \, dm = m_1 \vec{\mathbf{r}}_1 \tag{4.1.3}$$

$$\vec{\mathbf{d}} = \vec{\mathbf{d}}_0 + \vec{\mathbf{d}}_1 + m_1 \vec{\mathbf{b}}_1 \tag{4.1.4}$$

The second moment of inertia dyadics of body 0 and body 1 about O and H respectively are

$$\vec{\vec{\mathbf{J}}}_0 = \int_0 \left(\rho_0^2 \vec{\vec{\mathbf{1}}} - \vec{\boldsymbol{\rho}}_0 \vec{\boldsymbol{\rho}}_0 \right) dm \tag{4.1.5}$$

$$\vec{\vec{\mathbf{J}}}_1 = \int_0 \left(\rho_1^2 \vec{\vec{\mathbf{1}}} - \vec{\boldsymbol{\rho}}_1 \vec{\boldsymbol{\rho}}_1 \right) dm \tag{4.1.6}$$

Here $\vec{\vec{\mathbf{1}}}$ is the identity dyadic. The linear momenta of body 0 and body 1 are

$$\vec{\mathbf{p}}_0 = m_0 \vec{\mathbf{v}} - \vec{\mathbf{d}}_0 \times \vec{\boldsymbol{\omega}} \tag{4.1.7}$$

$$\vec{\mathbf{p}}_1 = m_1 \vec{\mathbf{v}}_H - \vec{\mathbf{d}}_1 \times \vec{\boldsymbol{\omega}}_1 \tag{4.1.8}$$

The expressions for the absolute velocity of point H and the angular rate of body 1 relative to body 0 can be inserted to yield the expression for the linear momentum of body 1;

$$\vec{\mathbf{p}}_1 = m_1 \vec{\mathbf{v}} - \left(m_1 \vec{\mathbf{b}}_1 + \vec{\mathbf{d}}_1 \right) \times \vec{\boldsymbol{\omega}} - \vec{\mathbf{d}}_1 \times \vec{\boldsymbol{\omega}}_{r1}. \tag{4.1.9}$$

Adding (4.1.7) and (4.1.9), we obtain the total linear momentum of the system

$$\vec{\mathbf{p}} = m_T \vec{\mathbf{v}} - \vec{\mathbf{d}} \times \vec{\boldsymbol{\omega}} - \vec{\mathbf{d}}_1 \times \vec{\boldsymbol{\omega}}_{r1} \tag{4.1.10}$$

The angular momentum is a bit more complicated. Let

$$\vec{\mathbf{h}}_0 \equiv \text{angular momentum of body 0 about } O \tag{4.1.11}$$

$$\vec{\mathbf{h}}_1 \equiv \text{angular momentum of body 1 about } H \tag{4.1.12}$$

For body 0, write

$$\vec{\mathbf{h}}_0 = \int \vec{\rho}_0 \times \left(\vec{\mathbf{v}} + \vec{\omega} \times \vec{\rho}_0\right) \, dm \tag{4.1.13}$$

By using (4.1.2) and (4.1.5) and the vector identity $\vec{\rho}_0 \times \left(\vec{\omega} \times \vec{\rho}_0\right) = \left(\vec{\rho}_0 \cdot \vec{\rho}_0\right)\vec{\omega} - \vec{\rho}_0\left(\vec{\rho}_0 \cdot \vec{\omega}\right)$ this becomes

$$\vec{\mathbf{h}}_0 = \vec{\mathbf{d}}_0 \times \vec{\mathbf{v}} + \vec{\mathbf{J}}_0 \cdot \vec{\omega} \tag{4.1.14}$$

The absolute angular momentum about H of body 1 can be obtained directly from (4.1.14), by changing the subscript to 1 and inserting the motion variables as seen at H:

$$\vec{\mathbf{h}}_1 = \vec{\mathbf{d}}_1 \times \vec{\mathbf{v}}_H + \vec{\mathbf{J}}_1 \cdot (\vec{\omega} + \vec{\omega}_{r1}) \tag{4.1.15}$$

Inserting the expression for the absolute velocity of point H from Table 4.1 into (4.1.15) and rearranging,

$$\vec{\mathbf{h}}_1 = \vec{\mathbf{d}}_1 \times \vec{\mathbf{v}} - \vec{\mathbf{d}}_1 \times \left(\vec{\mathbf{b}}_1 \times \vec{\omega}\right) + \vec{\mathbf{J}}_1 \cdot (\vec{\omega} + \vec{\omega}_{r1}) \tag{4.1.16}$$

The total angular momentum about O may be found as follows:

$$\vec{\mathbf{h}} = \vec{\mathbf{h}}_0 + \vec{\mathbf{h}}_1 + \vec{\mathbf{b}}_1 \times \vec{\mathbf{p}}_1 \tag{4.1.17}$$

After substituting from (4.1.9), (4.1.14), and (4.1.16), this can be expressed as

$$\begin{aligned}\vec{\mathbf{h}} = \left(\vec{\mathbf{d}}_0 + \vec{\mathbf{d}}_1\right) \times \vec{\mathbf{v}} + \left(\vec{\mathbf{J}}_0 + \vec{\mathbf{J}}_1\right) \cdot \vec{\omega} \\ - \vec{\mathbf{d}}_1 \times \left(\vec{\mathbf{b}}_1 \times \vec{\omega}\right) + \vec{\mathbf{J}}_1 \cdot \vec{\omega}_{r1} \\ + \vec{\mathbf{b}}_1 \times \left[m_1 \vec{\mathbf{v}} - \left(m_1 \vec{\mathbf{b}}_1 + \vec{\mathbf{d}}_1\right) \times \vec{\omega} - \vec{\mathbf{d}}_1 \times \vec{\omega}_{r1}\right]\end{aligned} \tag{4.1.18}$$

Using (4.1.4) and rearranging

$$\begin{aligned}\vec{\mathbf{h}} = \vec{\mathbf{d}} \times \vec{\mathbf{v}} + \left(\vec{\mathbf{J}}_0 + \vec{\mathbf{J}}_1\right) \cdot \vec{\omega} - \vec{\mathbf{d}}_1 \times \left(\vec{\mathbf{b}}_1 \times \vec{\omega}\right) \\ - \vec{\mathbf{b}}_1 \times \left(\vec{\mathbf{d}}_1 \times \vec{\omega}\right) - m_1 \vec{\mathbf{b}}_1 \times \vec{\mathbf{b}}_1 \times \vec{\omega} \\ + \vec{\mathbf{J}}_1 \cdot \vec{\omega}_{r1} - \vec{\mathbf{b}}_1 \times \left(\vec{\mathbf{d}}_1 \times \vec{\omega}_{r1}\right)\end{aligned} \tag{4.1.19}$$

4.2 Motion equations

By "motion equations" we mean the equations that are integrated to obtain $\vec{\mathbf{p}}$ and $\vec{\mathbf{h}}$. Let the total forces and torques on the system be given by $\vec{\mathbf{f}}$ and $\vec{\mathbf{g}}$ (about O), respectively. The components of these forces and torques are defined in Table 4.2.

Table 4.2 Force and torque components in pendulum system.

$\vec{\mathbf{f}}_0$	external force acting on body 0
$\vec{\mathbf{f}}_1$	external force acting on body 1
$\vec{\mathbf{g}}_0$	external torque acting on body 0 about O
$\vec{\mathbf{g}}_1$	external torque acting on body 1 about H
$\vec{\mathbf{g}}_{H1}$	interbody torque from body 0 acting on body 1

Thus $\vec{\mathbf{f}} = \vec{\mathbf{f}}_0 + \vec{\mathbf{f}}_1$ and $\vec{\mathbf{g}} = \vec{\mathbf{g}}_0 + \vec{\mathbf{g}}_1 + \vec{\mathbf{b}}_1 \times \vec{\mathbf{f}}_1$. The spring slosh model given in Chapter 3 used the matrix equations (3.2.11) and (3.2.12) for the rates of change of the system translational and angular momentum. The vector forms of these equations are given by:

$$\dot{\vec{\mathbf{p}}} = \vec{\mathbf{f}} \tag{4.2.1}$$

$$\dot{\vec{\mathbf{h}}} = -\vec{\mathbf{v}} \times \vec{\mathbf{p}} + \vec{\mathbf{g}} \tag{4.2.2}$$

The rate of change of angular momentum for body 1 can be obtained by simply adding appropriate subscripts to (4.2.2), and substituting the total torque applied to this body. Thus

$$\dot{\vec{\mathbf{h}}}_1 = -\vec{\mathbf{v}}_H \times \vec{\mathbf{p}}_1 + \vec{\mathbf{g}}_1 + \vec{\mathbf{g}}_{H1} \tag{4.2.3}$$

Using (4.1.8) this becomes

$$\dot{\vec{\mathbf{h}}}_1 = -\vec{\mathbf{v}}_H \times \left(m_1 \vec{\mathbf{v}}_H - \vec{\mathbf{d}}_1 \times \vec{\omega}_1 \right) + \vec{\mathbf{g}}_1 + \vec{\mathbf{g}}_{H1} \tag{4.2.4}$$

The first term on the RHS can be deleted since it is the cross product of parallel vectors. Thus we obtain

$$\dot{\vec{\mathbf{h}}}_1 = \vec{\mathbf{v}}_H \times \left(\vec{\mathbf{d}}_1 \times \vec{\omega}_1 \right) + \vec{\mathbf{g}}_1 + \vec{\mathbf{g}}_{H1} \tag{4.2.5}$$

We define the column matrices corresponding to the vectors $\vec{\mathbf{r}}_{cm}$, $\vec{\mathbf{r}}_0$, $\vec{\mathbf{r}}_1$ and in the following frames:

$$[\mathbf{r}_{cm}, \mathbf{r}_0] = \left[\vec{\mathbf{r}}_{cm}, \vec{\mathbf{r}}_0 \right] \text{ expressed in } \mathscr{F}_0 \tag{4.2.6}$$

$$\mathbf{r}_1 = \vec{\mathbf{r}}_1 \text{ expressed in } \mathscr{F}_1 \tag{4.2.7}$$

This scheme is followed generally herein, i.e., any variable with subscript 1 is in $\mathscr{F}_1$, otherwise it is in $\mathscr{F}_0$. The only exception is $\mathbf{b}_1$ which is expressed in $\mathscr{F}_0$. Thus $\mathbf{v}$ and $\boldsymbol{\omega}$ are column matrices representing the linear and angular velocity of body 0, expressed in the body 0 frame.

Moments of inertia

The first moments of inertia in matrix form are simply

$$\mathbf{d}_0 = m_0\mathbf{r}_0 \tag{4.2.8}$$

$$\mathbf{d}_1 = m_1\mathbf{r}_1 \tag{4.2.9}$$

The first moment of inertia of the entire system about O is given by

$$\mathbf{d} = m_T\mathbf{r}_{cm} = \mathbf{d}_0 + m_1\mathbf{b}_1 + \mathbf{C}_{01}\mathbf{d}_1 \tag{4.2.10}$$

$\mathbf{C}_{01}$ is the direction cosine matrix described at the outset of this section. Second moments of inertia (aka moments of inertia) are defined in the frame of their respective bodies. The overall system moment of inertia is defined in Frame 0. $\mathbf{J}_0$ and $\mathbf{J}_1$ are defined as the moment of inertia of bodies 0 and 1 about O and H, respectively, and the entire system inertia about O is given by $\mathbf{J}$.

Momentum in matrix form

From (4.1.10), the linear momentum becomes

$$\mathbf{p} = m_T\mathbf{v} - \mathbf{d}^\times\boldsymbol{\omega} - \mathbf{C}_{01}\mathbf{d}_1^\times\boldsymbol{\omega}_{r1} \tag{4.2.11}$$

For the body 1 angular momentum, all the Frame 0 vectors in (4.1.16) are expressed in Frame 1:

$$\mathbf{h}_1 = \mathbf{d}_1^\times(\mathbf{C}_{10}\mathbf{v}) + \mathbf{J}_1(\mathbf{C}_{10}\boldsymbol{\omega}) - \mathbf{d}_1^\times\left(\mathbf{C}_{10}\mathbf{b}_1\right)^\times(\mathbf{C}_{10}\boldsymbol{\omega}) + \mathbf{J}_1\boldsymbol{\omega}_{r1} \tag{4.2.12}$$

Using (4.1.1) it follows that

$$\left(\mathbf{C}_{10}\mathbf{b}_1\right)^\times = \mathbf{C}_{10}\mathbf{b}_1^\times\mathbf{C}_{01} \tag{4.2.13}$$

The rest of the parentheses in (4.2.12) are superfluous. Thus

$$\mathbf{h}_1 = \mathbf{d}_1^\times\mathbf{C}_{10}\mathbf{v} + \mathbf{J}_{10}\boldsymbol{\omega} + \mathbf{J}_1\boldsymbol{\omega}_{r1} \tag{4.2.14}$$

where

$$\mathbf{J}_{10} = \mathbf{J}_1\mathbf{C}_{10} - \mathbf{d}_1^{\times}\mathbf{C}_{10}\mathbf{b}_1^{\times} \tag{4.2.15}$$

The total angular momentum follows a similar development. From (4.1.19)

$$\begin{aligned}\mathbf{h} = \mathbf{d}^{\times}\mathbf{v} + \Big[\mathbf{J}_0 + \mathbf{C}_{01}\mathbf{J}_1\mathbf{C}_{10} - (\mathbf{C}_{01}\mathbf{d}_1)^{\times}\mathbf{b}_1^{\times} \\ -\mathbf{b}_1^{\times}(\mathbf{C}_{01}\mathbf{d}_1)^{\times} - m_1\mathbf{b}_1^{\times}\mathbf{b}_1^{\times}\Big]\boldsymbol{\omega} \\ + \mathbf{C}_{01}\mathbf{J}_1\boldsymbol{\omega}_{r1} - \mathbf{b}_1^{\times}(\mathbf{C}_{01}\mathbf{d}_1)^{\times}(\mathbf{C}_{01}\boldsymbol{\omega}_{r1})\end{aligned} \tag{4.2.16}$$

The two terms on the third line deserve some explanation. For the first term, which arises from $\vec{\mathbf{J}}_1 \cdot \vec{\omega}_{r1}$, the dot product is taken in $\mathscr{F}_1$ and the result is then expressed in $\mathscr{F}_0$. For the second term, all $\mathscr{F}_1$ column matrices are first expressed in $\mathscr{F}_0$ and then the cross product operations are performed. The result can be written as

$$\mathbf{h} = \mathbf{d}^{\times}\mathbf{v} + \mathbf{J}\boldsymbol{\omega} + \mathbf{J}_{01}\boldsymbol{\omega}_{r1} \tag{4.2.17}$$

where

$$\mathbf{J} = \mathbf{J}_0 + \mathbf{C}_{01}\mathbf{J}_1\mathbf{C}_{10} - m_1\mathbf{b}_1^{\times}\mathbf{b}_1^{\times} - (\mathbf{C}_{01}\mathbf{d}_1)^{\times}\mathbf{b}_1^{\times} - \mathbf{b}_1^{\times}(\mathbf{C}_{01}\mathbf{d}_1)^{\times} \tag{4.2.18}$$

and

$$\mathbf{J}_{01} \equiv \mathbf{C}_{01}\mathbf{J}_1 - \mathbf{b}_1^{\times}\mathbf{C}_{01}\mathbf{d}_1^{\times} \tag{4.2.19}$$

This last expression has been simplified by using (4.1.1). Note that

$$\mathbf{J}_{01} = \mathbf{J}_{10}^{T} \tag{4.2.20}$$

The present derivation gives results that appear in simpler forms than those given in the textbook by Hughes [6]. Most importantly, there are fewer appearances of $\mathbf{C}_{01}$ and $\mathbf{C}_{10}$. However, the results are equivalent.

Time derivatives of the system mass properties

For a single rigid body, the primary motivation for working in body coordinates is that the first and second moments of inertia remain constant. However, one complication of multibody dynamics is the presence of $\mathbf{C}_{01}$ and $\mathbf{C}_{10}$ in the system mass properties, which makes these elements vary with time. Derivatives of rotation matrices are given by the following kinematic equations;

$$\dot{\mathbf{C}}_{10} = -\boldsymbol{\omega}_{r1}^{\times}\mathbf{C}_{10} \tag{4.2.21}$$

$$\dot{\mathbf{C}}_{01} = \mathbf{C}_{01}\boldsymbol{\omega}_{r1}^{\times} \tag{4.2.22}$$

Using these expressions, the mass property derivatives are given by

$$\dot{\mathbf{d}} = \mathbf{C}_{01}\boldsymbol{\omega}_{r1}^{\times}\mathbf{d}_1 \tag{4.2.23}$$

$$\dot{\mathbf{J}}_{01} \equiv \mathbf{C}_{01}\boldsymbol{\omega}_{r1}^{\times}\mathbf{J}_1 - \mathbf{b}_1^{\times}\mathbf{C}_{01}\boldsymbol{\omega}_{r1}^{\times}\mathbf{d}_1^{\times} \tag{4.2.24}$$

$$\dot{\mathbf{J}}_{10} \equiv -\mathbf{J}_1\boldsymbol{\omega}_{r1}^{\times}\mathbf{C}_{10} + \mathbf{d}_1^{\times}\boldsymbol{\omega}_{r1}^{\times}\mathbf{C}_{10}\mathbf{b}_1^{\times} \tag{4.2.25}$$

$$\dot{\mathbf{J}} \equiv \mathbf{C}_{01}\boldsymbol{\omega}_{r1}^{\times}\mathbf{J}_1\mathbf{C}_{10} - \mathbf{C}_{01}\mathbf{J}_1\boldsymbol{\omega}_{r1}^{\times}\mathbf{C}_{10} - \left(\mathbf{C}_{01}\boldsymbol{\omega}_{r1}^{\times}\mathbf{d}_1\right)^{\times}\mathbf{b}_1^{\times} - \mathbf{b}_1^{\times}\left(\mathbf{C}_{01}\boldsymbol{\omega}_{r1}^{\times}\mathbf{d}_1\right)^{\times} \tag{4.2.26}$$

Forces and moments

The system forces and moments are given by

$$\mathbf{f} = \mathbf{f}_0 + \mathbf{C}_{01}\mathbf{f}_1 \tag{4.2.27}$$

$$\mathbf{g} = \mathbf{g}_0 + \mathbf{C}_{01}\mathbf{g}_1 + \mathbf{b}_1^{\times}\mathbf{C}_{01}\mathbf{f}_1 \tag{4.2.28}$$

The motion equations in matrix form come from (4.2.1), (4.2.2), and (4.2.5)

$$\dot{\mathbf{p}} = -\boldsymbol{\omega}^{\times}\mathbf{p} + \mathbf{f} \tag{4.2.29}$$

$$\dot{\mathbf{h}} = -\boldsymbol{\omega}^{\times}\mathbf{h} - \mathbf{v}^{\times}\mathbf{p} + \mathbf{g} \tag{4.2.30}$$

$$\dot{\mathbf{h}}_1 = \left[\mathbf{h}_1^{\times} + (\mathbf{C}_{10}\mathbf{v}_H)^{\times}\mathbf{d}_1^{\times}\right]\boldsymbol{\omega}_1 + \mathbf{g}_1 + \mathbf{g}_{H1} \tag{4.2.31}$$

where

$$\boldsymbol{\omega}_1 = \mathbf{C}_{10}\boldsymbol{\omega} + \boldsymbol{\omega}_{r1} \tag{4.2.32}$$

Using again the expression for the absolute velocity of point H, the last motion equation becomes

$$\dot{\mathbf{h}}_1 = \left[\mathbf{h}_1^{\times} + \mathbf{C}_{10}\left(\mathbf{v} + \boldsymbol{\omega}^{\times}\mathbf{b}_1\right)^{\times}\mathbf{C}_{01}\mathbf{d}_1^{\times}\right]\boldsymbol{\omega}_1 + \mathbf{g}_1 + \mathbf{g}_{H1} \tag{4.2.33}$$

This equation contains all the nonlinear torques on body 1, which can represent either an engine nozzle or a slosh pendulum when the equations are applied to a rocket.

Translation equation

Substituting (4.2.11) into (4.2.29), we obtain

$$m_T\dot{\mathbf{v}} - \dot{\mathbf{d}}^\times\boldsymbol{\omega} - \mathbf{d}^\times\dot{\boldsymbol{\omega}} - \dot{\mathbf{C}}_{01}\mathbf{d}_1^\times\boldsymbol{\omega}_{r1} - \mathbf{C}_{01}\mathbf{d}_1^\times\dot{\boldsymbol{\omega}}_{r1}$$
$$= -\boldsymbol{\omega}^\times\left(m\mathbf{v} - \mathbf{d}^\times\boldsymbol{\omega} - \mathbf{C}_{01}\mathbf{d}_1^\times\boldsymbol{\omega}_{r1}\right) + \mathbf{f} \quad (4.2.34)$$

The acceleration of O expressed in the body coordinate frame is

$$\mathbf{a}_b = \dot{\mathbf{v}} + \boldsymbol{\omega}\times\mathbf{v} \quad (4.2.35)$$

Using this expression along with (4.2.22) and (4.2.23), (4.2.34) becomes

$$m_T\mathbf{a}_b - \mathbf{d}^\times\dot{\boldsymbol{\omega}} - \mathbf{C}_{01}\mathbf{d}_1^\times\dot{\boldsymbol{\omega}}_{r1} = \left(\mathbf{C}_{01}\boldsymbol{\omega}_{r1}^\times\mathbf{d}_1\right)^\times\boldsymbol{\omega}$$
$$+ \mathbf{C}_{01}\boldsymbol{\omega}_{r1}^\times\mathbf{d}_1^\times\boldsymbol{\omega}_{r1} + \boldsymbol{\omega}^\times\mathbf{d}^\times\boldsymbol{\omega} + \boldsymbol{\omega}^\times\mathbf{C}_{01}\mathbf{d}_1^\times\boldsymbol{\omega}_{r1} + \mathbf{f} \quad (4.2.36)$$

The RHS of (4.2.36) can be written as the sum of a nonlinear term plus the external force

$$m_T\mathbf{a}_b - \mathbf{d}^\times\dot{\boldsymbol{\omega}} - \mathbf{C}_{01}\mathbf{d}_1^\times\dot{\boldsymbol{\omega}}_{r1} = \mathbf{f}_{NL} + \mathbf{f} \quad (4.2.37)$$

where

$$\mathbf{f}_{NL} = \boldsymbol{\omega}^\times\mathbf{d}^\times\boldsymbol{\omega} + 2\boldsymbol{\omega}^\times\mathbf{C}_{01}\mathbf{d}_1^\times\boldsymbol{\omega}_{r1} + \mathbf{C}_{01}\boldsymbol{\omega}_{r1}^\times\mathbf{d}_1^\times\boldsymbol{\omega}_{r1} \quad (4.2.38)$$

The first term in (4.2.38) is a term that appears whenever the origin is chosen at a point that differs from the system center of mass. The remaining two terms are the apparent Coriolis force and centrifugal force due to the motion of body 1 relative to body 0.

Rotation equation

Taking the derivative of (4.2.17), and using (4.2.23), the dynamics of rotation are given by

$$\dot{\mathbf{h}} = -\mathbf{v}^\times\mathbf{C}_{01}\boldsymbol{\omega}_{r1}^\times\mathbf{d}_1 + \mathbf{d}^\times\dot{\mathbf{v}} + \dot{\mathbf{J}}\boldsymbol{\omega} + \mathbf{J}\dot{\boldsymbol{\omega}} + \dot{\mathbf{J}}_{01}\boldsymbol{\omega}_{r1} + \mathbf{J}_{01}\dot{\boldsymbol{\omega}}_{r1} \quad (4.2.39)$$

An equivalent expression for $\dot{\mathbf{h}}$ can be obtained by substituting (4.2.11) and (4.2.17) into (4.2.30)

$$\dot{\mathbf{h}} = -\boldsymbol{\omega}^\times \left(\mathbf{d}^\times \mathbf{v} + \mathbf{J}\boldsymbol{\omega} + \mathbf{J}_{01}\boldsymbol{\omega}_{r1}\right) - \mathbf{v}^\times \left(m_T \mathbf{v} - \mathbf{d}^\times \boldsymbol{\omega} - \mathbf{C}_{01}\mathbf{d}_1^\times \boldsymbol{\omega}_{r1}\right) + \mathbf{g} \quad (4.2.40)$$

The fourth term on the RHS of this is zero. The sixth term matches the first term of (4.2.39), so these cancel when these two expressions are set equal to each other. Using the triple vector product, it follows that

$$-\boldsymbol{\omega}^\times \mathbf{d}^\times \mathbf{v} - \mathbf{v}^\times \boldsymbol{\omega}^\times \mathbf{d} = \mathbf{d}^\times \mathbf{v}^\times \boldsymbol{\omega} \quad (4.2.41)$$

Thus (4.2.39) and (4.2.40) can be equated and the result reduced to

$$\mathbf{d}^\times \dot{\mathbf{v}} + \dot{\mathbf{J}}\boldsymbol{\omega} + \mathbf{J}\dot{\boldsymbol{\omega}} + \dot{\mathbf{J}}_{01}\boldsymbol{\omega}_{r1} + \mathbf{J}_{01}\dot{\boldsymbol{\omega}}_{r1} = -\boldsymbol{\omega}^\times \left(\mathbf{J}\boldsymbol{\omega} + \mathbf{J}_{01}\boldsymbol{\omega}_{r1}\right) + \mathbf{d}^\times \mathbf{v}^\times \boldsymbol{\omega} + \mathbf{g} \quad (4.2.42)$$

Using (4.2.35), this expression can be arranged as

$$\mathbf{d}^\times \mathbf{a}_b + \mathbf{J}\dot{\boldsymbol{\omega}} + \mathbf{J}_{01}\dot{\boldsymbol{\omega}}_{r1} = \mathbf{g}_{NL} + \mathbf{g} \quad (4.2.43)$$

where the nonlinear torque is given by

$$\mathbf{g}_{NL} = -\dot{\mathbf{J}}\boldsymbol{\omega} - \dot{\mathbf{J}}_{01}\boldsymbol{\omega}_{r1} - \boldsymbol{\omega}^\times \left(\mathbf{J}\boldsymbol{\omega} + \mathbf{J}_{01}\boldsymbol{\omega}_{r1}\right) \quad (4.2.44)$$

Substituting from (4.2.19), (4.2.24), and (4.2.26), (4.2.44) becomes

$$\begin{aligned}\mathbf{g}_{NL} = &-\Big[\mathbf{C}_{01}\boldsymbol{\omega}_{r1}^\times \mathbf{J}_1 \mathbf{C}_{10} - \mathbf{C}_{01}\mathbf{J}_1 \boldsymbol{\omega}_{r1}^\times \mathbf{C}_{10} \\ &- \left(\mathbf{C}_{01}\boldsymbol{\omega}_{r1}^\times \mathbf{d}_1\right)^\times \mathbf{b}_1^\times - \mathbf{b}_1^\times \left(\mathbf{C}_{01}\boldsymbol{\omega}_{r1}^\times \mathbf{d}_1\right)^\times\Big]\boldsymbol{\omega} \\ &- \left[\mathbf{C}_{01}\boldsymbol{\omega}_{r1}^\times \mathbf{J}_1 - \mathbf{b}_1^\times \mathbf{C}_{01}\boldsymbol{\omega}_{r1}^\times \mathbf{d}_1^\times\right]\boldsymbol{\omega}_{r1} \\ &- \boldsymbol{\omega}^\times \left[\mathbf{J}\boldsymbol{\omega} + \left(\mathbf{C}_{01}\mathbf{J}_1 - \mathbf{b}_1^\times \mathbf{C}_{01}\mathbf{d}_1^\times\right)\boldsymbol{\omega}_{r1}\right]\end{aligned} \quad (4.2.45)$$

Define

$$\mathbf{c} \equiv \mathbf{C}_{01}\boldsymbol{\omega}_{r1}^\times \mathbf{d}_1 \quad (4.2.46)$$

Terms 3, 4, and 9 on the RHS of (4.2.45) combine into

$$-\mathbf{c}^\times \boldsymbol{\omega}^\times \mathbf{b}_1 + \mathbf{b}_1^\times \mathbf{c}^\times \boldsymbol{\omega} - \boldsymbol{\omega}^\times \mathbf{b}_1^\times \mathbf{c} = 2\mathbf{b}_1^\times \mathbf{c}^\times \boldsymbol{\omega} \quad (4.2.47)$$

Thus, the complete nonlinear expression is given by

$$
\begin{aligned}
\mathbf{g}_{NL} = -\boldsymbol{\omega}^{\times}\mathbf{J}\boldsymbol{\omega} - \boldsymbol{\omega}^{\times}\mathbf{C}_{01}\mathbf{J}_{1}\boldsymbol{\omega}_{r1} \\
- \mathbf{C}_{01}\boldsymbol{\omega}_{r1}^{\times}\mathbf{J}_{1}\left(\mathbf{C}_{10}\boldsymbol{\omega} + \boldsymbol{\omega}_{r1}\right) + \mathbf{C}_{01}\mathbf{J}_{1}\boldsymbol{\omega}_{r1}^{\times}\mathbf{C}_{10}\boldsymbol{\omega} \\
+ 2\mathbf{b}_{1}^{\times}\left(\mathbf{C}_{01}\boldsymbol{\omega}_{r1}^{\times}\mathbf{d}_{1}\right)^{\times}\boldsymbol{\omega} + \mathbf{b}_{1}^{\times}\mathbf{C}_{01}\boldsymbol{\omega}_{r1}^{\times}\mathbf{d}_{1}^{\times}\boldsymbol{\omega}_{r1}
\end{aligned} \quad (4.2.48)
$$

The first two lines in this expression represent various Euler coupling effects. The third line gives the torques on the system due to the Coriolis force and the centrifugal force on body 1. In the Coriolis term, it is helpful to reverse the order of the cross product within the parentheses, and then perform another reversal that gets rid of the parentheses. This yields

$$
\begin{aligned}
\mathbf{g}_{NL} = -\boldsymbol{\omega}^{\times}\mathbf{J}\boldsymbol{\omega} - \boldsymbol{\omega}^{\times}\mathbf{C}_{01}\mathbf{J}_{1}\boldsymbol{\omega}_{r1} \\
- \mathbf{C}_{01}\boldsymbol{\omega}_{r1}^{\times}\mathbf{J}_{1}\left(\mathbf{C}_{10}\boldsymbol{\omega} + \boldsymbol{\omega}_{r1}\right) + \mathbf{C}_{01}\mathbf{J}_{1}\boldsymbol{\omega}_{r1}^{\times}\mathbf{C}_{10}\boldsymbol{\omega} \\
+ 2\mathbf{b}_{1}^{\times}\boldsymbol{\omega}^{\times}\mathbf{C}_{01}\mathbf{d}_{1}^{\times}\boldsymbol{\omega}_{r1} + \mathbf{b}_{1}^{\times}\mathbf{C}_{01}\boldsymbol{\omega}_{r1}^{\times}\mathbf{d}_{1}^{\times}\boldsymbol{\omega}_{r1}
\end{aligned} \quad (4.2.49)
$$

In this form, it is easier to see the relationship to the Coriolis force in (4.2.38).

Pendulum equation

Taking the derivative of (4.2.14), we have

$$
\dot{\mathbf{h}}_{1} = \dot{\mathbf{d}}_{1}^{\times}\mathbf{C}_{10}\mathbf{v} + \mathbf{d}_{1}^{\times}\dot{\mathbf{C}}_{10}\mathbf{v} + \mathbf{d}_{1}^{\times}\mathbf{C}_{10}\dot{\mathbf{v}} + \dot{\mathbf{J}}_{10}\boldsymbol{\omega} + \mathbf{J}_{10}\dot{\boldsymbol{\omega}} + \dot{\mathbf{J}}_{1}\boldsymbol{\omega}_{r1} + \mathbf{J}_{1}\dot{\boldsymbol{\omega}}_{r1} \quad (4.2.50)
$$

Noting that $\dot{\mathbf{d}}_1 = \dot{\mathbf{J}}_1 = 0$ and using (4.2.21), this becomes

$$
\dot{\mathbf{h}}_{1} = -\mathbf{d}_{1}^{\times}\boldsymbol{\omega}_{r1}^{\times}\mathbf{C}_{10}\mathbf{v} + \mathbf{d}_{1}^{\times}\mathbf{C}_{10}\dot{\mathbf{v}} + \dot{\mathbf{J}}_{10}\boldsymbol{\omega} + \mathbf{J}_{10}\dot{\boldsymbol{\omega}} + \mathbf{J}_{1}\dot{\boldsymbol{\omega}}_{r1} \quad (4.2.51)
$$

Substituting (4.2.14) into (4.2.33) gives

$$
\begin{aligned}
\dot{\mathbf{h}}_{1} = \left(\mathbf{d}_{1}^{\times}\mathbf{C}_{10}\mathbf{v} + \mathbf{J}_{10}\boldsymbol{\omega} + \mathbf{J}_{1}\boldsymbol{\omega}_{r1}\right)^{\times}\boldsymbol{\omega}_{1} \\
+ \mathbf{C}_{10}\mathbf{v}^{\times}\mathbf{C}_{01}\mathbf{d}_{1}^{\times}\boldsymbol{\omega}_{1} \\
+ \mathbf{C}_{10}\left(\boldsymbol{\omega}^{\times}\mathbf{b}_{1}\right)^{\times}\mathbf{C}_{01}\mathbf{d}_{1}^{\times}\boldsymbol{\omega}_{1} + \mathbf{g}_{1} + \mathbf{g}_{H1}
\end{aligned} \quad (4.2.52)
$$

The first and fourth terms can be simplified by applying (4.1.1), then the triple vector product, then (4.2.32), such that

$$
\begin{aligned}
\left(\mathbf{d}_{1}^{\times}\mathbf{C}_{10}\mathbf{v}\right)^{\times}\boldsymbol{\omega}_{1} + \mathbf{C}_{10}\mathbf{v}^{\times}\mathbf{C}_{01}\mathbf{d}_{1}^{\times}\boldsymbol{\omega}_{1} &= -\boldsymbol{\omega}_{1}^{\times}\mathbf{d}_{1}^{\times}\mathbf{C}_{10}\mathbf{v} + \left(\mathbf{C}_{10}\mathbf{v}\right)^{\times}\mathbf{d}_{1}^{\times}\boldsymbol{\omega}_{1} \\
&= -\boldsymbol{\omega}_{1}^{\times}\mathbf{d}_{1}^{\times}\mathbf{C}_{10}\mathbf{v} - \left(\mathbf{C}_{10}\mathbf{v}\right)^{\times}\boldsymbol{\omega}_{1}^{\times}\mathbf{d}_{1}
\end{aligned}
$$

$$
\begin{aligned}
&= \mathbf{d}_1^{\times} (\mathbf{C}_{01}\mathbf{v})^{\times} \boldsymbol{\omega}_1 \\
&= -\mathbf{d}_1^{\times} (\mathbf{C}_{01}\boldsymbol{\omega})^{\times'} \mathbf{C}_{10}\mathbf{v} - \mathbf{d}_1^{\times}\boldsymbol{\omega}_{r1}^{\times}\mathbf{C}_{10}\mathbf{v}
\end{aligned} \tag{4.2.53}
$$

This expression can be used to replace the first and fourth terms in (4.2.52), giving

$$
\begin{aligned}
\dot{\mathbf{h}}_1 = (\mathbf{J}_{10}\boldsymbol{\omega} + \mathbf{J}_1\boldsymbol{\omega}_{r1})^{\times} \boldsymbol{\omega}_1 - \mathbf{d}_1^{\times} (\mathbf{C}_{01}\boldsymbol{\omega})^{\times'} \mathbf{C}_{10}\mathbf{v} \\
- \mathbf{d}_1^{\times}\boldsymbol{\omega}_{r1}^{\times}\mathbf{C}_{10}\mathbf{v} + \mathbf{C}_{10}(\boldsymbol{\omega}^{\times}\mathbf{b}_1)^{\times} \mathbf{C}_{01}\mathbf{d}_1^{\times}\boldsymbol{\omega}_1 + \mathbf{g}_1 + \mathbf{g}_{H1}
\end{aligned} \tag{4.2.54}
$$

By equating (4.2.51) and (4.2.54), the first term of (4.2.51) is canceled by the fourth term of (4.2.54). Also, the second term of (4.2.51) can be combined with the third term of (4.2.54) by using (4.2.35). The result is

$$
\begin{aligned}
\mathbf{d}_1^{\times}\mathbf{C}_{10}\mathbf{a}_b + \dot{\mathbf{J}}_{10}\boldsymbol{\omega} + \mathbf{J}_{10}\dot{\boldsymbol{\omega}} + \mathbf{J}_1\dot{\boldsymbol{\omega}}_{r1} = (\mathbf{J}_{10}\boldsymbol{\omega} + \mathbf{J}_1\boldsymbol{\omega}_{r1})^{\times} \boldsymbol{\omega}_1 \\
+ \mathbf{C}_{10}(\boldsymbol{\omega}^{\times}\mathbf{b}_1)^{\times} \mathbf{C}_{01}\mathbf{d}_1^{\times}\boldsymbol{\omega}_1 + \mathbf{g}_1 + \mathbf{g}_{H1}
\end{aligned} \tag{4.2.55}
$$

Using (4.1.1) yet again, this expression can be rearranged as

$$
\mathbf{d}_1^{\times}\mathbf{C}_{10}\mathbf{a}_b + \mathbf{J}_{10}\dot{\boldsymbol{\omega}} + \mathbf{J}_1\dot{\boldsymbol{\omega}}_{r1} = \mathbf{g}_1 + \mathbf{g}_{H1} + \mathbf{g}_{NL1} \tag{4.2.56}
$$

where

$$
\mathbf{g}_{NL1} = -\mathbf{J}_{10}\dot{\boldsymbol{\omega}} + (\mathbf{J}_{10}\boldsymbol{\omega} + \mathbf{J}_1\boldsymbol{\omega}_{r1})^{\times} \boldsymbol{\omega}_1 + (\mathbf{C}_{10}\boldsymbol{\omega}^{\times}\mathbf{b}_1)^{\times} \mathbf{d}_1^{\times}\boldsymbol{\omega}_1 \tag{4.2.57}
$$

System matrix equations

Eqs. (4.2.37), (4.2.43), and (4.2.56) may be expressed in a compact matrix model form as

$$
\mathcal{M}\ddot{\mathbf{x}} = \mathcal{F} \tag{4.2.58}
$$

where

$$
\ddot{\mathbf{x}} = \begin{bmatrix} \mathbf{a}_b \\ \dot{\boldsymbol{\omega}} \\ \dot{\boldsymbol{\omega}}_{r1} \end{bmatrix} \tag{4.2.59}
$$

$$
\mathcal{F} = \begin{bmatrix} \mathbf{f} + \mathbf{f}_{NL} \\ \mathbf{g} + \mathbf{g}_{NL} \\ \mathbf{g}_1 + \mathbf{g}_{NL1} + \mathbf{g}_{H1} \end{bmatrix} \tag{4.2.60}
$$

$$\mathcal{M} = \begin{bmatrix} m_T\mathbf{1} & -\mathbf{d}^\times & -\mathbf{C}_{01}\mathbf{d}_1^\times \\ \mathbf{d}^\times & \mathbf{J} & \mathbf{J}_{01} \\ \mathbf{d}_1^\times\mathbf{C}_{10} & \mathbf{J}_{10} & \mathbf{J}_1 \end{bmatrix} \tag{4.2.61}$$

As before, **1** is defined as the identity matrix. Extension to more than one pendulum is straightforward. The total mass becomes

$$m_T = m_0 + \sum m_i \tag{4.2.62}$$

Similar changes are made to the system first and second moments of inertia. For example, for a system with two pendulums the state vector would be written as

$$\ddot{\mathbf{x}} = \begin{bmatrix} \mathbf{a}_b \\ \dot{\boldsymbol{\omega}} \\ \dot{\boldsymbol{\omega}}_{r1} \\ \dot{\boldsymbol{\omega}}_{r2} \end{bmatrix} \tag{4.2.63}$$

and the mass matrix becomes

$$\mathcal{M} = \begin{bmatrix} m_T\mathbf{1}_{3\times3} & -\mathbf{d}^\times & -\mathbf{C}_{01}\mathbf{d}_1^\times & -\mathbf{C}_{02}\mathbf{d}_2^\times \\ & \mathbf{J} & \mathbf{J}_{01} & \mathbf{J}_{02} \\ & \cdots & \mathbf{J}_1 & \mathbf{O}_{3\times3} \\ & & & \mathbf{J}_2 \end{bmatrix} \tag{4.2.64}$$

where **O** is the null matrix and the mass matrix is necessarily symmetric.

Truncated motion equations

As shown in Fig. 4.1, the body 1 mass is treated as a point mass on a pendulum. The axes are set up such that the bar of each pendulum j lies along the x_j axis. For slosh, the nature of the fluid coupling is that there is no significant torque exerted about the pendulum x_j axis. That is, we could consider the slosh mass to be a mass of fluid with a finite moment of inertia about the pendulum axis, but since there are no significant torques about this axis, this moment of inertia has no effect. A similar, but not identical, situation applies to the nozzle, whose only degrees of freedom are about the y and z axes.

The attachment condition for a slosh mass, that no torque is transmitted about the pendulum bar axis, means that under equilibrium conditions the slosh mass rotation about the bar axis is disconnected from the rotation of the rocket about its x axis. On the other hand, for the nozzle the opposite is true – the nozzle roll is constrained to be the same as that of the rocket body. Recall from the discussion following (2.1.21) that the total moment of inertia $\mathbf{I}_T$ about the x axis does not include the inertia about the x axis of any of the fluid, either sloshing or non-sloshing. Thus, nothing further is required to implement the "disconnected" condition.

In either case, whether the pendulum model is used for a slosh mass or for a nozzle, we employ the same solution – delete the equation for rotation about the x axis. This approach is an approximation, and does result in the deletion of some nonlinear effects; however, for most boost vehicles this is an acceptable loss of fidelity. For the nozzle case, this is further justified by the fact that nozzle angles are typically quite small, so that the kinematic coupling from the y and z axes to the x axis is small. Alternatively, a more complete but complex and computationally intensive approach is the introduction of constraints as discussed in Chapter 10.

In the truncation method, the rows and columns of the mass matrix associated with the x angular degree of freedom are deleted. Two column matrices are defined corresponding to the unit vectors along the remaining axes:

$$\mathbf{n}_y \equiv \begin{bmatrix} 0 & 1 & 0 \end{bmatrix}^T \tag{4.2.65}$$

$$\mathbf{n}_z \equiv \begin{bmatrix} 0 & 0 & 1 \end{bmatrix}^T \tag{4.2.66}$$

From these, the following are defined:

$$\begin{aligned}
g_{jy} &\equiv \mathbf{n}_y^T \mathbf{g}_j = \text{component of } \mathbf{g}_j \text{ about the pitch axis (external torque)} \\
g_{jz} &\equiv \mathbf{n}_z^T \mathbf{g}_j = \text{component of } \mathbf{g}_j \text{ about the yaw axis (external torque)} \\
g_{Hjy} &\equiv \mathbf{n}_y^T \mathbf{g}_{Hj} = \text{component of } \mathbf{g}_{Hj} \text{ about the pitch axis (interbody torque)} \\
g_{Hjz} &\equiv \mathbf{n}_z^T \mathbf{g}_{Hj} = \text{component of } \mathbf{g}_{Hj} \text{ about the yaw axis (interbody torque)}
\end{aligned}$$

An overbar is used to represent a column matrix with only the last two elements. The relative angular rate can be defined as $\bar{\boldsymbol{\omega}}_{rj} = \begin{bmatrix} \omega_{rjy} & \omega_{rjz} \end{bmatrix}^T$ where the components are the relative pitch and yaw rates of body j, respectively. For example, the truncated acceleration vector for a system with two pendulums becomes

$$\ddot{\mathbf{x}} = \begin{bmatrix} \mathbf{a}_b & \dot{\boldsymbol{\omega}} & \dot{\bar{\boldsymbol{\omega}}}_{r1} & \dot{\bar{\boldsymbol{\omega}}}_{r2} \end{bmatrix} \tag{4.2.67}$$

The same truncation matrix that was defined in Chapter 2, Eq. (2.7.1) can be used, and the truncated mass matrix can be written

$$\mathcal{M}\ddot{\mathbf{x}} = \begin{bmatrix} m\mathbf{1}_{3\times 3} & -\mathbf{d}^{\times} & -\mathbf{C}_{01}\mathbf{d}_1^{\times}\bar{\mathbf{U}} & -\mathbf{C}_{02}\mathbf{d}_2^{\times}\bar{\mathbf{U}} \\ & \mathbf{J} & \mathbf{J}_{01}\bar{\mathbf{U}} & \mathbf{J}_{02}\bar{\mathbf{U}} \\ & \cdots & \bar{\mathbf{J}}_1 & \mathbf{O}_{2\times 2} \\ & & & \bar{\mathbf{J}}_2 \end{bmatrix} \begin{bmatrix} \mathbf{a}_b \\ \dot{\boldsymbol{\omega}} \\ \dot{\bar{\boldsymbol{\omega}}}_{r1} \\ \dot{\bar{\boldsymbol{\omega}}}_{r2} \end{bmatrix} \tag{4.2.68}$$

where $\bar{\mathbf{J}}_j$ is the 2, 3 submatrix of the inertia matrix of body j about its hinge point, given by

$$\bar{\mathbf{J}}_j = \bar{\mathbf{U}}^T \mathbf{J}_j \bar{\mathbf{U}}. \tag{4.2.69}$$

The equations for the momenta of body j about its associated pitch and yaw joint axes are

$$\dot{h}_{jy} = g_{jy} + g_{Hjy} + \mathbf{n}_y^T \mathbf{g}_{NLj} \tag{4.2.70}$$

$$\dot{h}_{iz} = g_{jz} + g_{Hjz} + \mathbf{n}_z^T \mathbf{g}_{NLj} \tag{4.2.71}$$

Obtaining the transformation matrices

If there are large deflections of the pendulums, any one of several parameterization methods can be used for generating the transformation matrices. The most popular methods are Euler angles and quaternions. Quaternions have the disadvantage that four parameters and a nonlinear constraint are required to describe the relative orientation. The main advantage of quaternions is that they avoid the problem of singularities. However, it can usually be assumed that at least one of the pendulum angles can be restricted to be less than 90 degrees, in which case singularities can be avoided. If this is the case, an Euler sequence is vastly preferable. Only two Euler angles are required, matching the two degrees of freedom of the pendulum. To illustrate this, let us use a 2-3-1 Euler Sequence to describe the transformation from Frame 0 to Frame 1. For a point mass pendulum, the last rotation, about the 1 axis (the x_1 axis) is irrelevant, since we have stipulated that the x_1 axis lies along the pendulum bar. Let the remaining Euler angles be γ_{1y} and γ_{1z}, corresponding to an initial rotation about the y axis followed by a rotation about the z axis. Then the transformation is

$$\mathbf{C}_{10} = \begin{bmatrix} \cos\gamma_{1z} & \sin\gamma_{1z} & 0 \\ -\sin\gamma_{1z} & \cos\gamma_{1z} & 0 \\ 0 & 0 & 1 \end{bmatrix} \begin{bmatrix} \cos\gamma_{1y} & 0 & -\sin\gamma_{1y} \\ 0 & 1 & 0 \\ \sin\gamma_{1y} & 0 & \cos\gamma_{1y} \end{bmatrix} \tag{4.2.72}$$

The Euler angles are obtained by integrating the Euler rates. To do this, we need an appropriate mapping from Euler rates to body rates. Such a mapping is called a projection matrix. The appropriate projection matrix for a two-axis rotation about y and z is given by

$$\begin{bmatrix} \omega_{r1x} \\ \omega_{r1y} \\ \omega_{r1z} \end{bmatrix} = \begin{bmatrix} \sin\gamma_{1z} & 0 \\ \cos\gamma_{1z} & 0 \\ 0 & 1 \end{bmatrix} \begin{bmatrix} \dot{\gamma}_{1y} \\ \dot{\gamma}_{1z} \end{bmatrix} \tag{4.2.73}$$

It is necessary to invert this expression. For the case at hand (a point mass pendulum), this is easy since the first row of the above equation can be ignored, leaving a square matrix. Inverting this we obtain

$$\begin{bmatrix} \dot{\gamma}_{1y} \\ \dot{\gamma}_{1z} \end{bmatrix} = \mathbf{\Gamma}_1 \begin{bmatrix} \omega_{r1y} \\ \omega_{r1z} \end{bmatrix} \tag{4.2.74}$$

where

$$\mathbf{\Gamma}_1 = \begin{bmatrix} \sec\gamma_{1z} & 0 \\ 0 & 1 \end{bmatrix} \tag{4.2.75}$$

One can see from this expression the exact nature of the singularity when γ_{1z} is 90 degrees. At this yaw angle, the Euler rate about the y axis may become infinite. Thus, with this choice of Euler angles, there is no restriction on the pitch angle γ_{1y}, but we must avoid the singularity region near $\gamma_{1z} = 90$ degrees.

If the pendulum angles are sufficiently small, as with a rocket nozzle, one could assume $\sec\gamma_{1z} = 1$ and obtain the Euler angles by simply integrating the ω_r's. After performing the integrations, one then uses Eq. (4.2.72). For both the slosh mass and the nozzle, the details of constraining the x axis rotation are avoided. For the slosh mass, there is no constraint and the x axis rotation can be ignored. For the nozzle, the small angle condition means that constraining the nozzle x rotation to be the same as that of the rocket is adequate.

Alternatively, Eq. (4.2.72) can be approximated by

$$\mathbf{C}_{10} \cong \begin{bmatrix} 1 & \gamma_{1z} & -\gamma_{1y} \\ -\gamma_{1z} & 1 & 0 \\ \gamma_{1y} & 0 & 1 \end{bmatrix} \tag{4.2.76}$$

This transformation has first-order accuracy. If greater accuracy is required, Appendix C provides a second-order transformation.

The decision as to whether to use the full kinematic transformation, the first-order approximation, or a second-order approximation depends on the application. If $\mathbf{C}_{10}$ is used to rotate an engine nozzle in simulation, the first-order rotations may not adequately compute the loss of axial thrust as an engine is rotated off-centerline. However, it is atypical for a TVC actuator model to include the detailed kinematic relationship of the actuator linkages, instead producing a torque or angle (for prescribed motion) about two mutually decoupled engine degrees of freedom. The most meticulous way to compute all the nonlinear effects would be to use a model that captures these linkage details. This would lead to an appropriate Euler sequence that in effect prioritizes one actuator degree of freedom over another. However, a simpler approach may be adequate. The second-order transformation of Appendix C essentially averages the rotations over all possible Euler sequences, and allows one to still capture the aforementioned axial thrust effect. In addition, the second-order rotation does not require numerical evaluation of any trigonometric functions. The use of the second-order transformation in computing the thrust forces is discussed further in Chapter 5.

4.3 Slosh dynamics using the pendulum model

Applying the pendulum model to the slosh problem requires the substitutions

$$m_j \Rightarrow m_{sj} \tag{4.3.1}$$

$$\mathbf{b}_j \Rightarrow \mathbf{b}_{Hj} \tag{4.3.2}$$

$$\mathbf{d} \Rightarrow \mathbf{s}_{TD} \tag{4.3.3}$$

where $\mathbf{s}_{TD}$ is the dynamic first moment of inertia, including the slosh motion, and $\mathbf{s}_T$ is the static first moment of inertia with slosh masses in equilibrium positions. The distinction between $\mathbf{s}_T$ and $\mathbf{s}_{TD}$ is discussed more thoroughly at the beginning of Section 3.2. Let

$$l_{pj} = \text{length of slosh } j \text{ pendulum}$$

The vector $\mathbf{b}_{Hj}$ for the hinge location is not the same as the $\mathbf{b}_{sj}$ of Chapter 2. If the pendulum is in its equilibrium position, such that Frame 0 coincides with Frame j $(\mathbf{C}_{0j} = \mathbf{1})$, then these are related by

$$\mathbf{b}_{sj} = \mathbf{b}_{Hj} + \mathbf{r}_j \tag{4.3.4}$$

where $\mathbf{r}_j$ corresponds to $\mathbf{r}_1$ of Fig. 4.1, i.e., the vector representing pendulum j.

$$\mathbf{r}_j = \begin{bmatrix} -l_{pj} & 0 & 0 \end{bmatrix}^T \tag{4.3.5}$$

The first moment of inertia of the pendulum is

$$\mathbf{d}_j = \begin{bmatrix} -m_{sj} l_{pj} & 0 & 0 \end{bmatrix}^T \tag{4.3.6}$$

The second moment of inertia is

$$\mathbf{J}_j = \begin{bmatrix} 0 & 0 & 0 \\ 0 & m_{sj} l_{pj}^2 & 0 \\ 0 & 0 & m_{sj} l_{pj}^2 \end{bmatrix} \tag{4.3.7}$$

The system first moment of inertia is obtained by extending the expression for the first moment to multiple pendulums and using (4.3.1) through (4.3.3) such that

$$\mathbf{s}_{TD} = \mathbf{d} = \mathbf{d}_0 + \sum \left(m_{sj} \mathbf{b}_{Hj} + \mathbf{C}_{0j} \mathbf{d}_j \right) \tag{4.3.8}$$

The system second moment of inertia and $\mathbf{J}_{0j}$ are obtained by applying a similar process to Eqs. (4.2.18) and (4.2.19), giving

$$\mathbf{J} = \mathbf{J}_0 + \sum \mathbf{C}_{0j} \mathbf{J}_j \mathbf{C}_{j0} - m_{sj} \mathbf{b}_{Hj}^{\times} \mathbf{b}_{Hj}^{\times} - \left(\mathbf{C}_{0j} \mathbf{d}_j \right) \mathbf{b}_{Hj}^{\times} - \mathbf{b}_{Hj}^{\times} \left(\mathbf{C}_{0j} \mathbf{d}_j \right)^{\times} \tag{4.3.9}$$

$$\mathbf{J}_{0j} \equiv \mathbf{C}_{0j} \mathbf{J}_j - \mathbf{b}_{Hj}^{\times} \mathbf{C}_{0j} \mathbf{d}_j^{\times} \tag{4.3.10}$$

This defines all of the variables necessary to complete the mass matrix. The only external force $\mathbf{f}_j$ on the slosh mass is that due to gravity. The only interbody torque $\mathbf{g}_{Hj}$ is that due to slosh damping, given by (4.3.13) and (4.3.14) below.

Mass matrix for pendulum slosh, prescribed engine motion, and flex

Using the notation defined in Chapter 2 for the flex parameters, the complete mass matrix for prescribed engine motion and including flexible modes is given in Eq. (4.3.11). This matrix assumes that the modes are obtained from an integrated-body flex model that includes all the slosh masses.

$$
\mathcal{M}\ddot{\mathbf{x}} \equiv
\begin{bmatrix}
m_T\mathbf{1} & -\mathbf{s}_{TD}^{\times} & -\mathrm{C}_{01}\mathbf{d}_1^{\times}\bar{\mathbf{U}} & -\mathrm{C}_{02}\mathbf{d}_2^{\times}\bar{\mathbf{U}} & \cdots & -\mathrm{C}_{0N}\mathbf{d}_N^{\times}\bar{\mathbf{U}} & \mathbf{0} & \mathbf{0} & \cdots & \mathbf{0} \\
\mathbf{s}_{TD}^{\times} & \mathbf{I}_T & \mathrm{J}_{01}\bar{\mathbf{U}} & \mathrm{J}_{02}\bar{\mathbf{U}} & \cdots & \mathrm{J}_{0N}\bar{\mathbf{U}} & \mathbf{0} & \mathbf{0} & \cdots & \mathbf{0} \\
\bar{\mathbf{U}}^T\mathbf{d}_1^{\times}\mathrm{C}_{10} & \bar{\mathbf{U}}^T\mathbf{J}_{10} & \mathbf{J}_{sub1} & \mathbf{O} & \cdots & \mathbf{O} & \bar{\mathbf{U}}^T\mathbf{d}_1^{\times}\mathrm{C}_{10}\boldsymbol{\psi}_{11} & \bar{\mathbf{U}}^T\mathbf{d}_1^{\times}\mathrm{C}_{10}\boldsymbol{\psi}_{12} & \cdots & \bar{\mathbf{U}}^T\mathbf{d}_1^{\times}\mathrm{C}_{10}\boldsymbol{\psi}_{1M} \\
\bar{\mathbf{U}}^T\mathbf{d}_2^{\times}\mathrm{C}_{20} & \bar{\mathbf{U}}^T\mathbf{J}_{20} & \mathbf{O} & \mathbf{J}_{sub2} & \cdots & \mathbf{O} & \bar{\mathbf{U}}^T\mathbf{d}_2^{\times}\mathrm{C}_{20}\boldsymbol{\psi}_{21} & \bar{\mathbf{U}}^T\mathbf{d}_2^{\times}\mathrm{C}_{20}\boldsymbol{\psi}_{22} & \cdots & \bar{\mathbf{U}}^T\mathbf{d}_2^{\times}\mathrm{C}_{20}\boldsymbol{\psi}_{2M} \\
\vdots & \vdots & \vdots & \vdots & \ddots & \vdots & \vdots & \vdots & \ddots & \vdots \\
\bar{\mathbf{U}}^T\mathbf{d}_N^{\times}\mathrm{C}_{N0} & \bar{\mathbf{U}}^T\mathbf{J}_{N0} & \mathbf{O} & \mathbf{O} & \cdots & \mathbf{J}_{subN} & \bar{\mathbf{U}}^T\mathbf{d}_N^{\times}\mathrm{C}_{N0}\boldsymbol{\psi}_{N1} & \bar{\mathbf{U}}^T\mathbf{d}_N^{\times}\mathrm{C}_{N0}\boldsymbol{\psi}_{N2} & \cdots & \bar{\mathbf{U}}^T\mathbf{d}_N^{\times}\mathrm{C}_{N0}\boldsymbol{\psi}_{NM} \\
\mathbf{0}^T & \mathbf{0}^T & -\boldsymbol{\psi}_{11}^T\mathrm{C}_{01}\mathbf{d}_1^{\times}\bar{\mathbf{U}} & -\boldsymbol{\psi}_{21}^T\mathrm{C}_{02}\mathbf{d}_2^{\times}\bar{\mathbf{U}} & \cdots & -\boldsymbol{\psi}_{N1}^T\mathrm{C}_{0N}\mathbf{d}_N^{\times}\bar{\mathbf{U}} & m_{B1} & 0 & \cdots & 0 \\
\mathbf{0}^T & \mathbf{0}^T & -\boldsymbol{\psi}_{12}^T\mathrm{C}_{01}\mathbf{d}_1^{\times}\bar{\mathbf{U}} & -\boldsymbol{\psi}_{22}^T\mathrm{C}_{02}\mathbf{d}_2^{\times}\bar{\mathbf{U}} & \cdots & -\boldsymbol{\psi}_{N2}^T\mathrm{C}_{0N}\mathbf{d}_N^{\times}\bar{\mathbf{U}} & 0 & m_{B2} & \cdots & 0 \\
\vdots & \vdots & \vdots & \vdots & \ddots & \vdots & \vdots & \vdots & \ddots & \vdots \\
\mathbf{0}^T & \mathbf{0}^T & -\boldsymbol{\psi}_{1M}^T\mathrm{C}_{01}\mathbf{d}_1^{\times}\bar{\mathbf{U}} & -\boldsymbol{\psi}_{2M}^T\mathrm{C}_{02}\mathbf{d}_2^{\times}\bar{\mathbf{U}} & \cdots & -\boldsymbol{\psi}_{NM}^T\mathrm{C}_{0N}\mathbf{d}_N^{\times}\bar{\mathbf{U}} & 0 & 0 & 0 & m_{BM}
\end{bmatrix}
\begin{bmatrix}
\mathbf{a}_b \\ \dot{\boldsymbol{\omega}} \\ \dot{\bar{\boldsymbol{\omega}}}_{r1} \\ \dot{\bar{\boldsymbol{\omega}}}_{r2} \\ \vdots \\ \dot{\bar{\boldsymbol{\omega}}}_{rN} \\ \ddot{\eta}_1 \\ \ddot{\eta}_2 \\ \vdots \\ \ddot{\eta}_M
\end{bmatrix}
\tag{4.3.11}
$$

Eq. (4.3.11): Mass matrix for prescribed motion with pendulum slosh

The RHS forcing function for the above matrix equation is given by (4.2.60), truncated as described below. The assumption of prescribed engine motion means that additional forcing terms are added to the RHS as described in Chapter 2. There are no such terms for the slosh equations. In (4.2.60), **f** and **g** are the external force and torque defined by the components described in Table 4.2. Actual computation of **f** and **g** is given in Chapter 5. The external torque on each pendulum is given by

$$\mathbf{g}_1 = g\mathbf{d}_1^{\times}\mathbf{C}_{10}\mathbf{u}_v, \quad \mathbf{g}_2 = g\mathbf{d}_2^{\times}\mathbf{C}_{20}\mathbf{u}_v, \quad \text{etc.} \tag{4.3.12}$$

where g (italic with no subscript) is the acceleration of gravity, and $\mathbf{u}_v$ is a unit vector in the downward vertical direction, expressed in the rocket body frame. This unit vector is also used in the system translation and rotation equations, as discussed when gravity is incorporated in Chapter 5.

In the pendulum model, there is no need for a torsional spring to return the pendulum to its equilibrium position. This centering tendency is built into the mass matrix, and is present whenever the thrust is sufficient to provide a forward acceleration. Thus in (4.2.60), the only interbody torque is that due to damping of the pendulum:

$$g_{H1y} = -m_{s1} l_{p1}^2 \left(2\zeta_{s1}\Omega_{s1}\right)\omega_{r1y} \tag{4.3.13}$$

$$g_{H1z} = -m_{s1} l_{p1}^2 \left(2\zeta_{s1}\Omega_{s1}\right)\omega_{r1z} \tag{4.3.14}$$

where $\Omega_{s1} = \sqrt{\frac{\|\mathbf{a}_b - g\mathbf{u}_v\|}{l_{p1}}}$ is the pendulum natural frequency, and ζ_{s1} is the damping ratio. Eq. (4.2.60) requires the computation of the nonlinear term $\mathbf{g}_{NL1}$ from (4.2.57). It is important to recognize that the $\boldsymbol{\omega}_1$ (with no r) of (4.2.57) is still the absolute angular rate. Since the subscript 1 of Section 4.2 translates to j here, it follows that

$$\boldsymbol{\omega}_j = \mathbf{C}_{j0}\boldsymbol{\omega} + \boldsymbol{\omega}_{rj}. \tag{4.3.15}$$

Method of solution for the slosh pendulum model

A somewhat contrived example will be used to illustrate the method of solution. Suppose we have a rocket with one slosh mass and one flexible mode. The first step is to compute the acceleration from

$$\ddot{\mathbf{x}} = \mathcal{M}^{-1}\bar{\mathcal{F}}$$

where

$$\ddot{\mathbf{x}} = \begin{bmatrix} \mathbf{a}_b \\ \dot{\boldsymbol{\omega}} \\ \dot{\bar{\boldsymbol{\omega}}}_{r1} \\ \ddot{\eta} \end{bmatrix} \tag{4.3.16}$$

and $\bar{\mathcal{F}}$ is the truncated version of $\mathcal{F}$. That is, all ten components of $\mathcal{F}$ are computed, and then the result is truncated (the slosh roll component is deleted) to fit into the matrix equation. In order to do this, it is necessary to compute $\mathbf{g}_{NL1}$ from (4.2.57). This is done using $\boldsymbol{\omega}$ and $\bar{\boldsymbol{\omega}}_{r1}$ from the previous time step. It is necessary to make up a value for the x component of $\boldsymbol{\omega}_{r1}$. The recommended approach is to set this component to zero. A more fastidious approach is to compute ω_{r1x} via (4.2.73). This approach results in consistency between the Euler rates and the angular rates of the slosh pendulum. It is not clear that this is necessary.

It turns out that all four elements of $\ddot{\mathbf{x}}$ must be integrated in a different way. The first element of (4.3.16), $\mathbf{a}_b$, is not holonomic. A detailed prescription for integrating the motion equations will be given in Chapter 10, but at this stage it is sufficient to state that the translation equation is solved by transforming from the body frame to the inertial frame

$$\dot{\mathbf{v}}_I = \mathbf{C}_{I0}\mathbf{a}_b \tag{4.3.17}$$

where $\mathbf{C}_{I0}$ is the transformation from body 0 frame to the inertial frame shown in Fig. 4.1. $\dot{\mathbf{v}}_I$ can be integrated to obtain the velocity of the origin O in the inertial frame. After the first integration, we are left with

$$\dot{\mathbf{x}} = \begin{bmatrix} \mathbf{v}_I \\ \boldsymbol{\omega} \\ \bar{\boldsymbol{\omega}}_{r1} \\ \dot{\eta} \end{bmatrix} \tag{4.3.18}$$

Element one of this can be integrated a second time to get position. However, elements two and three of this are not holonomic. The body angular velocity $\boldsymbol{\omega}$ must be transformed to quaternion rates, which can be integrated to get quaternions. This is also described in Chapter 10. The slosh angular velocity $\bar{\boldsymbol{\omega}}_{r1}$ is converted to Euler rates via (4.2.73). These can be integrated to get the Euler angles, which in turn are used to get $\mathbf{C}_{10}$. This allows computation of a new mass matrix. The flex amplitude acceleration $\ddot{\eta}$ is the only element of the state vector that can be integrated twice without any special treatment.

Linearized quasi-steady mass matrix for the pendulum model

The above mass matrix must be recomputed at each time step. For some applications, such as a frequency domain analysis tool, this is undesirable. This section describes how to create a quasi-steady mass matrix. We can write

$$\mathbf{C}_{10} = \mathbf{1} + \mathbf{T}_{10} \tag{4.3.19}$$

where

$$\mathbf{T}_{10} \equiv \mathbf{C}_{10} - \mathbf{1} \tag{4.3.20}$$

is the transformation perturbation, which equals the null matrix when the pendulum frame is lined up with the body frame. The word "secular" is defined as any term that remains finite even when there are no disturbances to the system. If the pendulum angles are small, we can use Eq. (4.2.76) to compute $\mathbf{C}_{10}$ from which we immediately see that $\mathbf{T}_{10}$ does not contain any secular terms.

We now substitute (4.3.19) into the mass matrix $\mathcal{M}$. In the acceleration vector $\ddot{\mathbf{x}}$ of (4.3.11), all elements are small perturbations except the first element $\mathbf{a}_b$, which contains the effect of steady-state thrust. The phrase "linearized analysis" means that we throw out all terms that contain products of two or more small perturbations. Since $\mathbf{a}_b$ is the only element with a secular term, it follows that for all columns except the first column of $\mathcal{M}$, every instance of $\mathbf{C}_{10}$ can be replaced by $\mathbf{1}$, that is, the $\mathbf{T}_{10}$ can be dropped since it will end up multiplying another small quantity. Thus the first column is the only one in which $\mathbf{T}_{10}$ has a first-order effect. This will be examined for the case of a single pendulum, from which the results can easily be extended to multiple pendulums. The element in the second row of the first column is $\mathbf{s}_{TD}^{\times} = \mathbf{d}^{\times}$. Substituting the transpose of (4.3.19) into (4.3.8) gives

$$\mathbf{d} = \mathbf{d}_0 + m_{s1}\mathbf{b}_{H1} + \mathbf{d}_1 + \mathbf{T}_{01}\mathbf{d}_1 \tag{4.3.21}$$

where $\mathbf{T}_{01} = \mathbf{T}_{10}^T = -\mathbf{T}_{10}$. This becomes

$$\mathbf{d} = \mathbf{d}_0 + m_{s1}\mathbf{b}_{H1} + \mathbf{d}_1 + \begin{bmatrix} 0 & -\gamma_{1z} & +\gamma_{1z} \\ +\gamma_{1z} & 0 & 0 \\ -\gamma_{1y} & 0 & 0 \end{bmatrix}\begin{bmatrix} -m_{s1}l_{p1} \\ 0 \\ 0 \end{bmatrix} \tag{4.3.22}$$

$$\mathbf{d} = \mathbf{d}_0 + m_{s1}\mathbf{b}_{H1} + \mathbf{d}_1 + \begin{bmatrix} 0 \\ -m_{s1}l_{p1}\gamma_{1z} \\ +m_{s1}l_{p1}\gamma_{1y} \end{bmatrix} \tag{4.3.23}$$

Using the relationship between $\mathbf{s}_T$ and $\mathbf{s}_{TD}$, it follows that

$$\mathbf{s}_{TD} = \mathbf{d} = \mathbf{s}_T + \begin{bmatrix} 0 \\ -m_{s1} l_{p1} \gamma_{1z} \\ +m_{s1} l_{p1} \gamma_{1y} \end{bmatrix} \tag{4.3.24}$$

where

$$\mathbf{s}_T = \mathbf{d}_0 + m_{s1} \mathbf{b}_{H1} + \mathbf{d}_1$$

Putting (4.3.24) into the rotation equation (2.5.2) produces

$$\mathbf{s}_T^{\times} \mathbf{a}_b + m_{s1} l_{p1} \begin{bmatrix} 0 \\ -\gamma_{1z} \\ +\gamma_{1y} \end{bmatrix}^{\times} \mathbf{a}_b + \mathbf{J}\dot{\boldsymbol{\omega}} + \mathbf{J}_{01} \dot{\boldsymbol{\omega}}_{r1} = \mathbf{g} \tag{4.3.25}$$

In the second term, it is only necessary to include the secular portion of $\mathbf{a}_b$:

$$\mathbf{s}_T^{\times} \mathbf{a}_b + m_{s1} l_{p1} \begin{bmatrix} 0 & \gamma_{1z} & +\gamma_{1z} \\ +\gamma_{1z} & 0 & 0 \\ -\gamma_{1y} & 0 & 0 \end{bmatrix} \begin{bmatrix} (F - D)/m_T \\ 0 \\ 0 \end{bmatrix} + \mathbf{J}\dot{\boldsymbol{\omega}} + \mathbf{J}_{01} \dot{\boldsymbol{\omega}}_{r1} = \mathbf{g} \tag{4.3.26}$$

After performing the above matrix vector multiplication, move the variable portion to the RHS:

$$\mathbf{s}_T^{\times} \mathbf{a}_b + \mathbf{J}\dot{\boldsymbol{\omega}} + \mathbf{J}_{01} \dot{\boldsymbol{\omega}}_{r1} = \frac{F - D}{m_T} m_{s1} l_{p1} \begin{bmatrix} 0 \\ -\gamma_{1z} \\ -\gamma_{1y} \end{bmatrix} + \mathbf{g} \tag{4.3.27}$$

Based on the arguments following (4.3.20), we can use the steady-state values of $\mathbf{J}$ and $\mathbf{J}_{01}$. Thus the left side of (4.3.27) is constant, or at least quasi-steady, meaning that it only varies slowly as propellant is depleted.

A similar procedure is applied to the slosh equation. We substitute Eq. (4.3.19) into Eq. (4.2.56) and then move the $\mathbf{T}_{10}$ term to the RHS.

$$\mathbf{d}_1^{\times} \mathbf{a}_b + \mathbf{J}_{10} \dot{\boldsymbol{\omega}} + \mathbf{J}_1 \dot{\boldsymbol{\omega}}_{r1} = -\mathbf{d}_1^{\times} \mathbf{T}_{10} \mathbf{a}_b + \mathbf{g}_{H1} + \mathbf{g}_1 \tag{4.3.28}$$

$$\mathbf{d}_1^{\times} \mathbf{a}_b + \mathbf{J}_{10} \dot{\boldsymbol{\omega}} + \mathbf{J}_1 \dot{\boldsymbol{\omega}}_{r1} = - \mathbf{d}_1^{\times} \begin{bmatrix} 0 & -\gamma_{1z} & +\gamma_{1z} \\ +\gamma_{1z} & 0 & 0 \\ -\gamma_{1y} & 0 & 0 \end{bmatrix} \begin{bmatrix} (F - D)/m_T \\ 0 \\ 0 \end{bmatrix} + \mathbf{g}_{H1} + \mathbf{g}_1 \tag{4.3.29}$$

$$\mathbf{d}_1^\times \mathbf{a}_b + \mathbf{J}_{10}\dot{\boldsymbol{\omega}} + \mathbf{J}_1\dot{\boldsymbol{\omega}}_{r1} = \\ -\frac{F-D}{m_T}\begin{bmatrix} 0 & 0 & 0 \\ 0 & 0 & m_{s1}l_{p1} \\ 0 & -m_{s1}l_{p1} & 0 \end{bmatrix}\begin{bmatrix} 0 \\ -\gamma_{1z} \\ \gamma_{1y} \end{bmatrix} + \mathbf{g}_{H1} + \mathbf{g}_1 \quad (4.3.30)$$

$$\mathbf{d}_1^\times \mathbf{a}_b + \mathbf{J}_{10}\dot{\boldsymbol{\omega}} + \mathbf{J}_1\dot{\boldsymbol{\omega}}_{r1} = -\frac{F-D}{m_T}m_{s1}l_{p1}\begin{bmatrix} 0 \\ -\gamma_{1z} \\ -\gamma_{1y} \end{bmatrix} + \mathbf{g}_{H1} + \mathbf{g}_1 \quad (4.3.31)$$

$\mathbf{g}_1$ and $\mathbf{g}_{H1}$ are given by (4.3.12) through (4.3.14). The first term on the RHS is the "spring term" – the tendency of the slosh pendulum to develop a torque that returns it to the zero position, assuming the thrust is greater than the drag. The slosh equations for all the other slosh masses are similarly changed. Doing so moves the variable portion of each of the remaining elements in the first column of the mass matrix to the RHS. This results in a linearized mass matrix that is quasi-steady.

It may be desirable to convert the translation equation from body coordinates to inertial coordinates. An explanation of how to do this for a linearized analysis is provided in Chapter 7.

Relation of the linearized pendulum slosh model to the spring slosh model

Eq. (2.5.4) is the slosh equation for the spring model. The slosh excitation is the sum of a spring term, a damping term, and a gravity term:

$$\mathbf{f}_{sj} = m_{sj}\left(-\Omega_{sj}^2\boldsymbol{\delta}_{sj} - 2\zeta_{sj}\Omega_{sj}\dot{\boldsymbol{\delta}}_{sj} + g\mathbf{u}_v\right)$$

The slosh equation becomes

$$m_{sj}\mathbf{a}_b - m_{sj}\mathbf{r}_{sj}^\times\dot{\boldsymbol{\omega}} + m_{sj}\ddot{\boldsymbol{\delta}}_{sj} + m_{sj}\sum_i \boldsymbol{\psi}_{ji}\ddot{\eta}_i \\ = m_{sj}\left(-\Omega_{sj}^2\boldsymbol{\delta}_{sj} - 2\zeta_{sj}\Omega_{sj}\dot{\boldsymbol{\delta}}_{sj} + g\mathbf{u}_v\right) \quad (4.3.32)$$

In this form, every term has units of force. To convert to torques about the pendulum hinge, the whole equation is multiplied by $\mathbf{r}_j^\times$, where $\mathbf{r}_j$ is the pendulum vector defined in Fig. 4.1. This is not the same as the $\mathbf{r}_{sj}$ of Chapter 2, which is the distance from the system origin to the slosh mass. The relation between $\dot{\boldsymbol{\delta}}_{sj}$ from the spring model and $\boldsymbol{\omega}_{rj}$ of the pendulum model is

$$\dot{\boldsymbol{\delta}}_{sj} = \boldsymbol{\omega}_{rj}^\times\mathbf{r}_j \quad (4.3.33)$$

The derivative and integral of this are

$$\ddot{\boldsymbol{\delta}}_{sj} = \dot{\boldsymbol{\omega}}_{rj}^{\times}\mathbf{r}_j = -\mathbf{r}_j^{\times}\dot{\boldsymbol{\omega}}_{rj} \tag{4.3.34}$$

$$\boldsymbol{\delta}_{sj} = -\mathbf{r}_j^{\times}\int \boldsymbol{\omega}_{rj} = -\mathbf{r}_j^{\times}\boldsymbol{\gamma}_{sj} \tag{4.3.35}$$

This expression can be inserted into (4.3.32) and the result is multiplied by $\mathbf{r}_j^{\times}$. We then use

$$\mathbf{r}_j^{\times} = \mathbf{d}_j^{\times}/m_{sj} \tag{4.3.36}$$

The whole process gives

$$\begin{aligned}\mathbf{d}_j^{\times}\mathbf{a}_b - \mathbf{d}_j^{\times}\mathbf{r}_{sj}^{\times}\dot{\boldsymbol{\omega}} - \mathbf{d}_j^{\times}\mathbf{r}_j^{\times}\dot{\boldsymbol{\omega}}_{rj} + \mathbf{d}_j^{\times}\sum_i \boldsymbol{\psi}_{ji}\ddot{\eta}_i \\ = -\Omega_{sj}^2\mathbf{d}_j^{\times}\mathbf{r}_j^{\times}\boldsymbol{\gamma}_{sj} - 2\zeta_{sj}\Omega_{sj}\mathbf{d}_j^{\times}\mathbf{r}_j^{\times}\boldsymbol{\omega}_{rj} + g\mathbf{d}_j^{\times}\mathbf{u}_v\end{aligned} \tag{4.3.37}$$

Thus the linearized mass matrix can be converted from the spring model to the pendulum model via the following steps:

1. Pre-multiply each slosh row by $\mathbf{r}_j^{\times}$.
2. Post-multiply each slosh column by $-\mathbf{r}_j^{\times}$.
3. Eliminate $\mathbf{r}_j$ using Eq. (4.3.36).

The spring model uses the vector $\mathbf{b}_{sj}$ giving the equilibrium position of the slosh mass, whereas the pendulum model uses the vector $\mathbf{b}_{Hj}$ giving the hinge position; see (4.3.4). Both of these vectors are defined relative to the body origin. It is necessary to show that (4.3.31), from the pendulum model, is equivalent to (4.3.37), from the spring model. The two are not exactly the same, since (4.3.31) does not include flex. We immediately see that the first terms of these expressions match. Skipping for the moment to the third terms, note that the factor multiplying $\dot{\boldsymbol{\omega}}_{rj}$ in (4.3.37) is the second moment of inertia:

$$-\mathbf{d}_j^{\times}\mathbf{r}_j^{\times} = -m_{sj}\mathbf{r}_j^{\times}\mathbf{r}_j^{\times} = \mathbf{J}_j \tag{4.3.38}$$

Thus the third term in (4.3.37) matches the third term in (4.3.31). Turning now to the second term in (4.3.31), if $\mathbf{C}_{10} = \mathbf{1}$, we also have, from (4.2.15), (4.3.2), and (4.3.38)

$$\mathbf{J}_{j0} = \mathbf{J}_j - \mathbf{d}_j^{\times}\mathbf{b}_{Hj}^{\times} = -\mathbf{d}_j^{\times}\left(\mathbf{r}_j^{\times} + \mathbf{b}_{Hj}^{\times}\right) \tag{4.3.39}$$

Using (4.3.4), this becomes

$$\mathbf{J}_{j0} = -\mathbf{d}_j^{\times}\mathbf{b}_{sj}^{\times} \tag{4.3.40}$$

This is the same as the term multiplying $\dot{\boldsymbol{\omega}}$ in (4.3.37), since $\mathbf{r}_{sj} = \mathbf{b}_{sj}$ under equilibrium conditions. Thus the first three terms of (4.3.37) can be made to match the first three terms of (4.3.31). The fourth term of (4.3.37) shows how flex is included. The terms on the RHS of (4.3.31) can also be shown to match those on the RHS of (4.3.37). Thus it turns out that all the relevant terms in (4.3.31) are the same as those of (4.3.37).

A quasi-steady mass matrix for the pendulum model may be required if a linearized frequency domain analysis is needed to verify closed-loop stability. Its value in a time-domain simulation is not obvious. For small slosh displacements, the pendulum model is the same as the spring model. The whole point of using the pendulum model is to obtain better model fidelity for the case of large slosh displacements. We can define a "large" displacement as one for which the pendulum angle γ differs significantly from $\sin\gamma$. The process of setting $\sin\gamma = \gamma$ erases any distinction between the models. Unless the sine effects are captured, there is no benefit to the pendulum model. A rigorous implementation of the pendulum model, in which the mass matrix is updated at every time step, will capture these effects. It is not clear how to capture these effects in a quasi-steady model. The main benefit of Eqs. (4.3.19) through (4.3.40) is to show the relationship between the models. It is also worth noting that if a spring slosh model has been implemented and the desire is to convert this implementation to a pendulum model, these equations can provide a means of cross checking.

4.4 Nozzle dynamics using the pendulum model

Throughout this book, the terms engine and nozzle are used interchangeably. One thing that distinguishes the nozzle dynamics from the slosh model is that the nozzle rotation about its x axis is constrained. Thus there is a finite constraint torque about this axis. Note from (4.2.73) that the body rate about the x axis is not in general zero even if the Euler rate about the x axis is zero. For the case of a gimbal-mounted rocket engine, setting the Euler rate about x to zero would be a good representation of the gimbal mechanism. There exist general approaches to this class of problem in the dynamics literature. These methods require that we recast Eqs. (4.2.58) through (4.2.61) in terms of gimbal angle rates (essentially Euler rates) rather than $\boldsymbol{\omega}_r$'s, using projection matrices. The projection matrices and their time derivatives create more complicated nonlinear terms, with an additional computational burden. Because of the fact that rocket nozzle motion is restricted to very small angles, typically less than 10 degrees, this additional computation is usually unwarranted.

The simplest rocket dynamic analyses assume prescribed engine motion, in which case the torques on the RHS of Eq. (4.2.56) are irrelevant. This simplification is justified by the fact that the nozzle actuators are "stiff," i.e., to first order, they do what they are commanded to do in spite of any motion of the rocket body. Removing the assumption that the nozzle motion is prescribed and including the first-order *linear* effects of rocket body motion on the nozzle motion thus represents a degree of accuracy that goes beyond this. Going further, and including the small *nonlinear* torques on the nozzle motion is not required unless there are very unusual circumstances. However, there is less difficulty computing the nonlinear effects of nozzle motion on the rocket body itself, which can be done using the same equations as for the slosh pendulum.

In view of the above arguments, the following is recommended:

1. At the first stage of integration (i.e., while integrating the accelerations), use $\boldsymbol{\omega}_{r1}$ in the state vector, as is done herein, rather than using gimbal angle rates, a.k.a. joint angle rates.
2. Do not include nonlinear terms in the nozzle equation.
3. Include the nonlinear terms from Eqs. (4.2.38) and (4.2.48) in the rotation and translation equations.

Conversion to the notation of Chapter 2 requires the following substitutions:

$$\mathbf{J} = \mathbf{I}_{TD} \tag{4.4.1}$$

$$\mathbf{d} = \mathbf{s}_{TD} \tag{4.4.2}$$

$$\mathbf{b}_1 = \mathbf{r}_G \tag{4.4.3}$$

$$\mathbf{J}_1 = \mathbf{I}_E \tag{4.4.4}$$

$$\mathbf{d}_1 = \mathbf{s}_E \tag{4.4.5}$$

$$\boldsymbol{\omega}_{r1} = \boldsymbol{\omega}_E \tag{4.4.6}$$

The vector $\mathbf{r}_G$ is shown in Fig. 4.1. Here, $\mathbf{s}_E$ and $\mathbf{I}_E$ are the first and second moments of inertia of the engine, expressed in the engine frame. Recall that the body frame analysis in Chapter 2 uses $\mathbf{s}_{Eb}$ and $\mathbf{I}_{Eb}$. The subscript b was attached to these parameters to indicate they are expressed in the body frame. The conversion between frames is given by

$$\mathbf{s}_{Eb} = \mathbf{C}_{01}\mathbf{s}_E \tag{4.4.7}$$

$$\mathbf{I}_{Eb} = \mathbf{C}_{01}\mathbf{I}_E\mathbf{C}_{10} \tag{4.4.8}$$

$$\boldsymbol{\omega}_{Eb} = \mathbf{C}_{01}\boldsymbol{\omega}_E \tag{4.4.9}$$

It is natural to think of engine mass properties expressed in the engine frame as the input parameters for the dynamic problem. Thus the above equations are presented as conversions from the engine frame to the body frame. Even if it is decided to leave the engine equations in the engine frame, it may still be necessary to have engine mass properties in the body frame in order to compute the engine/flex coupling vectors $\mathbf{c}_{EFi}$ given by (2.4.22). Also, if a reduced body model is used for the FEM, the parameters defined in Eqs. (2.6.18) through (2.6.19) require mass properties in the body frame. Some simplification can result from the assumption that the nozzle is axisymmetric about the nozzle x axis, and that the only significant rotation represented in $\mathbf{C}_{10}$ is about the body x axis.

These assumptions are aimed at the situation in which the gimbal axis frame is rotated by some fixed angle (typically 45 degrees) about the x axis. The nozzle deflections about the y and z axes are assumed to be small enough that they can be ignored. It is common to refer to these rotated gimbal axes as rock and tilt axes rather than y (pitch) and z (yaw) axes. Under the above assumptions we have

$$\mathbf{I}_{Eb} = \mathbf{I}_E \tag{4.4.10}$$

and

$$\mathbf{s}_{Eb} = \mathbf{s}_E \tag{4.4.11}$$

That is, the engine first and second moments of inertia are unchanged by rotation about the x axis. If the gimbal frame is canted, i.e., the nominal value of $\mathbf{C}_{10}$ represents a rotation about an axis that is not parallel to the body x axis, then this simplification is not applicable.

Eq. (4.2.19) becomes

$$\mathbf{J}_{01} = \mathbf{C}_{01}\mathbf{I}_E - \mathbf{r}_G^{\times}\mathbf{C}_{01}\mathbf{s}_E^{\times} \tag{4.4.12}$$

By comparing to (2.4.23) we see that if $\mathbf{C}_{01} = \mathbf{1}$, then $\mathbf{J}_{01} = \mathbf{I}_{TWD}$. However, even if both of the above simplifications are applicable, and the rotation due to nozzle deflections is neglected, it is still necessary to include $\mathbf{C}_{01}$ in (4.4.12) if the engine frame is rotated about the x axis. Eqs. (4.2.38) and (4.2.48) become

$$\mathbf{f}_{NL} = \boldsymbol{\omega}^{\times}\mathbf{d}^{\times}\boldsymbol{\omega} + 2\boldsymbol{\omega}^{\times}\mathbf{C}_{01}\mathbf{s}_E^{\times}\boldsymbol{\omega}_E + \mathbf{C}_{01}\boldsymbol{\omega}_E^{\times}\mathbf{s}_E^{\times}\boldsymbol{\omega}_E \tag{4.4.13}$$

$$\mathbf{g}_{NL} = -\boldsymbol{\omega}^{\times}\mathbf{I}_{TD}\boldsymbol{\omega} - \boldsymbol{\omega}^{\times}\mathbf{C}_{01}\mathbf{I}_{E}\boldsymbol{\omega}_{E} - \mathbf{C}_{01}\boldsymbol{\omega}_{E}^{\times}\mathbf{I}_{E}\left(\mathbf{C}_{10}\boldsymbol{\omega} + \boldsymbol{\omega}_{E}\right) + \mathbf{C}_{01}\mathbf{I}_{E}\boldsymbol{\omega}_{E}^{\times}\mathbf{C}_{10}\boldsymbol{\omega} + 2\mathbf{r}_{G}^{\times}\boldsymbol{\omega}^{\times}\mathbf{C}_{01}\mathbf{s}_{E}^{\times}\boldsymbol{\omega}_{E} + \mathbf{r}_{G}^{\times}\mathbf{C}_{01}\boldsymbol{\omega}_{E}^{\times}\mathbf{s}_{E}^{\times}\boldsymbol{\omega}_{E} \quad (4.4.14)$$

These are the desired nonlinear effects used in Chapter 5 for the translation and rotation equations. Through a similar process, one could develop an expression for the nonlinear torque for the engine equation, but as discussed at the beginning of this section, under normal circumstances such effects are too small to be significant.

The engine equation is not complete until the effects of flexibility are included. These effects are derived in Chapter 2. Pre-multiplying (2.5.3) by $\mathbf{C}_{10}$ gives

$$\mathbf{C}_{10}\mathbf{s}_{Eb}^{\times}\mathbf{a}_{b} + \mathbf{C}_{10}\mathbf{I}_{TWD}^{T}\dot{\boldsymbol{\omega}} + \mathbf{C}_{10}\mathbf{I}_{Eb}\dot{\boldsymbol{\omega}}_{Eb} + \mathbf{C}_{10}\sum \mathbf{c}_{EFi}\ddot{\eta}_{i} = \mathbf{C}_{10}\mathbf{g}_{Eb} = \mathbf{g}_{E} \quad (4.4.15)$$

One can either make a careful substitution of (4.4.7) through (4.4.9) into this, or use (4.2.56) to replace the first three terms on the LHS. Either process gives the same result, although the latter approach is more straightforward and less likely to lead to error. If nonlinear torques are neglected, the driving torques on the RHS are related by

$$\mathbf{g}_{E} = \mathbf{g}_{1} + \mathbf{g}_{H1} \quad (4.4.16)$$

Here, $\mathbf{g}_1$ is the external torque on the engine due to gravity plus aerodynamic forces. The latter can be ignored for most rocket configurations. Thus $\mathbf{g}_1$ can be computed from (4.3.12) and (4.4.5) such that

$$\mathbf{g}_{1} = g\mathbf{s}_{E}^{\times}\mathbf{C}_{10}\mathbf{u}_{v} \quad (4.4.17)$$

The interbody torque is that due to the TVC system, expressed in the engine frame

$$\mathbf{g}_{H1} = \mathbf{g}_{TVC} \quad (4.4.18)$$

Using (4.2.56) as described above, plus (4.4.4) through (4.4.6), we obtain

$$\mathbf{s}_{E}^{\times}\mathbf{C}_{10}\mathbf{a}_{b} + \mathbf{J}_{10}\dot{\boldsymbol{\omega}} + \mathbf{I}_{E}\dot{\boldsymbol{\omega}}_{E} + \mathbf{C}_{10}\sum \mathbf{c}_{EFi}\ddot{\eta}_{i} = g\mathbf{s}_{E}^{\times}\mathbf{C}_{10}\mathbf{u}_{v} + \mathbf{g}_{TVC} \quad (4.4.19)$$

For the illustrative case of a single engine and no slosh masses, the mass matrix with the engine equation in the engine frame is given by (4.4.20).

$$
\begin{bmatrix}
m_T\mathbf{1} & -\mathbf{s}_{TD}^{\times} & -\mathbf{C}_{01}\mathbf{s}_E^{\times} & \mathbf{0} & \mathbf{0} & \cdots & \mathbf{0} \\
\mathbf{s}_{TD}^{\times} & \mathbf{I}_{TD} & \mathbf{J}_{10} & \mathbf{0} & \mathbf{0} & \cdots & \mathbf{0} \\
\mathbf{s}_E^{\times}\mathbf{C}_{10} & \mathbf{J}_{01} & \mathbf{I}_E & \mathbf{C}_{10}\mathbf{c}_{EF1} & \mathbf{C}_{10}\mathbf{c}_{EF2} & \cdots & \mathbf{C}_{10}\mathbf{c}_{EFM} \\
\mathbf{0}^T & \mathbf{0}^T & \mathbf{c}_{EF1}^T\mathbf{C}_{01} & m_{B1} & 0 & \cdots & 0 \\
\mathbf{0}^T & \mathbf{0}^T & \mathbf{c}_{EF2}^T\mathbf{C}_{01} & 0 & m_{B2} & \cdots & 0 \\
\vdots & \vdots & \vdots & \vdots & \vdots & \ddots & \vdots \\
\mathbf{0}^T & \mathbf{0}^T & \mathbf{c}_{EFM}^T\mathbf{C}_{01} & 0 & 0 & \cdots & m_{BM}
\end{bmatrix}
\begin{bmatrix}
\mathbf{a}_b \\ \dot{\boldsymbol{\omega}} \\ \dot{\boldsymbol{\omega}}_E \\ \ddot{\eta}_1 \\ \ddot{\eta}_2 \\ \vdots \\ \ddot{\eta}_M
\end{bmatrix}
=
\begin{bmatrix}
\tilde{\mathbf{f}} \\ \tilde{\mathbf{g}} \\ \tilde{\mathbf{g}}_E \\ \tilde{f}_{B1} \\ \tilde{f}_{B2} \\ \vdots \\ \tilde{f}_{BM}
\end{bmatrix}
\tag{4.4.20}
$$

Eq. (4.4.20): Mass matrix for pendulum engine model

For simulation purposes, whether this version of the matrix equation is preferred over (2.5.20) depends on the details of the implementation. If all of the elements of the mass matrix must be computed at every time step, and a large number of modes are included, (4.4.20) can result in excessive computation time. On the other hand, if it can be assumed that the nozzle deflections can be ignored in the computation of the mass matrix, i.e. the only reason for $\mathbf{C}_{10}$ is to capture the fixed rotation between the body frame and the gimbal frame, then it becomes acceptable to compute the elements involving $\mathbf{C}_{10}$ only as often as the coupling vectors $\mathbf{c}_{\mathrm{EFi}}$ change. Note that since (2.5.20) produces $\boldsymbol{\omega}_{Eb}$ in the body frame, there must be a conversion both ways, i.e., $\boldsymbol{\omega}_{Eb}$ must be transformed into the engine frame (or more accurately the gimbal frame) for use in the TVC model, and the TVC torques must be transformed from the gimbal frame into the body frame for use in the dynamics. Both of these conversions are avoided by using (4.4.20).

CHAPTER 5

Forces and torques

The preceding chapters developed the equations of motion for a variety of rocket configurations, and focused on identifying internal (interbody or inertial) forces and torques that depend on the state variables of the rocket or their derivatives. In this chapter, the remaining external forces needed for a simulation will be provided, and combined with the nonlinear terms that were developed in Chapters 3 and 4 to produce all of the terms on the right hand side of the equations. The pendulum model of Chapter 4 provides nonlinear terms for the nozzle motion. Chapter 4 also provides the required nonlinear terms if the pendulum model is used for slosh. If the spring slosh model is used, nonlinear slosh terms are obtained from Chapter 3.

5.1 External forces and torques

The primary external forces and moments on a rocket are due to thrust (including moments due to engine gimbaling), aerodynamics, and secondary control effectors such as reaction control thrusters. The gravitational acceleration on the rocket will also be discussed, along with the forces and torques that appear due to fluid motion known as *jet damping*. There are other forces and torques can be included in a rocket simulation but are usually not important unless the vehicle is in coast or orbit for long durations outside the atmosphere. These include gravity gradient torques, magnetic dipole torque, solar radiation pressure, and so on. For these models, the reader is referred to standard spacecraft dynamics texts such as Hughes [6].

Thrust forces and torques

The thrust can be divided into a fixed component F_0 and a gimbaled component F_R, where the total thrust $F = F_0 + F_R$. This is the case when some, but not all, of the engines can be gimbaled. For the case when the nominal thrust vector is aligned to the body x axis, the thrust forces corresponding to each of these scalars are defined as follows:

$$\mathbf{f}_0 = \begin{bmatrix} F_0 & 0 & 0 \end{bmatrix}^T \tag{5.1.1}$$

Dynamics and Simulation of Flexible Rockets
https://doi.org/10.1016/B978-0-12-819994-7.00010-8

$$\mathbf{f}_R = \begin{bmatrix} F_R & 0 & 0 \end{bmatrix}^T \tag{5.1.2}$$

Fig. 5.1 shows how this thrust produces lateral forces when there is an engine pitch β_{Ey} plus elastic rotation from a number of modes $\sum \sigma_{y\beta i}$. For ease of illustration, this figure has been drawn for the case of a single engine on the centerline. In Chapter 4, a distinction was made between the engine frame in which the engine mass properties are defined, a gimbal frame about which the rotations of the engine are about local y and z axes, and a body frame in which the equations of motion are integrated (see Chapter 4, Eqs. (4.4.7)-(4.4.9)). To simplify the present discussion, these frames are assumed to be nominally aligned and β_{Ey} is the rotation of the engine about the body y axis. However, these results can be easily extended to handle arbitrary engine orientations by defining (5.1.1) and (5.1.2) using a nominal unit vector in the body frame.

The gimbal location in the body frame is given by

$$\mathbf{r}_G = \begin{bmatrix} X_G & 0 & 0 \end{bmatrix}^T \tag{5.1.3}$$

The locations X along the body x axis are positive if forward of the origin and negative if aft of the origin, thus X_G is usually a negative quantity. The subscript β indicates that the modal parameters are to be taken at the location of the gimbal.

$$\psi_{z\beta i} \equiv \psi_{zi}(\mathbf{r}_G) \tag{5.1.4}$$

$$\sigma_{y\beta i} \equiv \sigma_{yi}(\mathbf{r}_G) \tag{5.1.5}$$

These allow computation of the z component of thrust, as shown in Fig. 5.1. The y component is derived similarly.

The changes in the thrust angle due to flexibility are always assumed to be small rotations. The effects of rotating the thrust vector with the engine gimbal depend on the relationship of the engine to the body frame and the mechanism that moves the thrust vector. Under the aforementioned assumption that the engine, gimbal, and body frames are nominally aligned, the rotated thrust vector is given by (using Chapter 4, Eq. (4.2.72))

$$\mathbf{f}_R' = F_R \begin{bmatrix} \cos\beta_{Ey}\cos\beta_{Ez} \\ \sin\beta_{Ez} \\ -\cos\beta_{Ez}\sin\beta_{Ey} \end{bmatrix} \tag{5.1.6}$$

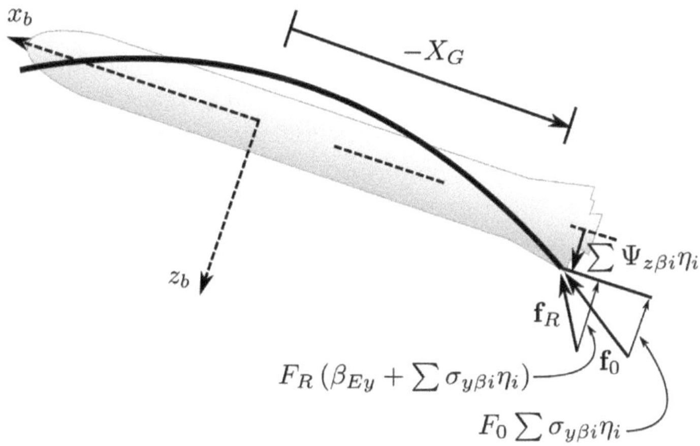

Figure 5.1 Generalized forces in the pitch plane resulting from thrust.

If the gimbal angles are also small and independent, the approximations $\cos\beta_{Ez} = 1$ and $\sin\beta_{Ey} = \beta_{Ey}$ can be used. The three components of the thrust force are therefore

$$f_{x\ \text{thrust}} = F_0 + F_R = F \tag{5.1.7}$$

$$f_{y\ \text{thrust}} = F_R\beta_{Ez} + F\sum \sigma_{z\beta i}\eta_i \tag{5.1.8}$$

$$f_{z\ \text{thrust}} = -F_R\beta_{Ey} - F\sum \sigma_{y\beta i}\eta_i \tag{5.1.9}$$

The corresponding vector expression is

$$\mathbf{f}_{\text{thrust}} = \mathbf{f}_0 + \mathbf{f}_R + \left(\sum \eta_i \boldsymbol{\sigma}_{\beta i}\right)^{\times} (\mathbf{f}_0 + \mathbf{f}_R) + \boldsymbol{\beta}_E^{\times}\mathbf{f}_R \tag{5.1.10}$$

The external torque arising from this thrust is

$$\mathbf{g}_{\text{thrust}} = \left(\mathbf{r}_G + \sum \eta_i \boldsymbol{\psi}_{\beta i}\right)^{\times} \mathbf{f}_{\text{thrust}} \tag{5.1.11}$$

This is the torque about whatever origin is used to define $\mathbf{r}_G$.

Steering loss

It was suggested in Chapter 4 that a small angle approximation can produce unacceptable errors in simulations if the gimbal angles are large. This problem can also occur if the gimbal angles have a steady state value, for example, due to aerodynamic trim. The problem arises because the true axial component of thrust, $f_x = F_1 \cos\beta_{Ey} \cos\beta_{Ez}$, is rotated away from the direction that accelerates the rocket along its trajectory. This force is wasted to counter winds or aerodynamic moments, or to cant the engines and reduce sensitivity to center of mass uncertainty. The deflected force does not contribute to the rocket's performance. In launch vehicle performance analysis, this term is called *steering loss*. A similar effect also occurs due to flexibility, although it is usually very small.

If it is necessary to capture the effect of rotations on the axial thrust, and improve the fidelity of the thrust force expression over a larger range of angles, a second-order small rotation can be used. The details of this transformation are discussed in Appendix C. The thrust force is then given by

$$\begin{aligned}\mathbf{f}_{\text{thrust}} &= \mathbf{f}_0 + \mathbf{f}_R \\ &\quad + \left[\left(\sum \eta_i \boldsymbol{\sigma}_{\beta i}\right)^{\times} - \frac{1}{2}\left(\sum \eta_i \boldsymbol{\sigma}_{\beta i}\right)^{\times}\left(\sum \eta_i \boldsymbol{\sigma}_{\beta i}\right)^{\times}\right]\mathbf{f}_0 \\ &\quad + \left[\left(\boldsymbol{\beta}_E + \sum \eta_i \boldsymbol{\sigma}_{\beta i}\right)^{\times} - \frac{1}{2}\left(\boldsymbol{\beta}_E + \sum \eta_i \boldsymbol{\sigma}_{\beta i}\right)^{\times}\left(\boldsymbol{\beta}_E + \sum \eta_i \boldsymbol{\sigma}_{\beta i}\right)^{\times}\right]\mathbf{f}_R\end{aligned} \tag{5.1.12}$$

where

$$\boldsymbol{\beta}_E = \begin{bmatrix} 0 & \beta_{Ey} & \beta_{Ez} \end{bmatrix}^T. \tag{5.1.13}$$

The computation of the moment (5.1.11) is unchanged. This expression is a better approximation that balances fidelity and complexity, since it does not require any kinematic equations.

Jet damping

There is a torque that develops if a rocket has a finite rotation rate and mass is being expelled. The term *jet damping* arises from the fact that this force usually acts to decrease the rotation rate. This is discussed by Thomson [22] and Wertz [23].

Fig. 5.2 introduces the concept of a "pipe." This is meant to convey the simple notion that propellant mass originates in a tank and exits out the back. Only the mass flow rate $\dot{m}$ matters, so a simple pipe is adequate to describe the flow. The analysis can ignore the details of pumps, valves, area changes in the nozzle, or even the startling changes in density and velocity that occur in the combustion chamber. $\dot{m}$ is the rate at which the rocket mass changes, so this is a negative quantity. The forward end of the pipe is

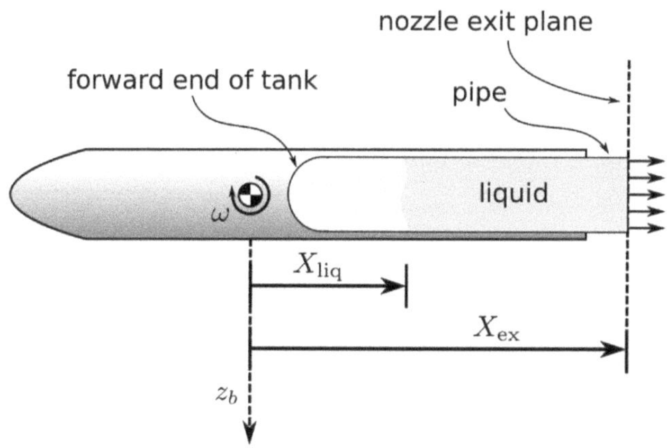

Figure 5.2 Rotating liquid rocket ejecting propellant.

at the liquid surface. The distance from the center of mass of the rocket to the forward end of the pipe is given by X_{liq}, as shown.

The rocket in Fig. 5.2 has a pitch angular rate ω. This causes the propellant gas to leave the rocket at the nozzle exit plane with a velocity perpendicular to the longitudinal axis. There is thus a continuous momentum change in the z_b direction given by

$$\dot{p}_{\text{gas}} = -\omega \dot{m} X_{\text{ex}} \tag{5.1.14}$$

where X_{ex} is the distance from the center of mass to the nozzle exit plane. (Unlike X_G in Fig. 5.1, X_{liq} and X_{ex} are positive as shown in Fig. 5.2.) This momentum is continually being added to the exit gases and must therefore come from the rocket itself. The corresponding angular momentum change to the rocket about its center of mass is

$$\dot{h} = \omega \dot{m} X_{ex}^2 \tag{5.1.15}$$

Likewise, the rate of change of angular momentum is related to the angular rate and angular acceleration by

$$\dot{h} = I_{yy}\dot{\omega} + \dot{I}_{yy}\omega \tag{5.1.16}$$

where I_{yy} is the pitch moment of inertia about the center of mass.

A suitable approximation for the moment of inertia of the propellant mass is given by

$$I_{yyprop} = \int_{X_{liq}}^{X_{ex}} x^2 \, dm \tag{5.1.17}$$

Here, dm is an element of propellant mass. This expression ignores the z contribution to the integral. Letting ρ be the density of the propellant,

$$dm = \rho A \, dx \tag{5.1.18}$$

where A is the cross-sectional area of the pipe or tank. Thus

$$I_{yyprop} = \frac{\rho A}{3}\left(X_{ex}^3 - X_{liq}^3\right) \tag{5.1.19}$$

Since the non-propellant portion of the rocket has a fixed moment of inertia as the propellant is depleted, the rate at which (5.1.19) changes is the same as the rate at which the total inertia changes. Taking the derivative of (5.1.19),

$$\dot{I}_{yy} = \dot{I}_{yyprop} = -\rho A X_{liq}^2 \dot{X}_{liq} \tag{5.1.20}$$

The speed at which the liquid surface moves is related to the mass flow rate by

$$\rho A \dot{X}_{liq} = -\dot{m} \tag{5.1.21}$$

Thus

$$\dot{I}_{yy} = \dot{m} X_{liq}^2 \tag{5.1.22}$$

This expression justifies the extreme simplicity of the pipe model. Only the location of the liquid surface matters. Having the correct cross-sectional area of the pipe is not necessary, since A does not appear in (5.1.22). It is not even necessary that the volume of the pipe be equal to the volume of the propellant, since Eq. (5.1.17) itself is not used, only its derivative. Combining (5.1.15), (5.1.16), and (5.1.22) yields

$$I\dot{\omega} = \omega \dot{m}\left(X_{ex}^2 - X_{liq}^2\right) \tag{5.1.23}$$

Jet damping can be computed equivalently in terms of the Coriolis forces on the propellant mass as it moves toward the nozzle exit plane. Consider the force on a small length dx of the pipe. The Coriolis acceleration of the propellant is

$$a_{cor} = 2\omega V \tag{5.1.24}$$

where $V = \dot{X}_{\text{liq}}$ is the rearward velocity of the propellant in the pipe. The incremental force $\mathrm{d}F$ acting on the length $\mathrm{d}x$ of the pipe acts in the direction opposite to the Coriolis acceleration of the propellant. Using Eq. (5.1.18), it follows that

$$\mathrm{d}F = -2\omega V\ \mathrm{d}m = -2\omega V\rho A\ \mathrm{d}x \tag{5.1.25}$$

However,

$$V\rho A = -\dot{m}, \tag{5.1.26}$$

thus

$$\mathrm{d}F = 2\omega\dot{m}\ \mathrm{d}x. \tag{5.1.27}$$

The torque about the center of mass, integrated over the length of the pipe, becomes

$$M_{\text{cor}} = 2\int_{X_{\text{liq}}}^{X_{\text{ex}}} \omega\dot{m}x\ \mathrm{d}x \tag{5.1.28}$$

$$M_{\text{cor}} = \omega\dot{m}\left(X_{\text{ex}}^2 - X_{\text{liq}}^2\right) \tag{5.1.29}$$

The RHS of this agrees with (5.1.23). This is the jet damping torque. Since $\dot{m}$ is negative, this torque is negative, i.e., opposite to the direction of ω, whenever $X_{\text{ex}} > X_{\text{liq}}$. It is possible to contrive a situation in which the propellant originates in a tank that is so far forward of the cm that $|X_{\text{liq}}| > |X_{\text{ex}}|$, in which case the Coriolis forces act to increase the rotation rate ω (negative damping). Other than that situation, we can expect that the jet damping will be, in fact, damping.

The analysis of jet damping for solid rocket motors differs in a couple of respects from that of liquid motors. Fig. 5.3 again shows a "pipe" of solid propellant that is used to simplify the analysis. At the beginning of the burn, the pipe starts out with a small inner diameter, so that the walls are very thick, and the propellant burns radially outward as time progresses. The forward end of the pipe remains at a fixed location in the rocket, at a distance X_1 behind the center of mass. The figure shows the aft end of the pipe at the same location as the nozzle exit plane. Obviously, this is not quite accurate, and the analysis could be refined, if desired, to recognize the fact that there must be some distance between the aft end of the propellant and the exit plane. Letting m_p be the mass of propellant, the mass per unit length is m_p/L, where

$$L = X_{\text{ex}} - X_1 \tag{5.1.30}$$

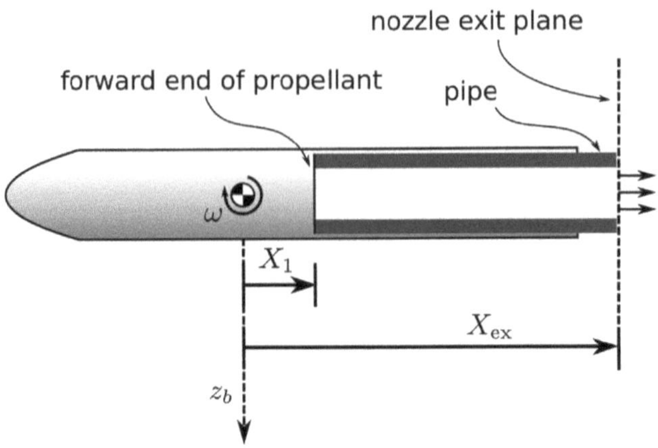

Figure 5.3 Rotating solid rocket ejecting propellant.

Using the same simplification that was used in (5.1.17) for the liquid rocket, the moment of inertia of the solid propellant is given by

$$I_{yyprop} = \int_{X_1}^{X_{ex}} \left(\frac{m_p}{L}\right) x^2 \,\mathrm{d}x = \frac{m_p}{3L}\left(X_{ex}^3 - X_1^3\right) \tag{5.1.31}$$

Taking the derivative of this quantity with respect to time,

$$\dot{I}_{yy} = \dot{I}_{yyprop} = \frac{\dot{m}}{3L}\left(X_{ex}^3 - X_1^3\right) \tag{5.1.32}$$

after substituting $\dot{m}_p = \dot{m}$. The angular momentum being carried off by propellant gas is still given by Eq. (5.1.15). Using (5.1.30), Eq. (5.1.15) can be written as

$$\dot{h} = \omega \dot{m} X_{ex}^2 \frac{(X_{ex} - X_1)}{L} \tag{5.1.33}$$

Combining (5.1.16), (5.1.32), and (5.1.33) the expression for the jet damping torque due to a solid motor burn appears on the RHS;

$$I_{yy}\dot{\omega} = \frac{\omega \dot{m}}{3L}\left(2X_{ex}^3 - 3X_1 X_{ex}^2 + X_1^3\right) \tag{5.1.34}$$

The quantity within parentheses is more or less constant, although there is some variation as the overall center of mass moves in response to propellant consumption.

It is possible to again compute the jet damping effect using Coriolis forces on a length $\mathrm{d}x$ of the pipe, for comparison with the expression in

Eq. (5.1.34). Following the same steps that led to (5.1.27), the differential force is given by

$$dF = 2\omega V\rho A\ dx = 2\omega\dot{\mu}\ dx \tag{5.1.35}$$

where V, ρ, and A are the velocity, density, and area of the gaseous flow in the pipe, which together are replaced by $\dot{\mu}(x)$, defined here as the local mass flow rate at a position x. At the forward end of the pipe, this is zero, and at the aft end this equals $\dot{m}$. It is often reasonable to assume a linear variation in between, which simplifies the integral; that is,

$$\dot{\mu}(x) = \left(\frac{x - X_1}{L}\right)\dot{m} \quad X_1 < x < X_{ex} \tag{5.1.36}$$

It is not necessary to define $\dot{\mu}(x)$ outside the indicated range. The moment due to Coriolis forces becomes

$$M_{cor} = \int_{X_1}^{X_{ex}} 2\omega\dot{\mu}x\ dx = \frac{2\omega\dot{m}}{L}\int_{X_1}^{X_{ex}} (x - X_1)\,x\ dx \tag{5.1.37}$$

$$M_{cor} = \frac{2\omega\dot{m}}{L}\left[\frac{x^3}{3} - \frac{X_1 x^2}{2}\right]_{X_1}^{X_{ex}} \tag{5.1.38}$$

The simplified expression,

$$M_{cor} = \frac{\omega\dot{m}}{3L}\left(2X_{ex}^3 - 3X_1X_{ex}^2 + X_1^3\right) \tag{5.1.39}$$

matches the RHS of (5.1.34). Thus it has been confirmed that the contribution to $\dot{\omega}$ from jet damping can be computed either by integrating the Coriolis forces or from the momentum carried away in the exhaust gases, for both liquid and solid propellants. The jet damping torque becomes part of $\mathbf{g}$. Using Eqs. (5.1.23) and (5.1.34), the jet damping torques are

$$\mathbf{g}_{jet} = \dot{m}\left(X_{ex}^2 - X_{liq}^2\right)\begin{bmatrix} 0 & \omega_y & \omega_z \end{bmatrix}^T \tag{5.1.40}$$

for a liquid rocket, and

$$\mathbf{g}_{jet} = \frac{\dot{m}}{3L}\left(2X_{ex}^3 - 3X_1X_{ex}^2 + X_1^3\right)\begin{bmatrix} 0 & \omega_y & \omega_z \end{bmatrix}^T \tag{5.1.41}$$

for a solid rocket. These expressions are valid for a rocket having axial symmetry. The analysis must be extended if the rocket engine and tank are arbitrarily located in the body.

Aerodynamic forces and moments

The subject of aerodynamic forces on a rocket is a field of study in itself. Specialized analyses may include static and dynamic aero-elastic forces, control surface moments, plume interaction effects, and moments due to interactions of gimbaled engines with the freestream. Since aerodynamic bending moments are a key driver of the structural design for large rockets, these and the subject of unsteady flow and shock interactions often receive great scrutiny in rocket development. These effects are often assessed using specialized analyses.

Lift L and drag D are defined as the forces perpendicular to and parallel to the relative wind vector, respectively. They can be thought of as the forces in the "wind frame" or "wind tunnel frame." They have nondimensional coefficients C_L and C_D. If the sideslip angle β is zero, these can be converted to body frame coefficients as follows:

$$C_X = -C_D \cos\alpha + C_L \sin\alpha \tag{5.1.42}$$

$$C_Z = -C_D \sin\alpha - C_L \cos\alpha \tag{5.1.43}$$

where α is the angle of attack. Using the conventional orientation of the body axes (x forward, z down), both of these body axis coefficients are negative for positive α. Analysts therefor find it more convenient to use the normal force coefficient $C_N \equiv -C_Z$ and the axial force coefficient $C_A \equiv -C_X$. For the nominal state in which the angle of attack is zero, the axial force is the same as the drag force. Even with nonzero α, the difference is very small for missile-shaped vehicles. For stability and control analysis, C_A can almost always be taken to be as the same as C_D. The only application for which the difference between drag and axial force might be significant would be a precision guidance analysis.

It is usually sufficient for a preliminary simulation environment to introduce quasi-steady aerodynamic forces acting on the rigid body as a linear function of the aerodynamic angles;

$$f_{x\,\text{aero}} = -C_A \bar{q} S_{ref} \tag{5.1.44}$$

$$f_{y\,\text{aero}} = \bar{q} S_{ref} \left(C_{Y\beta_0} + C_{Y\beta}\beta \right) \tag{5.1.45}$$

$$f_{z\,\text{aero}} = -\bar{q} S_{ref} \left(C_{N\alpha_0} + C_{N\alpha}\alpha \right) \tag{5.1.46}$$

In (5.1.44) through (5.1.46), $\bar{q}$ is the dynamic pressure, S_{ref} is the reference area, $C_{Y\beta}$ and $C_{N\alpha}$ and are the side and normal force coefficient slopes. The aerodynamic force bias, e.g., force at zero incidence angles, is expressed by

$C_{Y\beta_0}$ and $C_{N\alpha_0}$. This expression does not include any aerodynamic moment effects.

In a general purpose simulation, the angles of attack and sideslip are determined from the rocket velocity relative to the atmosphere

$$\mathbf{v}_{\text{rel}} = \mathbf{v} - \mathbf{v}_{\text{atm}} = \begin{bmatrix} u & v & w \end{bmatrix}^T \tag{5.1.47}$$

where $\mathbf{v}_{\text{atm}}$ is the local velocity of the atmosphere expressed in the body frame. This must take into account the rotation of the earth and winds. In Chapters 3 and 4, considerable effort has been devoted to eliminating $\mathbf{v}$ from the dynamic equations. Thus it is necessary to compute $\mathbf{v}$ by transforming $\mathbf{v}_I$ into the body coordinate frame.

The definition of the aerodynamic angles depends in part on the assumptions used in generating the aerodynamic data for simulation. Using typical aircraft and missile body frame conventions, the aerodynamic angles are obtained from

$$\alpha = \tan^{-1}(w/u) \tag{5.1.48}$$

$$\beta = \sin^{-1}(v/V) \tag{5.1.49}$$

where $V = \|\mathbf{v}_{\text{rel}}\|$. The aerodynamic angles and signs of the aerodynamic coefficients are shown in Fig. 5.4. Note that the sideslip angle β is defined using a left-handed rotation about the z axis. This is discussed further in Chapter 7.

In many rocket programs, particularly for those vehicles having a degree of axial symmetry, the aerodynamic forces and moments may be defined in terms of a total angle of attack α_T and an aerodynamic roll angle ϕ_a, defined as

$$\alpha_T = \sqrt{\alpha^2 + \beta^2} \tag{5.1.50}$$

$$\phi_a = \tan^{-1}(v/w)\,. \tag{5.1.51}$$

The total angle of attack is useful since its product with the dynamic pressure, $\bar{q}\alpha_T$, is a metric for aerodynamic loads and bending moments.

When using higher-fidelity aerodynamic tables, the coefficient slopes $C_{Y\beta}$ and $C_{N\alpha}$ are replaced by the general aerodynamic coefficients shown in Table 5.1. In practical designs, the normal forces and aerodynamic moments may not be well-behaved (e.g., linear) about zero angle of attack, or the incidence angles may extend beyond the linear range, necessitating a more general model.

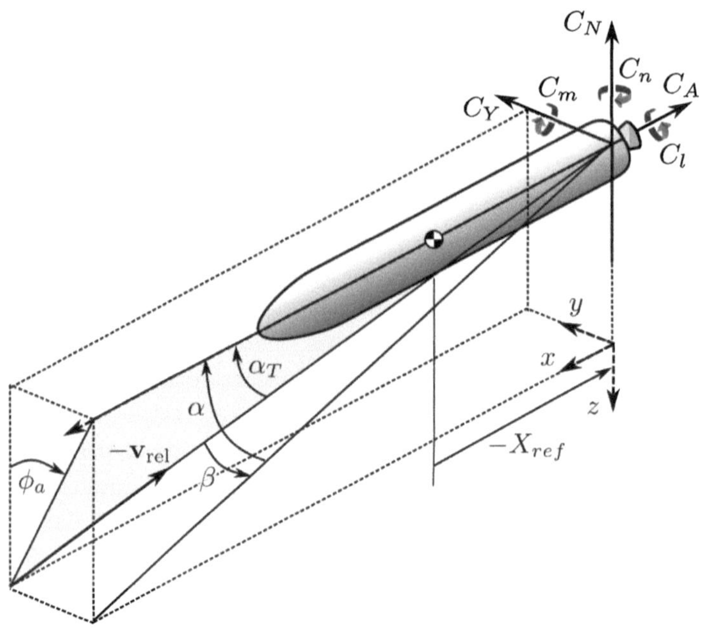

Figure 5.4 Aerodynamic angles and coefficients.

Table 5.1 Aerodynamic coefficient nomenclature.

C_N	Normal force coefficient (-C_Z)
C_Y	Side force coefficient
C_A	Axial force coefficient (-C_X)
C_l	Rolling moment coefficient
C_m	Pitching moment coefficient
C_n	Yawing moment coefficient

Aerodynamic moment coefficients are defined with respect to an aerodynamic reference point X_{ref}, which must be supplied as part of the aerodynamic data. This may be the forward-most tip of the rocket or some location near the aft structure, such as a gimbal point.[1] In the following, the primes indicate moments relative to this point. These aerodynamic moments are given by

[1] The gimbal point and a wind tunnel sting location are sometimes coincident, simplifying the scaling and computation of forces and moments from wind tunnel tests.

$$g'_{x\text{ aero}} = \bar{q} S_{ref} b C_l \tag{5.1.52}$$

$$g'_{y\text{ aero}} = \bar{q} S_{ref} \bar{c} C_m \tag{5.1.53}$$

$$g'_{z\text{ aero}} = \bar{q} S_{ref} b C_n \tag{5.1.54}$$

where b and $\bar{c}$ are the reference span and mean aerodynamic chord, respectively. For rockets, it is common to let $b = \bar{c} = L_{ref}$, the aerodynamic reference length, usually taken to be the diameter of the rocket. The aerodynamic forces are similarly given by

$$f'_{x\text{ aero}} = -\bar{q} S_{ref} C_A \tag{5.1.55}$$

$$f'_{y\text{ aero}} = \bar{q} S_{ref} C_Y \tag{5.1.56}$$

$$f'_{z\text{ aero}} = -\bar{q} S_{ref} C_N \tag{5.1.57}$$

Unprimed symbols indicate forces and moments about the body frame origin. Assuming that X_{ref} is expressed in the body coordinate frame as

$$\mathbf{r}_{ref} = \begin{bmatrix} -X_{ref} & 0 & 0 \end{bmatrix}^T \tag{5.1.58}$$

the aerodynamic moment at the origin of the body frame can be written as

$$\mathbf{g}_{\text{aero}} = \mathbf{g}'_{\text{aero}} + \mathbf{r}^{\times}_{ref} \mathbf{f}_{\text{aero}} \tag{5.1.59}$$

or

$$g_{x\text{ aero}} = \bar{q} S_{ref} L_{ref} C_l \tag{5.1.60}$$

$$g_{y\text{ aero}} = \bar{q} S_{ref} \left(L_{ref} C_m - X_{ref} C_N \right) \tag{5.1.61}$$

$$g_{z\text{ aero}} = \bar{q} S_{ref} \left(L_{ref} C_n - X_{ref} C_Y \right) \tag{5.1.62}$$

For axisymmetric rockets, it is the aerodynamic designers' intention to keep the rolling moment as small as possible so as to minimize the need for auxiliary roll control devices. Rolling moments can be produced by protuberances such as propellant feedlines. These protuberances cause a rolling moment when the rocket is at a nonzero angle of attack or sideslip. Unlike most aircraft, the rolling moment due to sideslip is sometimes unstable.

An alternate way to represent the pitch and yaw moments is to compute centers of pressure;

$$X_{cp\text{ pitch}} = X_{ref} - \frac{C_m}{C_N} L_{ref} \tag{5.1.63}$$

$$X_{cp\ \text{yaw}} = X_{ref} - \frac{C_n}{C_Y} L_{ref} \tag{5.1.64}$$

For an axisymmetric vehicle, one would expect that $X_{cp\ \text{pitch}} = X_{cp\ \text{yaw}}$. In each plane (pitch or yaw), the center of pressure is the point of application of the lateral force, i.e., the point about which the moment is zero. The significance of this may be much easier to grasp than the coefficient values, since it directly indicates the degree of static stability of the airframe. If the center of pressure is forward of the center of mass, the rocket is aerodynamically unstable, and the distance between these points is an indication of how much flight control feedback gain will be required for stabilization.

At low altitudes, the aerodynamic coefficients are in general nonlinear functions of the flight condition, as in

$$C_N = C_N(\phi_a, \alpha_T, M) \tag{5.1.65}$$

where M is the freestream Mach number. Given these inputs, the coefficients are usually obtained in simulation using lookup tables. For most rockets, secondary effects that depend on body angular rates (short period damping), rates of change of angle of attack, and so on are usually insignificant and are not included in lookup tables. Highly maneuverable rockets, rockets without active control, and rockets undergoing separation dynamics (e.g., staging) may have supplemental lookup tables or polynomial approximations that must be included in simulations. Stengel [24] provides a detailed discussion of many of these effects in the context of aircraft dynamics, which applies analogously to some rocket configurations.

Although seldom used in large launch vehicles for performance reasons, many smaller rockets use aerosurface controls such as actuated fins. The detailed treatment of these effects on the aerodynamics is beyond the scope of the present text, but simulations often implement the effects of aerosurface deflections using aerosurface increment lookup tables. The aerodynamic coefficients are given as a linear combination of Eq. (5.1.65) and an increment due to deflection angles;

$$C_N = C_{N0}(\phi_a, \alpha_T, M) + C_{N\Delta}(\phi_a, \alpha_T, M, \mathbf{\Delta}) \tag{5.1.66}$$

where $\mathbf{\Delta}$ is a vector of control surface deflection angles. If the rocket has engines or plumes that interact with the freestream, $\mathbf{\Delta}$ may also include the TVC gimbal angles.

For the modeling of aerodynamics at higher altitudes, Regan and Anandakrishnan [25] provide an excellent discussion of how to deal with the aerodynamics when the mean free path between molecules becomes large. In this regime, it is appropriate for aerodynamic tables to be based on altitude rather than Mach number.

Gravity

Let the force of gravity in body coordinates be

$$\mathbf{f}_{\text{grav}} = \begin{bmatrix} f_{x,\,\text{grav}} & f_{y,\,\text{grav}} & f_{z,\,\text{grav}} \end{bmatrix}^T \tag{5.1.67}$$

This can be computed by knowing the transformation between the body frame and whatever frame is used to define the local vertical. Chapter 4 of Greensite [1] discusses this problem in detail. For our present purposes, this is done by defining a vector $\mathbf{u}_v$ in the body frame that is oriented in the downward vertical direction; thus,

$$\mathbf{f}_{\text{grav}} = m_T g\, \mathbf{u}_v \tag{5.1.68}$$

The torque about the body origin due to gravity is

$$\mathbf{g}_{\text{grav}} = g\mathbf{s}_{TD}^{\times}\mathbf{u}_v \tag{5.1.69}$$

If the origin is chosen to coincide with the center of mass, then the gravity torque is zero.

The value of the apparent gravity magnitude, g, is a function of the position of the rocket with respect to the earth. Basis function methods are used to model the spatial dependency of the gravity field over the earth's surface, using a combination of spherical harmonics to express a potential function U. For example, the simplest spherical potential model is given by

$$g = U_{\bigcirc}\frac{1}{r_I} \tag{5.1.70}$$

where $U_{\bigcirc} = \mu/r_I$ is the potential function, μ is the earth gravitational constant, and r_I is the magnitude of the earth-relative inertial position vector. The most common gravity models used in rocket simulations include additional terms in U, such as J_2, which accounts for earth oblateness effects. The details and implementation of these models are given in Vallado [26].

Finally, the precise definition of $\mathbf{u}_v$ depends on the frame in which the simulation defines "vertical." For example, $\mathbf{u}_v$ may be coincident with a so-called navigation plumbline, which is the true orientation of the gravity vector at a fixed point on the earth's surface. Conversely, $\mathbf{u}_v$ may be defined antinormal to the reference geoid of the earth's surface, in which case the actual gravitational acceleration vector will differ in direction from $\mathbf{u}_v$ by a small amount. Except in precision navigation and guidance analyses, these differences are negligible. When using more advanced gravitational models, $\mathbf{u}_v$ is simply a unit vector toward the center of the earth. Its value is computed in simulation using

$$\mathbf{u}_v = -\mathbf{C}_{bI}\frac{\mathbf{r}_I}{r_I} \tag{5.1.71}$$

where $\mathbf{C}_{bI}$ is the inertial to body kinematic transformation discussed in Chapter 2.

RCS forces and torques

If a rocket is configured with a reaction control system (RCS), the force from the RCS thrusters is also part of the external force $\mathbf{f}$. Define the RCS vector $\mathbf{u}_{\mathrm{RCS}}$ as an array of 1's and 0's indicating which jets are firing. That is,

$$\mathbf{u}_{\mathrm{RCS}} = \begin{bmatrix} u_{\mathrm{RCS}1} & u_{\mathrm{RCS}2} & \cdots & u_{\mathrm{RCS}K} \end{bmatrix}^T \tag{5.1.72}$$

where each element $u_{\mathrm{RCS}k}$ is either 1 or 0, and K is the number of thrusters. Further define $\mathbf{f}_{\mathrm{RCS}k}$ as the force from thruster k, and $\mathbf{f}_{\mathrm{RCS}} \equiv \begin{bmatrix} f_{\mathrm{RCS}x} & f_{\mathrm{RCS}y} & f_{\mathrm{RCS}z} \end{bmatrix}^T$ as the total force from all the thrusters. Thus we have

$$\mathbf{f}_{\mathrm{RCS}} = \sum_{k=1}^{K} u_{\mathrm{RCS}k}\mathbf{f}_{\mathrm{RCS}k} \tag{5.1.73}$$

There is also a torque from the RCS thrusters that contributes to the external torque $\mathbf{g}$. Letting $\mathbf{g}_{\mathrm{RCS}} \equiv \begin{bmatrix} g_{\mathrm{RCS}x} & g_{\mathrm{RCS}y} & g_{\mathrm{RCS}z} \end{bmatrix}^T$ represent the resultant torque from all RCS thrusters, for K thrusters

$$\mathbf{g}_{\mathrm{RCS}} = \sum_{k=1}^{K} u_{\mathrm{RCS}k}\mathbf{r}^{\times}_{\mathrm{RCS}k}\mathbf{f}_{\mathrm{RCS}k} \tag{5.1.74}$$

where $\mathbf{r}_{\mathrm{RCS}k}$ is a vector from the origin to thruster k.

For particularly flexible rockets, one should also include the effects of elastic displacement and rotation on the RCS thrust vector. If this is the case, the RCS thrust forces and torques can be calculated using expressions similar to (5.1.10) and (5.1.11), or (5.1.12).

5.2 Fuel and nozzle offset torques

The symbol $\mathbf{s}_T$ represents the first moment of inertia with the slosh masses in their equilibrium positions. The actual first moment of inertia varies dynamically because of fuel slosh and nozzle motion:

$$\mathbf{s}_{TD} = \mathbf{s}_T + \sum_j m_{sj}\boldsymbol{\delta}_{sj} + \boldsymbol{\beta}_{Eb}^{\times}\mathbf{s}_{Eb} \tag{5.2.1}$$

As discussed in Chapter 3, instead of putting the dynamic first moment of inertia in the mass matrix, it may be more convenient to capture the dynamic effect as part of the external torque, and to use a quasi-steady mass matrix. The effect on the system rotation equation is given by the last term of (3.2.34). Extending this to multiple tanks and adding an engine term gives

$$\mathbf{g}_{\text{offset}} = \sum_j m_{sj}\mathbf{a}_{b0}^{\times}\boldsymbol{\delta}_{sj} + \mathbf{a}_{b0}^{\times}\boldsymbol{\beta}_{Eb}^{\times}\mathbf{s}_{Eb} = \begin{bmatrix} 0 & g_{\text{offset } y} & g_{\text{offset } z} \end{bmatrix} \tag{5.2.2}$$

where $\mathbf{a}_{b0}$ is defined in Eq. (3.2.33). With $\boldsymbol{\beta}_{Eb} = \begin{bmatrix} 0 & \beta_{Ey} & \beta_{Ez} \end{bmatrix}^T$, $\boldsymbol{\delta}_{sj} = \begin{bmatrix} 0 & y_{sj} & z_{sj} \end{bmatrix}^T$, and $\mathbf{s}_{Eb} = \begin{bmatrix} S_{Ex} & 0 & 0 \end{bmatrix}^T$ this yields the offset terms

$$g_{\text{offset } y} \cong \frac{(F-D)}{m_T}\left[-\sum_j m_{sj}z_{sj} + \beta_{Ey}S_{Ex}\right] \tag{5.2.3}$$

$$g_{\text{offset } z} \cong \frac{(F-D)}{m_T}\left[\sum_j m_{sj}y_{sj} + \beta_{Ez}S_{Ex}\right] \tag{5.2.4}$$

Note that S_{Ex} is normally a negative quantity when expressed in the body frame.

The quasi-steady approach implies that $\mathbf{s}_T$ and $\mathbf{I}_T$ are used in the mass matrix. Also, $\mathbf{b}_{sj}$, the undisplaced slosh mass location, is substituted for $\mathbf{r}_{sj}$. It is worth emphasizing that the fuel offset terms must not be included on the right hand side if the mass matrix uses $\mathbf{s}_{TD}$ and $\mathbf{I}_{TD}$ and is updated at every time step.

There is also a fuel offset force in the translation equation. Indeed, in the development of the slosh dynamics in Chapter 3, the dynamic part of $\mathbf{s}_{TD}$ was shifted to the RHS, giving rise to a similar fuel offset force, which can be identified as the term $m_{sj}\boldsymbol{\delta}_{sj}^{\times}\dot{\boldsymbol{\omega}}$ in Eq. (3.2.21). Note that in the rotation equation, the offset torque is proportional to the secular term $\mathbf{a}_{b0}$, with the result that it becomes part of the linear equations. The term "secular" is defined after Eq. (4.3.20). In the translation equation, however, the offset force becomes proportional to the product of two small quantities. It is thus a nonlinear effect which becomes part of $\mathbf{f}_{NL}$. In a numerical simulation, this term must be computed from a previous time step. It can thus lead to numerical instability. As discussed in Chapter 3, the most reliable options are to compute a new mass matrix at each time step, or to leave out all terms involving accelerations and accept a loss in simulation accuracy.

5.3 Slosh, engine, and bending excitation

In the system translation and rotation equations, the interbody forces have no net effect. However, this is not true for the remaining degrees of freedom. It becomes useful to revisit the rearranged version of Lagrange's equation;

$$\frac{\mathrm{d}}{\mathrm{d}t}\left(\frac{\partial T}{\partial \dot{q}_i}\right) = -\frac{\partial V}{\partial q_i} - \frac{\partial D}{\partial \dot{q}_i} + Q_i \tag{5.3.1}$$

When the phrases "right hand side" (RHS) or "left hand side" (LHS) are used below, this is the arrangement that is envisioned. In Chapter 2, a mass matrix is developed from the LHS. The first two terms on the RHS can be used to compute interbody forces. Thus in the slosh equation a spring potential energy term and dissipation term can be inserted, and likewise for the bending equations. The important point is that for the present bookkeeping scheme, Q_i represents excitation *from external forces only*.

Slosh

If the spring model is used for slosh, then the force acting on the slosh mass j is given by Eq. (2.5.4)

$$m_{sj}\left(\mathbf{a}_b - \mathbf{r}_{sj}^{\times}\dot{\boldsymbol{\omega}} + \ddot{\boldsymbol{\delta}}_{sj} + \sum \boldsymbol{\psi}_{ji}\ddot{\eta}_i\right) = \mathbf{f}_{sj} + \mathbf{f}_{sj,\,\mathrm{NL}} \tag{5.3.2}$$

$\mathbf{f}_{sj}$ is the combination of the interbody forces and the external excitation appearing on the RHS of (5.3.1). The slosh spring potential V and dissipation D are given by

$$V_{sj} = \frac{1}{2} k_{sj} \boldsymbol{\delta}_{sj}^T \boldsymbol{\delta}_{sj} \tag{5.3.3}$$

$$D_{sj} = \frac{1}{2} C_{sj} \dot{\boldsymbol{\delta}}_{sj}^T \dot{\boldsymbol{\delta}}_{sj} \tag{5.3.4}$$

where k_{sj} and C_{sj} are the slosh spring and damping constants. Using the derivative with respect to a column matrix described in Chapter 1,

$$\frac{\partial V_{sj}}{\partial \boldsymbol{\delta}_{sj}} = k_{sj} \boldsymbol{\delta}_{sj} \tag{5.3.5}$$

$$\frac{\partial D_{sj}}{\partial \boldsymbol{\delta}_{sj}} = C_{sj} \dot{\boldsymbol{\delta}}_{sj} \tag{5.3.6}$$

Because of the form of (5.3.5) and (5.3.6), the spring and damping terms only appear in the jth slosh equation – they do not appear in the equations for any other degree of freedom. For example, if one uses $\partial V/\partial q_i$ to compute the slosh spring force in the y direction, this force only appears in the slosh y equation. In contrast, an external force such as thrust will produce Q_i's that appear in the rotation, translation, and bending equations, and possibly also in the engine equation. That simple observation is the primary motivation for presenting Eq. (5.3.1). By specifying that the potential energy V and dissipation D are only used for interbody forces, the bookkeeping becomes simplified.

The only external excitation of slosh is that of gravity. Thus the RHS of (5.3.1) becomes

$$\mathbf{f}_{sj} = -k_{sj} \boldsymbol{\delta}_{sj} - C_{sj} \dot{\boldsymbol{\delta}}_{sj} + m_{sj} g \, \mathbf{u}_v \tag{5.3.7}$$

The relations to the natural frequency and damping ratio are given by

$$\Omega_{sj}^2 \equiv \frac{k_{sj}}{m_{sj}} \tag{5.3.8}$$

$$\zeta_{sj} \equiv \frac{C_{sj}}{2 m_{sj} \Omega_{sj}} \tag{5.3.9}$$

These can be used to create an alternate version of (5.3.7)

$$\mathbf{f}_{sj} = m_{sj} \left(-\Omega_{sj}^2 \boldsymbol{\delta}_{sj} - 2 \zeta_{sj} \Omega_{sj} \dot{\boldsymbol{\delta}}_{sj} + g \, \mathbf{u}_v \right) \tag{5.3.10}$$

The nonlinear effect on the slosh equation is given by (3.2.15). If the pendulum model is used for slosh, refer to Eqs. (4.3.12) through (4.3.14) and the associated discussion. Note that in Eqs. (5.3.7) and (5.3.10), the vector $\mathbf{u}_\nu$ in the body frame must be continually computed as the rocket orientation changes in time.

Engine

Engine excitation forces only have to be considered if the engine motion is not prescribed. Let the total moment on the engine about the gimbal point be given by $\mathbf{g}_E$. A subscript b is attached to this if the engine equation is written in the body frame, otherwise it is in the engine frame (see Section 4.4). The main component of this is the moment from the TVC system, which requires a TVC model. There is also a torque due to gravity, which takes one of the following forms, depending on whether the engine equation is written in the engine frame or the body frame;

$$\mathbf{g}_{Eb,\,\text{grav}} = g\,\mathbf{s}_{Eb}^{\times}\mathbf{u}_\nu \tag{5.3.11}$$

$$\mathbf{g}_{E,\,\text{grav}} = g\,\mathbf{s}_{E}^{\times}\mathbf{C}_{10}\mathbf{u}_\nu \tag{5.3.12}$$

It is possible that there is an aerodynamic moment on the nozzle, or a torque due to the Coriolis forces on the exhaust gases. The latter would require an analysis similar to the jet damping computation in Section 5.1, only confined to the length of the nozzle. If either of these are significant, they would become part of $\mathbf{g}_E$.

To understand the nature of this term, consider the engine equation (2.5.3) with $\dot{\boldsymbol{\omega}} = \ddot{\eta} = 0$. Further, divide the body acceleration into a sensed portion and a remainder as in (3.2.32). The result is

$$\mathbf{s}_{Eb}^{\times}\left(\mathbf{a}_{b0} + \mathbf{a}_b'\right) + \mathbf{I}_{Eb}\dot{\boldsymbol{\omega}}_{Eb} = \mathbf{g}_{Eb} \tag{5.3.13}$$

Now consider the benign situation in which there are no perturbations, only the steady effects of thrust and gravity. In that case, the remainder acceleration is solely due to gravity, that is $\mathbf{a}_b' = g\,\mathbf{u}_\nu$, and the only engine torque is the gravity torque from (5.3.11) plus the TVC torque. Thus, (5.3.13) becomes

$$\mathbf{s}_{Eb}^{\times}\left(\mathbf{a}_{b0} + g\,\mathbf{u}_\nu\right) + \mathbf{I}_{Eb}\dot{\boldsymbol{\omega}}_{Eb} = g\mathbf{s}_{Eb}^{\times}\mathbf{u}_\nu + \mathbf{g}_{TVC} \tag{5.3.14}$$

Here we see that the gravity torque on the right side of the equation is simply a balancing term whose function is to cancel out a similar term on the left. Thus, as expected, gravity does not really have any effect on the motion of the engine.

$$\mathbf{s}_{Eb}^{\times}\mathbf{a}_{b0} + \mathbf{I}_{Eb}\dot{\boldsymbol{\omega}}_{Eb} = \mathbf{g}_{TVC} \tag{5.3.15}$$

Assuming the engine frame lines up with the body frame under equilibrium conditions, and that the engine deflections are very small, the transformation between engine frame and body frame can be written using a transformation of the same form as (2.1.2). To first order, the engine first moment of inertia in the body frame is

$$\mathbf{s}_{Eb} = \left(\mathbf{1} + \boldsymbol{\beta}^{\times}\right)\mathbf{s}_E \tag{5.3.16}$$

Thus

$$\left(\mathbf{s}_E + \boldsymbol{\beta}^{\times}\mathbf{s}_E\right)^{\times}\mathbf{a}_{b0} + \mathbf{I}_{Eb}\dot{\boldsymbol{\omega}}_{Eb} = \mathbf{g}_{TVC} \tag{5.3.17}$$

This can be rearranged as

$$\mathbf{s}_{Eb}^{\times}\mathbf{a}_{b0} + \mathbf{I}_{Eb}\dot{\boldsymbol{\omega}}_{Eb} = \mathbf{g}_{TVC} - \mathbf{a}_{b0}^{\times}\mathbf{s}_E^{\times}\boldsymbol{\beta} \tag{5.3.18}$$

If it is further assumed that $\mathbf{a}_{b0}$ and $\mathbf{s}_E$ are each parallel to the x axis, the two cross product matrices on the right side multiply to create a diagonal matrix. We see that the sensed acceleration creates a stiffness effect – an engine deflection creates a centering torque. In order to return this to a form that uses the total acceleration rather than the sensed acceleration, we add back the gravity torque on both sides:

$$\mathbf{s}_E^{\times}\mathbf{a}_b + \mathbf{I}_{Eb}\dot{\boldsymbol{\omega}}_{Eb} = \mathbf{g}_{TVC} - \mathbf{a}_{b0}^{\times}\mathbf{s}_E^{\times}\boldsymbol{\beta} + g\,\mathbf{s}_E^{\times}\mathbf{u}_v \tag{5.3.19}$$

It turns out that all of the equations, and not just the engine equations, require a term on the right due to gravity. A way to compute a single gravity vector that takes care of the entire matrix equation is presented at the end of this section.

Bending

The bending equation (2.5.5) is repeated here for convenience;

$$\mathbf{c}_{EFi}^{T}\dot{\boldsymbol{\omega}}_{Eb} + \sum_j m_{sj}\boldsymbol{\psi}_{ji}^{T}\ddot{\boldsymbol{\delta}}_{sj} + m_{Bi}\ddot{\eta} = f_{Bi} \tag{5.3.20}$$

The LHS of this corresponds to the LHS of (5.3.1). Equating the RHS of (5.3.1) and (5.3.20), and replacing q_i with η_i, we obtain

$$f_{Bi} = -\frac{\partial V}{\partial \eta_i} - \frac{\partial D}{\partial \dot{\eta}} + Q_{\eta i} \tag{5.3.21}$$

The first two terms on the RHS become the spring and damping terms.

The generalized force $Q_{\eta i\,\text{thrust}}$ arises from thrust. The thrust is divided into gimbaled and ungimbaled components as shown by (5.1.1) and (5.1.2). It is usually adequate to only consider the components of the gimbaled thrust that are parallel to the y and z lateral deflections.

$$Q_{\eta i\,\text{thrust}} = F_R\left(\beta_{Ez}\psi_{y\beta i} - \beta_{Ey}\psi_{z\beta i}\right) \tag{5.3.22}$$

For a more accurate but more demanding computation, the vector thrust force with a second order transformation (Eq. (5.1.12)) can be computed, and then applied to the bending dynamics;

$$Q_{\eta i,\,\text{thrust}} = \boldsymbol{\psi}_{\beta i}^{T}\mathbf{f}_{\text{thrust}}. \tag{5.3.23}$$

Note that this expression inherently includes a *force following* term, which may affect the accuracy of the bending frequencies as discussed in Chapter 6.

The generalized force due to aerodynamics is computed from

$$Q_{\eta i\,\text{aero}} = \bar{q}S_{ref}\left(C_{Y\eta i}\beta - C_{N\eta i}\alpha\right) \tag{5.3.24}$$

where

$$C_{Y\eta i} = \int \frac{\partial C_{Y\beta}}{\partial x}\psi_{yi}\,\mathrm{d}x \tag{5.3.25}$$

$$C_{N\eta i} = \int \frac{\partial C_{N\alpha}}{\partial x}\psi_{zi}\,\mathrm{d}x \tag{5.3.26}$$

A term such as $\partial C_{N\alpha}/\partial x$ represents the spatial derivative of the normal force coefficient along the length of the rocket, determined via test or computational analysis. This distribution is known as the *line load*. The expression in Eq. (5.3.24) represents a simplified effect of wind loads on bending. In this equation, it is assumed that the angles of attack and sideslip are the same everywhere on the rocket body. That is, the local angle of attack at station x is the same as the average angle of attack of the entire structure. A more detailed approach to distributed aerodynamics, better suited to linearized models, is introduced in Chapter 7.

If there is an active RCS, computation of the flex mode excitation requires the product of the thrust generated by each individual thruster and the translational mode shape of each mode at that thruster node. Let $\boldsymbol{\psi}_{Rki}$

be the displacement of mode i at the node for thruster k. Using (5.1.72) and (5.1.73),

$$Q_{\eta i\,\mathrm{RCS}} = \sum_k u_{\mathrm{RCS}k} \mathbf{f}_{\mathrm{RCS}k}^T \boldsymbol{\psi}_{Rki}. \tag{5.3.27}$$

If any of the slosh masses have a lateral offset, this has the effect of inducing a torque on the structure, which can cause bending. The generalized excitation of each mode i due to the slosh offset plus engine rotation is given by

$$Q_{\eta i\,\mathrm{offset}} = \sum_j m_{sj} \boldsymbol{\sigma}_{ji}^T \left(\mathbf{a}_{b0}^{\times} \boldsymbol{\delta}_{sj}\right) + \boldsymbol{\sigma}_{\beta}^T \mathbf{a}_{b0}^{\times} \boldsymbol{\beta}_{Eb}^{\times} \mathbf{s}_{Eb} \tag{5.3.28}$$

where $\boldsymbol{\sigma}_{ji}$ is a modal rotation that is defined in the same way as the parameter $\boldsymbol{\psi}_{ji}$ from (2.3.13). Note the relationship to Eq. (5.2.2). It is somewhat curious that Frosch and Vallely [4] do not include any such terms in their equations, even though equivalent terms are included by Garner [3]. In the case of slosh, one could argue that the above representation applies the slosh offset torque in a concentrated manner at the particular location of the slosh point mass, whereas in reality this torque is distributed over a wide region of the tank walls. Thus, the above equation may have a tendency to overestimate the effect. If deemed significant, this term should be included even if the mass matrix is updated at every time step.

The generalized force from all components is

$$Q_{\eta i} = Q_{\eta i\,\mathrm{thrust}} + Q_{\eta i\,\mathrm{aero}} + Q_{\eta i\,\mathrm{RCS}} + Q_{\eta i\,\mathrm{offset}} + Q_{\eta i\,\mathrm{grav}} \tag{5.3.29}$$

The gravity term is discussed below. The RHS force term for bending also includes the flex spring and damping terms, which are computed in exact parallel to the corresponding terms in the slosh equation

$$f_{Bi} = Q_{\eta i} - m_{Bi} \left(\Omega_{Bi}^2 \eta_i + 2\zeta_{Bi} \Omega_{Bi} \dot{\eta}_i\right) \tag{5.3.30}$$

Note that (5.3.29) does not contain any term representing the torque from the TVC system. The bending excitation due to TVC torques comes entirely from cross coupling in the mass matrix, i.e., the $\mathbf{c}_{EFi}^T$ terms in (2.6.21). If we have a reduced body model in which the FEM does not include the mass of the engine, one could obtain a bending equation that explicitly includes the TVC torques, in a manner that is analogous to the way the bending equation can be written with excitation on the RHS from the slosh spring and damping forces. For more information on this approach, the reader is referred to the discussion in Section 2.6.

Fuel offset effect – slosh pendulum model versus slosh spring model

It has been repeatedly stated that for small motions, the slosh pendulum model should give results that are equivalent to the slosh spring model. This raises an apparent contradiction, since the hinge point of the pendulum can only transmit forces to the elastic dynamics, and the offset excitation (5.3.28) represents a generalized torque that is applied to the bending equation. How is this torque transmitted in the case of the pendulum model? The answer to this question provides some insight into the nature of bending excitation due to slosh for both models.

In order to explore this issue, the following simplifications are made:

1. The engine gimbal angles are held fixed, i.e., $\boldsymbol{\beta}_E = \boldsymbol{\omega}_E = \dot{\boldsymbol{\omega}}_E = \mathbf{0}$.
2. There is no flex excitation other than slosh motion.
3. The rocket mass m_T and moment of inertia $\mathbf{I}_T$ are so large compared to the slosh mass that the rocket only has an axial acceleration from the thrust, with no significant lateral acceleration or angular acceleration.
4. Gravity and slosh damping are ignored.

We also stipulate that there is only one slosh mass, so the summation in (5.3.20) can be dropped:

$$m_{sj}\boldsymbol{\psi}_{Sji}^T\ddot{\boldsymbol{\delta}}_{sj} + m_{Bi}\ddot{\eta}_i = f_{Bi} \tag{5.3.31}$$

Here, the subscript S has been added to $\boldsymbol{\psi}$, to emphasize that this is the $\boldsymbol{\psi}$ that is being used in the spring model. This is not exactly the same as the $\boldsymbol{\psi}_p$ that is used for the pendulum force, which is applied at the hinge point rather than the location of the slosh mass. On the RHS we must include the excitation from (5.3.19) plus the flex spring and damping terms:

$$f_{Bi} = Q_{\eta i\ \text{offset}} - m_{Bi}\left(2\zeta_{Bi}\Omega_{Bi}\dot{\eta}_i + \Omega_{Bi}^2\eta_i\right) \tag{5.3.32}$$

Substituting (5.3.28) and (5.3.32) into (5.3.31) and rearranging gives

$$m_{Bi}\left(\ddot{\eta} + 2\zeta_{Bi}\Omega_{Bi}\dot{\eta}_i + \Omega_{Bi}^2\eta_i\right) = m_{sj}\left(-\boldsymbol{\psi}_{Sji}^T\ddot{\boldsymbol{\delta}}_{sj} + \boldsymbol{\sigma}_{Sij}^T\mathbf{a}_{b0}^{\times}\boldsymbol{\delta}_{sj}\right) \tag{5.3.33}$$

Simplification 3 above means that there is no significant difference between the slosh relative acceleration $\ddot{\boldsymbol{\delta}}_{sj}$ and the absolute acceleration that comes from dividing the slosh force by the slosh mass. This in turn means we can write (5.3.33) as

$$m_{Bi}\left(\ddot{\eta} + 2\zeta_{Bi}\Omega_{Bi}\dot{\eta}_i + \Omega_{Bi}^2\eta_i\right) = \boldsymbol{\psi}_{Sji}^T\bar{\mathbf{f}}_{sj} + \boldsymbol{\sigma}_{Sji}^T\bar{\mathbf{g}}_{sj} \tag{5.3.34}$$

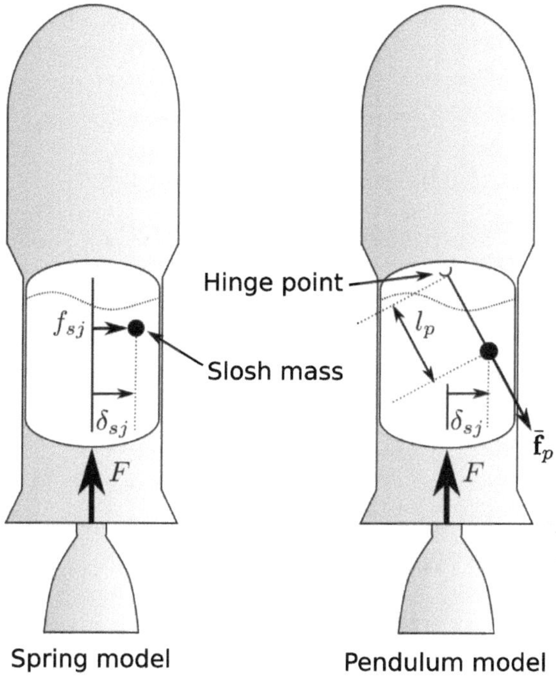

Figure 5.5 Comparison of bending excitation due to slosh offset for the spring model with that for the pendulum model.

where $\bar{\mathbf{f}}_{sj}$ and $\bar{\mathbf{g}}_{sj}$ are the force and torque being applied to the structure at the location of the slosh mass. $\bar{\mathbf{f}}_{sj}$ is equal and opposite to $\mathbf{f}_{sj}$, the force on the slosh mass. There is no torque on the slosh mass, but there is a torque on the structure:

$$\bar{\mathbf{g}}_{sj} = m_{sj}\mathbf{a}_{b0}^{\times}\boldsymbol{\delta}_{sj} \tag{5.3.35}$$

This alternate representation of the bending excitation, which is only accurate if the slosh mass is negligible compared to the rocket mass, is presented for the sole purpose of simplifying the following discussion.

In Fig. 5.5, the pendulum force $\bar{\mathbf{f}}_p$ is shown acting on a line that passes through the hinge point. Again, the overbar indicates this is the force acting on the structure, equal and opposite to the force acting on the slosh mass. The lateral components of this constitute the slosh excitation

$$\bar{\mathbf{f}}_{sj} \cong f_p \begin{bmatrix} 0 & \sin\gamma_{sjy} & \sin\gamma_{sjz} \end{bmatrix}^T \tag{5.3.36}$$

where $f_p = \left|\mathbf{f}_p\right|$, and γ_{sj} is the slosh pendulum angle. We have, for small angles,

$$f_p = \left(\frac{F-D}{m_T}\right) m_s = a_{b0} m_s \tag{5.3.37}$$

The equivalent of (5.3.34) for the pendulum model is

$$m_{Bi}\left(\ddot{\eta} + 2\zeta_{Bi}\Omega_{Bi}\dot{\eta}_i + \Omega_{Bi}^2 \eta_i\right) = \boldsymbol{\psi}_{pji}^T \bar{\mathbf{f}}_{sj} \tag{5.3.38}$$

This differs from (5.3.34) in two ways – there is no torque excitation, and we use $\boldsymbol{\psi}_{pji}$, defined as the modal parameter at the hinge location. To establish that the spring and pendulum models provide the same excitation, we must show that the right sides are equal:

$$\boldsymbol{\psi}_{Sji}^T \bar{\mathbf{f}}_{sj} + \boldsymbol{\sigma}_{Sji}^T \bar{\mathbf{g}}_{sj} = \boldsymbol{\psi}_{pji}^T \bar{\mathbf{f}}_{sj} \tag{5.3.39}$$

In order to reduce the notational clutter, we now drop the subscripts i and j. The spring force becomes

$$\bar{\mathbf{f}}_s = \bar{\mathbf{f}}_{sj} = k_s \boldsymbol{\delta}_s = m_s \Omega_s^2 \boldsymbol{\delta}_s \tag{5.3.40}$$

The slosh model and the pendulum model both have the same natural frequency:

$$\Omega_s^2 = \frac{a_{b0}}{l_p} \tag{5.3.41}$$

Substituting (5.3.35) and (5.3.40) into (5.3.39) and using Eq. (1.8),

$$\left(\frac{a_{b0}}{l_p}\right) m_s \boldsymbol{\psi}_S^T \boldsymbol{\delta}_s + m_s \left(\boldsymbol{\sigma}_S^\times \mathbf{a}_{b0}\right)^T \boldsymbol{\delta}_s = \left(\frac{a_{b0}}{l_p}\right) m_s \boldsymbol{\psi}_p^T \boldsymbol{\delta}_s \tag{5.3.42}$$

Given that $\mathbf{a}_{b0}$ has no y or z component, the second term multiplies out to

$$\left(\boldsymbol{\sigma}_S^\times \mathbf{a}_{b0}\right)^T = a_{b0}\begin{bmatrix} 0 & \sigma_{Sz} & -\sigma_{Sy} \end{bmatrix} \tag{5.3.43}$$

Thus (5.3.42) becomes

$$\boldsymbol{\psi}_S^T + l_p \begin{bmatrix} 0 & \sigma_{Sz} & -\sigma_{Sy} \end{bmatrix} = \boldsymbol{\psi}_p^T \tag{5.3.44}$$

One can also take an independent approach, by using a first-order Taylor series expansion to extrapolate $\boldsymbol{\psi}_S$ forward a distance l_p from the slosh location in order to get another estimate of $\boldsymbol{\psi}_p$.

$$\psi_{py} = \psi_{Sy} + \psi'_{Sy} l_p = \psi_{Sy} + \sigma_{Sz} l_p \tag{5.3.45}$$

$$\psi_{pz} = \psi_{Sz} + \psi'_{Sz} l_p = \psi_{Sz} - \sigma_{Sy} l_p \tag{5.3.46}$$

Here the primes represent the slopes of the modal parameter, i.e., the derivative with respect to x, which have essentially the same numerical magnitudes as the elements of the modal rotations $\boldsymbol{\sigma}_s$. These two equations are equivalent and thus validate Eq. (5.3.44). This in turn establishes the validity of Eq. (5.3.39).

It can be seen that the torsional input from the fuel offset term is equivalent to moving the point of application of the slosh force forward a distance l_p. Unless there is reason to believe that there is a substantial difference in the modal parameters at the two locations, it can be argued that this term is small. This may explain why Frosch and Vallely [4] leave out this term in their bending equation.

It must be emphasized that the fuel offset torque should only be applied if the spring slosh model is being used. If the pendulum model is used, there is no such offset torque. The reader can show that the same rigid-body motion is produced either by applying the combination of slosh force and offset torque at the slosh location, or by applying the slosh force alone at the hinge location.

Effect of gravity on bending

Because the bending equations use free-free modes, there is no gravity term if an integrated body model is used. This may seem surprising. Suppose a rocket travels horizontally through the atmosphere while being supported by aerodynamic forces, as is done by some air-launched rockets and missiles. Just to make a more graphic hypothetical case, suppose there are aerodynamic surfaces that supply concentrated lift forces at both ends. Fig. 5.6 shows the ultimate extreme of this thought experiment. The structure looks like a beam supported at both ends. The most natural way to determine the beam deflection in this hypothetical case would be to use pinned-pinned modes and apply gravity along the length of the beam. If this approach were taken, it would not be necessary to specify the forces at the ends, since the effect of these forces is built into the pinned-pinned boundary conditions. However, the shape of the beam in this case could also be determined using free-free modes. This would require many modes to get the same accuracy as would be obtained from just one or two of the pinned-pinned modes, but nonetheless it can be done.

With free-free modes, we have the reverse of the pinned-pinned situation. It is essential to impose the forces at the ends, and gravity can be

$$
\begin{bmatrix}
m_T\mathbf{1} & -\mathbf{s}_{TD}^{\times} & -\mathbf{s}_E^{\times} & \mathbf{p}_{\psi 1} & \mathbf{p}_{\psi 2} & \cdots & \mathbf{p}_{\psi M} \\
\mathbf{s}_{TD}^{\times} & \mathbf{I}_{TD} & \mathbf{I}_{TWD} & \mathbf{h}_{\psi 1} & \mathbf{h}_{\psi 2} & \cdots & \mathbf{h}_{\psi M} \\
\mathbf{s}_E^{\times} & \mathbf{I}_{TWD}^T & \mathbf{I}_{EG} & \mathbf{c}_{EF1} & \mathbf{c}_{EF2} & \cdots & \mathbf{c}_{EFM} \\
\mathbf{p}_{\psi 1}^T & \mathbf{h}_{\psi 1}^T & \mathbf{c}_{EF1}^T & m_{B1} + d_{\psi 11} & d_{\psi 12} & \cdots & d_{\psi 1M} \\
\mathbf{p}_{\psi 2}^T & \mathbf{h}_{\psi 2}^T & \mathbf{c}_{EF2}^T & d_{\psi 21} & m_{B2} + d_{\psi 22} & \cdots & d_{\psi 2M} \\
\vdots & \vdots & \vdots & \vdots & \vdots & \ddots & \vdots \\
\mathbf{p}_{\psi M}^T & \mathbf{h}_{\psi M}^T & \mathbf{c}_{EFM}^T & d_{\psi M1} & d_{\psi M2} & \cdots & m_{BM} + d_{\psi MM}
\end{bmatrix}
\begin{bmatrix}
\mathbf{a}_b \\ \dot{\boldsymbol{\omega}} \\ \dot{\boldsymbol{\omega}}_E \\ \ddot{\eta}_1 \\ \ddot{\eta}_2 \\ \vdots \\ \ddot{\eta}_M
\end{bmatrix}
= \mathcal{F}_{\text{grav}} \qquad (5.3.47)
$$

Eq. (5.3.47): Reduced-body equation for gravity alone

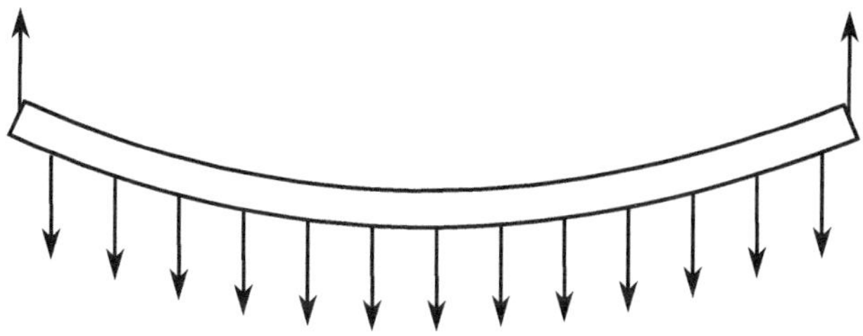

Figure 5.6 Beam with concentrated forces at the ends.

left out. The key to understanding this is to recognize that with free-free modes there is no coupling between the rigid-body motion and the flexible modes. The forces at the ends determine the beam deflection. Without gravity, the beam will accelerate upward with a value of g. Adding gravity will only affect the rigid-body motion; that is, it will cancel the upward acceleration so that the beam remains stationary. Gravity has no effect on the shape. We can imagine conducting this experiment inside an elevator. The shape of the beam must be the same whether the elevator is in free space accelerating in a direction normal to the beam or sitting stationary on Earth in a gravity field. This is consistent with Einstein's observation that without some kind of external reference, it is impossible to distinguish between an upward acceleration and a uniform gravity field.

With a reduced-body model, however, a gravity term is required on the RHS of the flex equations. Let us consider the case of no slosh, and no excitation other than gravity. By inserting reduced body terms into (2.5.20), the mass matrix in (5.3.47) can be obtained. In this simplified equation, $\mathcal{F}_{\text{grav}}$ is a vector containing all the generalized forces due to gravity;

$$\mathcal{F}_{\text{grav}} = \text{col}\left(\mathbf{f}_{\text{grav}}, \mathbf{g}_{\text{grav}}, \mathbf{g}_{E\text{grav}}, f_{B1\text{grav}}, f_{B2\text{grav}}, \dots, f_{BM\text{grav}}\right) \tag{5.3.48}$$

Examination of (5.1.68), (5.1.69) and (5.3.10) suggests that all the elements of $\mathcal{F}_{\text{grav}}$ can be obtained by multiplying the first column of the mass matrix in (5.3.47) by $g\,\mathbf{u}_v$, the gravity acceleration in the body frame. If this is done, it is readily confirmed that the solution of the system of Eqs. (5.3.47) is $\mathbf{a}_b = g\,\mathbf{u}_v$, with all the other elements of the acceleration vector being zero. With this solution, the rotation equation ends up with $\mathbf{s}_{TD}^{\times} g\,\mathbf{u}_v$ on both sides, the engine equation ends up with $\mathbf{s}_E^{\times} g\mathbf{u}_v$ on both sides, and most pertinent to the present discussion, all the bending equations are similarly in balance. This solution is exactly what we expect if the only excitation is gravity, which would result in translational acceleration but no acceleration of any other degree of freedom. Thus we could write

$$f_{Bi\,\text{grav}} = \mathbf{p}_{\psi i}^{T} g\,\mathbf{u}_{\nu} \tag{5.3.49}$$

although it would seem to be unnecessary to compute this explicitly – it makes more sense to compute the entire vector $\mathcal{F}_{\text{grav}}$ all at once as just described. Using a notation analogous to that used in *MATLAB*®, one can write

$$\mathcal{F}_{\text{grav}} = \mathcal{M}_{[\ldots,\,1\cdots3]}\, g\,\mathbf{u}_{\nu} \tag{5.3.50}$$

The first factor on the left represents all rows of the first three columns of the mass matrix. Nothing in the above discussion is altered if slosh equations are added. $\mathcal{F}_{\text{grav}}$ becomes part of the overall excitation vector $\mathcal{F}$ as described below.

The above analysis suggests an alternate approach to the issue of gravity. Instead of having the first element of the state vector be the total acceleration $\mathbf{a}_b$, we instead use the sensed acceleration $\mathbf{a}_{b\,\text{sen}}$, which is the acceleration with gravity not included. The gravity terms are then left out of the forces and moments on the right side of every equation of the dynamic system. This procedure will introduce a change in the translation equation, but a careful examination of the above discussion reveals that no net change is introduced to any of the other equations. Computing the correct translation requires special treatment, but this is required anyway. As will be described in Chapter 10, it is necessary to transform $\mathbf{a}_b$ into the inertial frame before integrating the translation equation. If instead we translate $\mathbf{a}_{b\,\text{sen}}$ into the inertial frame, then it is a straightforward matter to add to this the gravitational acceleration (also expressed in the inertial frame) to get the total acceleration in the inertial frame $\mathbf{a}_I$. This can be integrated to get the inertial velocity. This alternate approach is both simpler and computationally more efficient.

5.4 Summary of excitation terms

Let $\mathcal{F}$ be the vector formed from the right hand sides of (2.5.1) through (2.5.5):

$$\mathcal{F} = \text{col}\left(\tilde{\mathbf{f}}, \tilde{\mathbf{g}}, \tilde{\mathbf{g}}_E, \tilde{\mathbf{f}}_{s1}, \tilde{\mathbf{f}}_{s2}, \cdots, \tilde{\mathbf{f}}_{sN}, \tilde{f}_{B1}, \tilde{f}_{B2}, \cdots, \tilde{f}_{BM}\right) \tag{5.4.1}$$

If the engine motion is prescribed, then $\tilde{\mathbf{g}}_E$ is truncated out of this expression. This section provides a roadmap to all of these terms. Throughout this section, if the alternate approach to gravity described at the end of the previous section is used, then all the gravity terms included below (subscript grav) should be left out.

It is useful to define the phrase "nonlinear $\dot{\boldsymbol{\omega}}$ terms." Eq. (3.2.31) and the discussion thereafter provides a good illustration – the phrase refers to any situation in which $\dot{\boldsymbol{\omega}}$ is multiplied by either a slosh deflection or an engine deflection. The following summary assumes either a quasi-steady mass matrix, with nonlinear $\dot{\boldsymbol{\omega}}$ terms left out, or that the mass matrix is updated at each time step. If there is a desire to pursue the option of using a quasi-steady mass matrix and putting nonlinear $\dot{\boldsymbol{\omega}}$ terms on the RHS, then the discussion following (3.2.31) should be reviewed in detail. If the pendulum model is being used for slosh, the discussion following (4.3.40) should be reviewed.

System translation equation

From (2.5.7),

$$\tilde{\mathbf{f}} = \mathbf{f} + \mathbf{f}_{NL} + \mathbf{f}_{\text{presc}} \tag{5.4.2}$$

$\mathbf{f}_{presc}$ is given by (2.5.11), and the external force $\mathbf{f}$ is given by

$$\mathbf{f} = \mathbf{f}_{\text{thrust}} + \mathbf{f}_{\text{aero}} + \mathbf{f}_{\text{grav}} + \mathbf{f}_{\text{RCS}} \tag{5.4.3}$$

Each of these contributions is described in Section 5.1. The system nonlinear forces may be broken down as follows:

$$\mathbf{f}_{NL} = \mathbf{f}_{NL\ \text{locked}} + \mathbf{f}_{NL\ \text{nozzle}} + \mathbf{f}_{NL\ \text{slosh}} \tag{5.4.4}$$

where

$$\mathbf{f}_{NL\ \text{locked}} = \boldsymbol{\omega}^{\times} \mathbf{s}_T^{\times} \boldsymbol{\omega} \tag{5.4.5}$$

This is the term that shows up even if the nozzle and slosh mass are locked in position. It is straightforward to show that the difference between $\boldsymbol{\omega}^{\times}\mathbf{s}_T^{\times}\boldsymbol{\omega}$ and $\boldsymbol{\omega}^{\times}\mathbf{s}_{TD}^{\times}\boldsymbol{\omega}$ is third order and is usually not significant. Subtracting this term from the RHS of (4.4.13) (in order to avoid double counting) gives the nonlinear nozzle force:

$$\mathbf{f}_{NL\ \text{nozzle}} = 2\boldsymbol{\omega}^{\times}\mathbf{C}_{01}\mathbf{s}_E^{\times}\boldsymbol{\omega}_E + \mathbf{C}_{01}\boldsymbol{\omega}_E^{\times}\mathbf{s}_E^{\times}\boldsymbol{\omega}_E \tag{5.4.6}$$

For the spring slosh model, the nonlinear slosh forces are given by the following terms from the right side of (3.2.20):

$$\mathbf{f}_{NL\,\text{slosh}} = \sum_j m_{sj}\left(\boldsymbol{\omega}^\times \boldsymbol{\delta}_{sj}^\times \boldsymbol{\omega} - 2\boldsymbol{\omega}^\times \dot{\boldsymbol{\delta}}_{sj}\right) \tag{5.4.7}$$

The nonlinear term $\boldsymbol{\delta}_{sj}^\times \dot{\boldsymbol{\omega}}$ has been left out of this expression. This term will arise if a quasi-steady mass matrix is assumed, i.e. the term $\dot{\mathbf{s}}_{TD}$ in (3.2.19) must be moved to the RHS. This falls in the category of "nonlinear $\dot{\boldsymbol{\omega}}$ terms" discussed at the outset of this section.

For the pendulum model, the nonlinear slosh terms come from (4.2.38):

$$\mathbf{f}_{NL\,\text{slosh}} = \sum_j (2\boldsymbol{\omega}^\times \mathbf{C}_{0j}\mathbf{d}_j^\times \boldsymbol{\omega}_{rj} + \mathbf{C}_{0j}\boldsymbol{\omega}_{rj}^\times \mathbf{d}_j^\times \boldsymbol{\omega}_{rj}) \tag{5.4.8}$$

System rotation equation

Adding the fuel offset torques to (2.5.8) gives

$$\tilde{\mathbf{g}} = \mathbf{g} + \mathbf{g}_{NL} + \mathbf{g}_{\text{presc}} + \mathbf{g}_{\text{offset}} \tag{5.4.9}$$

The external torque $\mathbf{g}$ is given by

$$\mathbf{g} = \mathbf{g}_{\text{thrust}} + \mathbf{g}_{\text{aero}} + \mathbf{g}_{\text{grav}} + \mathbf{g}_{\text{RCS}} + \mathbf{g}_{\text{jet}} \tag{5.4.10}$$

Each of the contributions in (5.4.10) is described in Section 5.1. $\mathbf{g}_{\text{offset}}$ is given in Section 5.2, and $\mathbf{g}_{presc}$ is given by (2.5.12). The nonlinear torque is broken down in the same way as the nonlinear force:

$$\mathbf{g}_{NL} = \mathbf{g}_{NL\,\text{locked}} + \mathbf{g}_{NL\,\text{nozzle}} + \mathbf{g}_{NL\,\text{slosh}} \tag{5.4.11}$$

where

$$\mathbf{g}_{NL\,\text{locked}} = -\boldsymbol{\omega}^\times \mathbf{I}_T \boldsymbol{\omega} \tag{5.4.12}$$

The arguments leading up to (3.2.34) make the point that the difference between $\boldsymbol{\omega}^\times \mathbf{I}_T^\times \boldsymbol{\omega}$ and $\boldsymbol{\omega}^\times \mathbf{I}_{TD}^\times \boldsymbol{\omega}$ is also third order and thus not significant. We subtract this term from the RHS of (4.4.14), since we do not wish to include it twice. This gives the nonlinear nozzle torque:

$$\begin{aligned}\mathbf{g}_{NL\,\text{nozzle}} = &-\boldsymbol{\omega}^\times \mathbf{C}_{01}\mathbf{I}_E\boldsymbol{\omega}_E - \mathbf{C}_{01}\boldsymbol{\omega}_E^\times \mathbf{I}_E(\mathbf{C}_{10}\boldsymbol{\omega} + \boldsymbol{\omega}_E) + \mathbf{C}_{01}\mathbf{I}_E\boldsymbol{\omega}_E^\times \mathbf{C}_{10}\boldsymbol{\omega} \\ &+ 2\mathbf{r}_G^\times \boldsymbol{\omega}^\times \mathbf{C}_{01}\mathbf{s}_E^\times \boldsymbol{\omega}_E + \mathbf{r}_G^\times \mathbf{C}_{01}\boldsymbol{\omega}_E^\times \mathbf{s}_E^\times \boldsymbol{\omega}_E\end{aligned} \tag{5.4.13}$$

For the spring model, the nonlinear slosh term is given by the following from the right side of (3.2.34). Recall from the discussion leading up to (3.2.34) that a number of higher-order terms proportional to $\dot{\boldsymbol{\omega}}$ have been left out.

$$\mathbf{g}_{NL\,\text{slosh}} = -2\sum_j m_{sj}\mathbf{r}_{sj}^{\times}\boldsymbol{\omega}^{\times}\dot{\boldsymbol{\delta}}_{sj} \tag{5.4.14}$$

For the pendulum model, the nonlinear slosh terms come from (4.2.49):

$$\begin{aligned}\mathbf{g}_{NL,\,\text{slosh}} = +\sum_j -\boldsymbol{\omega}^{\times}\mathbf{C}_{0j}\mathbf{J}_j\boldsymbol{\omega}_{rj} - \mathbf{C}_{0j}\boldsymbol{\omega}_{rj}^{\times}\mathbf{J}_j\left(\mathbf{C}_{j0}\boldsymbol{\omega}+\boldsymbol{\omega}_{rj}\right) + \mathbf{C}_{0j}\mathbf{J}_j\boldsymbol{\omega}_{rj}^{\times}\mathbf{C}_{j0}\boldsymbol{\omega} \\ +\sum_j 2\mathbf{b}_j^{\times}\left(\mathbf{C}_{0j}\boldsymbol{\omega}_{rj}^{\times}\mathbf{d}_j\right)^{\times} + \mathbf{b}_j^{\times}\mathbf{C}_{0j}\boldsymbol{\omega}_{rj}^{\times}\mathbf{d}_j^{\times}\boldsymbol{\omega}_{rj}\end{aligned} \tag{5.4.15}$$

Slosh equation

$$\tilde{\mathbf{f}}_{sj} = \mathbf{f}_{sj} + \mathbf{f}_{sj\,NL} \tag{5.4.16}$$

The slosh force $\mathbf{f}_{sj}$ is obtained from (5.3.10). Note that this term contains the gravity force. If the spring model is used, the nonlinear terms for each slosh equation come from (3.2.15):

$$\mathbf{f}_{sj\,NL} = m_{sj}\left(\boldsymbol{\omega}^{\times}\mathbf{r}_{sj}^{\times}\boldsymbol{\omega} - 2\boldsymbol{\omega}^{\times}\dot{\boldsymbol{\delta}}_{sj}\right) \tag{5.4.17}$$

Following the discussion after (3.2.16), the $\boldsymbol{\delta}_{sj}^{\times}\dot{\boldsymbol{\omega}}$ term has been dropped. To be clear, $\mathbf{f}_{sj\,NL}$ is the term that appears in the slosh equation for slosh mass j, and $\mathbf{f}_{NL\,\text{slosh}}$ is the term that appears in the system translation equation. The first term in (5.4.17) is similar but not identical to that in (5.4.7). The reason for the difference is explained in the discussion following (3.2.20).

For the slosh pendulum model, the slosh equation has torques rather than forces, and the terms come from (4.2.57):

$$\mathbf{g}_{sj\,NL} = -\dot{\mathbf{J}}_{j0}\boldsymbol{\omega} + \left(\mathbf{J}_{j0}\boldsymbol{\omega}+\mathbf{J}_j\boldsymbol{\omega}_{rj}\right)^{\times}\boldsymbol{\omega}_j + \left(\mathbf{C}_{j0}\boldsymbol{\omega}^{\times}\mathbf{b}_j\right)^{\times}\mathbf{d}_j^{\times}\boldsymbol{\omega}_j \tag{5.4.18}$$

Engine equation

An engine equation is only included in the set of dynamic equations if the engine motion is not prescribed. In that case, we have

$$\tilde{\mathbf{g}}_E = \mathbf{g}_E + \mathbf{g}_{E\,NL} \tag{5.4.19}$$

The torque on the engine $\mathbf{g}_E$ must be obtained from a Thrust Vector Control (TVC) model.

In practice, there is very little evidence that nonlinear engine torques would have a noticeable influence on the nozzle motion. That is, the torque given by (5.4.11) may be significant for the overall rocket, but it is not necessary to include second-order terms in the nozzle motion. The effect on the *absolute* angular velocity of the rocket may be significant, but the effect on the relative angular velocity $\boldsymbol{\omega}_E$ of the nozzle is not, since the nozzle actuator is presumably quite stiff.

Bending equation

The total external excitation in bending is given by

$$\tilde{f}_{Bi} = f_{Bi} + f_{Bi\,\mathrm{presc}} \tag{5.4.20}$$

These are obtained from (5.3.30) and (2.5.13). For a reduced body model, there is an additional excitation due to gravity that is described in Section 5.3.

CHAPTER 6

Engine interactions

Deflecting the thrust via the TVC system is a principal means of controlling a rocket's attitude. For rockets that can be considered rigid, the rotational dynamics in response to engine deflections are relatively straightforward. The following section discusses one particularly simple example. Subsequent sections discuss possible complications that arise as the rocket becomes more flexible.

6.1 The tail-wags-dog (TWD) zero

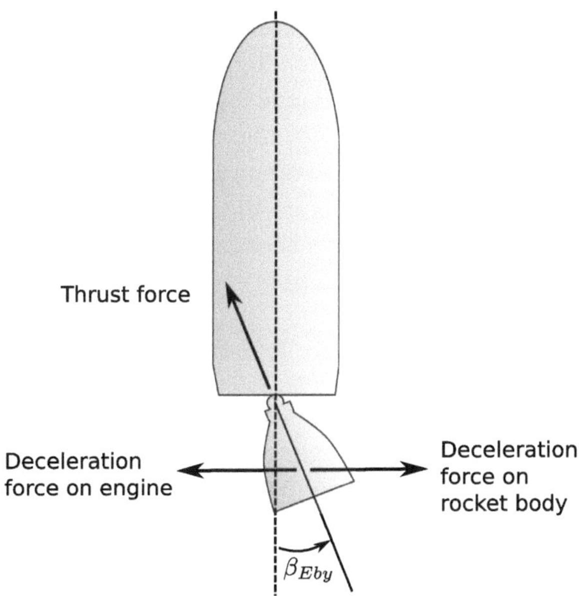

Figure 6.1 Rocket with engine undergoing sinusoidal motion.

Consider the case of sinusoidal engine pitch motion:

$$\beta_{Eby} = A \sin \omega_a t \tag{6.1.1}$$

Dynamics and Simulation of Flexible Rockets
https://doi.org/10.1016/B978-0-12-819994-7.00011-X

where ω_a is the applied frequency, A is the amplitude of the motion, and β_{Eby} is the engine deflection about the body y axis, expressed in the body frame. Fig. 6.1 shows the particular instant at which the engine is at its maximum deflection. This is also the instant at which the angular deceleration of the nozzle is maximized, having an instantaneous value of $\ddot{\beta}_{Eby} = -\omega_a^2 A$. The loads from the TVC system can be resolved as a combination of force and torque acting at the center of mass of the engine. For illustrative purposes, consider just the force, which can be called the "deceleration force" at this instant. This is an interbody force, so the force from the rocket body acting on the engine is equal and opposite to the force from the engine acting on the rocket body. As shown in Fig. 6.1, the deceleration force on the rocket body acts toward the right. At this instant, the lateral component of the thrust force acts to the left. The possibility therefore exists that the lateral component of the thrust force is canceled by the deceleration force. This can only happen at one frequency. Even if the above simplifications are removed and the effect of the local torque about the engine center of mass is considered, there is still a frequency at which there is no net torque about the rocket center of mass. The applied frequency at which this happens is called the "tail-wags-dog" (TWD) frequency. At this frequency, a sinusoidal motion of the engine produces no net effect on the rocket motion.

To show how this frequency is computed, consider a rocket with no slosh and no flexible modes, and locate the origin at the overall center of mass, so that $\mathbf{s}_T = \mathbf{0}$. The relationship between the engine and body angular acceleration, using Eq. (2.5.2), becomes

$$\mathbf{I}_T\dot{\boldsymbol{\omega}} = \mathbf{g} - \mathbf{I}_{TWD}\dot{\boldsymbol{\omega}}_{Eb} \tag{6.1.2}$$

To further simplify, assume the gimbal is on the centerline of the rocket body, and the engine center of mass is on the engine centerline. Thus the y and z components of $\mathbf{r}_G$ and $\mathbf{s}_E$ are zero;

$$\mathbf{r}_G = \begin{bmatrix} X_G & 0 & 0 \end{bmatrix}^T \tag{6.1.3}$$

$$\mathbf{s}_E = \begin{bmatrix} m_E l_E & 0 & 0 \end{bmatrix}^T \tag{6.1.4}$$

where l_E is the distance of the center of mass of the engine from the gimbal point. Using the same convention as X_G, this value is negative if the engine CM is aft of the gimbal (Fig. 6.5); thus, it is important to note that for conventional rockets, both X_G and l_E are negative quantities.

With these simplifications, Eqs. (5.1.10) and (5.1.11) can be combined to give

$$\mathbf{g} = \mathbf{r}_G^{\times} \boldsymbol{\beta}_{Eb}^{\times} \mathbf{f}_R \tag{6.1.5}$$

This expression can be expanded as

$$\mathbf{g} = \begin{bmatrix} 0 \\ X_G \beta_{Eby} F_R \\ X_G \beta_{Ebz} F_R \end{bmatrix} \tag{6.1.6}$$

Assuming the moment of inertia of the engine is diagonal,

$$\mathbf{I}_{Eb} = \begin{bmatrix} I_{Exx} & 0 & 0 \\ 0 & I_{EG} & 0 \\ 0 & 0 & I_{EG} \end{bmatrix}. \tag{6.1.7}$$

Consider the y component of Eq. (6.1.2), which provides the pitch dynamics:

$$I_{Tyy} \dot{\omega}_y = X_G \beta_{Eby} F_R - I_{TWDyy} \ddot{\beta}_{Eby} \tag{6.1.8}$$

Eq. (2.4.23) gives

$$I_{TWDyy} = I_{EG} + X_G m_E l_E \tag{6.1.9}$$

If the engine motion is sinusoidal, Eq. (6.1.1) can be differentiated twice to yield

$$\ddot{\beta}_{Eby} = -\omega_a^2 \beta_{Eby} \tag{6.1.10}$$

At the TWD zero, $\omega_a = \omega_{TWD}$, and $\dot{\omega}_y = 0$, and the two terms on the RHS of (6.1.8) add up to zero:

$$X_G \beta_{Eby} F_R + \left(I_{EG} + X_G m_E l_E \right) \omega_{TWD}^2 \beta_{Eby} = 0$$

This can be solved for the desired TWD frequency:

$$\omega_{TWD}^2 = \frac{-F_R X_G}{I_{EG} + X_G m_E l_E} \tag{6.1.11}$$

In the case of a rigid rocket, this frequency is of some importance for the design of the control system, and is called the "TWD zero." For a flexible rocket, the zero will become shifted. In that case, (6.1.11) may still provide a rough guide to a region in the frequency domain where an important phase shift will occur. It is important to remember that the X_G used here must be defined relative to an origin at the center of mass.

6.2 Engine/flex interaction

Using prescribed engine motion in the dynamics amounts to the assumption that the TVC actuators are so stiff that the engine nozzle angle β_E is not affected by the flexible dynamics. With such an assumption, the inertial effect of the engine angular acceleration can affect the motion of the rocket, but the elastic motion of the rocket does not affect β_E. The stiff actuator assumption allows the dynamicist to avoid the details of mutual coupling between the engine motion and the flex motion. In later stages of development, it becomes necessary to recognize that the actuators are not infinitely stiff. In this case, β_E can be affected by the motion of the gimbal point. This is known as the "dog-wags-tail" (DWT) effect. In order to capture this effect, an equation describing the engine dynamics is included in the overall set of dynamic equations, and β_E becomes an element of the state vector, as described at the end of Section 2.5. The present section focuses on the interaction between this engine equation and the flexible modes.

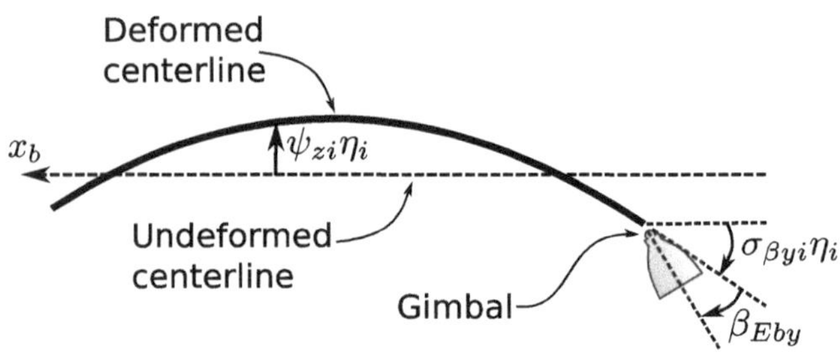

Figure 6.2 Mode shape for the lowest frequency pitch bending mode ($i = 1$), and definition of engine deflection.

Recall from Section 2.2 how the typical rocket analysis requires a FEM that supplies a large "$\mathbf{\Phi}$ matrix" containing eigenvectors for each node of the structure. For purposes of the present discussion, we can assume that a $\mathbf{\Phi}$ matrix is provided that contains just the centerline nodes. We can extract the eigenvector of mode i for the node at the gimbal point (subscript β) and represent it as follows:

$$\boldsymbol{\phi}_{\beta i} = \begin{bmatrix} \psi_{\beta xi} & \psi_{\beta yi} & \psi_{\beta zi} & \sigma_{\beta xi} & \sigma_{\beta yi} & \sigma_{\beta zi} \end{bmatrix}^T \qquad (6.2.1)$$

Quantities with a node subscript (for a total of three subscripts) are scalars. Quantities without the node subscript are long vectors containing values for all the centerline nodes. The mode shape in the pitch plane is obtained by plotting ψ_{zi} versus x, as shown in Fig. 6.2. This figure is also used to define the engine deflection β_{Eby}, which is not part of the mode shape.

In order to illustrate the DWT dynamics, we assume a rocket with no rigid-body rotation or translation, no slosh, no thrust, and no external excitation of any kind. With these simplifications, the engine and bending equations (2.5.3) and (2.5.5) become, with the aid of Eq. (5.3.30),

$$\mathbf{I}_{Eb}\dot{\boldsymbol{\omega}}_{Eb} + \sum \mathbf{c}_{EFi}\ddot{\eta}_i = \mathbf{g}_{Eb} \tag{6.2.2}$$

$$\mathbf{c}_{EFi}^T\dot{\boldsymbol{\omega}}_{Eb} + m_{Bi}\ddot{\eta}_i = -m_{Bi}\left(\Omega_{Bi}^2\eta_{Bi} + 2\zeta_{Bi}\Omega_{Bi}\dot{\eta}_i\right) \tag{6.2.3}$$

There is a separate bending equation for each mode i. Here, the only excitation of bending is that from the engine angular acceleration relative to the deformed centerline. Repeating all the simplifying assumptions of the previous section (diagonal inertia matrices, center of mass and gimbals on the centerline, etc.), the engine/flex coupling vector defined by (2.4.22) becomes simplified. Define the engine/flex coupling scalar as the pitch component of this expression;

$$\Gamma_i \equiv (\mathbf{c}_{EFi})_y = I_{Ebyy}\sigma_{y\beta i} - m_E l_E \psi_{z\beta i} \tag{6.2.4}$$

This is defined in the body frame, so the relationships that follow must also be consistently expressed in the body frame. Assuming that the only motion is in the pitch plane, Eqs. (6.2.2) and (6.2.3) become

$$I_{Ebyy}\ddot{\beta}_{Eby} + \sum \Gamma_i\ddot{\eta}_i = g_{Eby} \tag{6.2.5}$$

$$\Gamma_i\ddot{\beta}_{Eby} + m_{Bi}\ddot{\eta}_i = -m_{Bi}\left(\Omega_{Bi}^2\eta_i + 2\zeta_{Bi}\Omega_{Bi}\dot{\eta}_i\right) \tag{6.2.6}$$

The summation in (6.2.5) includes the effect of all the modes on the rotation and translation of the gimbal point.

Fig. 6.3 shows the aft structure of a hypothetical rocket. For purposes of the present discussion, the actuator should be thought of as a simple linear spring of stiffness K_A, as shown. Here we use upper case K to represent a linear spring stiffness, and lower case k represents the same element converted to torsional stiffness. The conversion is simply

$$k_A = d_{arm}^2 K_A \tag{6.2.7}$$

With this, the actuator torque can be written as

$$g_{Eby} = -k_A\beta_{Eby} \tag{6.2.8}$$

It is perhaps illuminating to write (6.2.8) as $\beta_{Eby} = -g_{Eby}/k_A$. In this form, one can see that the effect of infinite stiffness is not to create infinite torques but to drive β_{Eby} to zero. Using (6.2.8), Eq. (6.2.5) becomes

$$I_{Eb}\ddot{\beta}_{Eby} + \sum \Gamma_i \ddot{\eta}_i = -k_A \beta_{Eby} \tag{6.2.9}$$

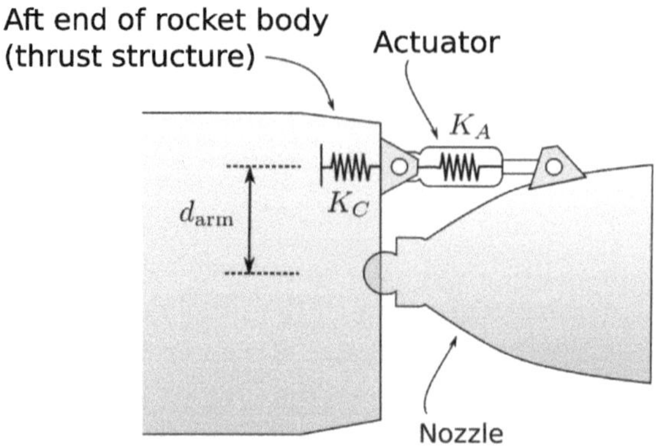

Figure 6.3 Schematic of the connection between a rocket body and its nozzle. K_C = attach point stiffness, K_A = actuator stiffness.

In order to reduce the problem to its simplest form we now assume that only the first mode of the FEM is included. Thus $i = 1$. To simplify the notation, let

$$\beta \equiv \beta_{Eby} \tag{6.2.10}$$

$$\psi = \psi_{z\beta 1} \tag{6.2.11}$$

$$\sigma = \sigma_{y\beta 1} \tag{6.2.12}$$

The nozzle equation (6.2.9) reduces to

$$I_{Eb}\ddot{\beta} + \Gamma_1 \ddot{\eta}_1 = -k_A \beta \tag{6.2.13}$$

In the bending equation (6.2.6), assume the damping is zero and use the following for the bending stiffness:

$$k_{B1} \equiv m_{B1}\Omega_{B1}^2 \tag{6.2.14}$$

Then the system of Eqs. (6.2.5) and (6.2.13) becomes

$$\begin{bmatrix} I_{Eb} & \Gamma_1 \\ \Gamma_1 & m_{B1} \end{bmatrix} \begin{bmatrix} \ddot{\beta} \\ \ddot{\eta}_1 \end{bmatrix} + \begin{bmatrix} k_A & 0 \\ 0 & k_{B1} \end{bmatrix} \begin{bmatrix} \beta \\ \eta_1 \end{bmatrix} = \mathbf{0} \tag{6.2.15}$$

One point that was not mentioned previously is that the engine equation (6.2.5) is only valid if the FEM does not include the actuator compliance $1/k_A$. This restriction becomes more evident when the equations are assembled as in (6.2.15). If the FEM includes k_A, then this system of equations includes k_A twice.

The first matrix in (6.2.15) is the mass matrix for this problem. The determinant of the mass matrix is

$$D = I_{Eb} m_{B1} - \Gamma_1^2 \tag{6.2.16}$$

If more bending modes are included, it becomes convenient to mass-normalize the bending mode shapes, such that $m_{Bi} = 1$ for all i. If three modes are included, the mass matrix becomes

$$\begin{bmatrix} I_{Eb} & \Gamma_1 & \Gamma_2 & \Gamma_3 \\ \Gamma_1 & 1 & 0 & 0 \\ \Gamma_2 & 0 & 1 & 0 \\ \Gamma_3 & 0 & 0 & 1 \end{bmatrix} \tag{6.2.17}$$

It is not difficult to evaluate the determinant of this using expansion by minors:

$$D = I_{Eb} - \Gamma_1^2 - \Gamma_2^2 - \Gamma_3^2 \tag{6.2.18}$$

One can see an unfortunate trend here. As more and more modes are added, the determinant of the mass matrix becomes smaller and smaller. If enough modes are included, the determinant may become negative and the mass matrix will not be positive definite. Thus there is a limit to how many modes can be included. This is discussed further in the next section.

In the finite element model, it is possible to specify a rotary joint at the gimbal point. This results in an "inboard node" attached to the rocket, and an "outboard node" attached to the nozzle. These two nodes would be co-located. Returning now to the analysis of a single flexible mode, define

$$\Delta\sigma = \sigma_{out} - \sigma_{in} \tag{6.2.19}$$

The angle between the inboard rocket centerline and the nozzle centerline is the sum of effects from the attach point stiffness and the actuator stiffness. This angle is apportioned as follows:

$$\beta = \text{angle due to } k_A \tag{6.2.20}$$

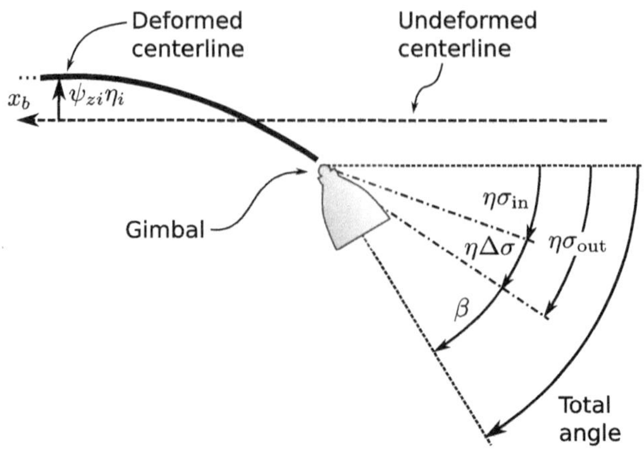

Figure 6.4 Detail of inboard and outboard modal rotations.

$$\eta\Delta\sigma = \text{angle due to } k_C \tag{6.2.21}$$

Fig. 6.4 shows these angles. To get the angle between the *undeformed* rocket centerline and the nozzle centerline, it is necessary to also include $\eta\sigma_{in} =$ inboard angle due to rocket body flex rotation at the gimbal. Thus the total angle between these two centerlines is

$$\text{Total angle} = \beta + \eta\sigma_{out} \tag{6.2.22}$$

The attach point stiffness may be allocated to the finite element model, or it may become part of the TVC model. If no attach point or actuator compliance is contained in the FEM, $\Delta\sigma$ is zero and there is no distinction between the inboard and the outboard node. For that approach, the two springs of Fig. 6.3 would be added in series and subsumed into the TVC model. If the FEM does include the attach point compliance, then the modal rotation σ in Eq. (6.2.12), and the $\sigma_{\beta yi}$ in Fig. 6.2, must be that for the outboard node, as indicated by (6.2.22).

Case 1 – zero attach point stiffness

The assumption that the attach point stiffness k_C is zero is examined here in order to provide some insight into how the math agrees with the physics, even in this limiting case. Assume the engine is modeled as a point mass on a rigid, massless rod. The engine first and second moments of inertia are

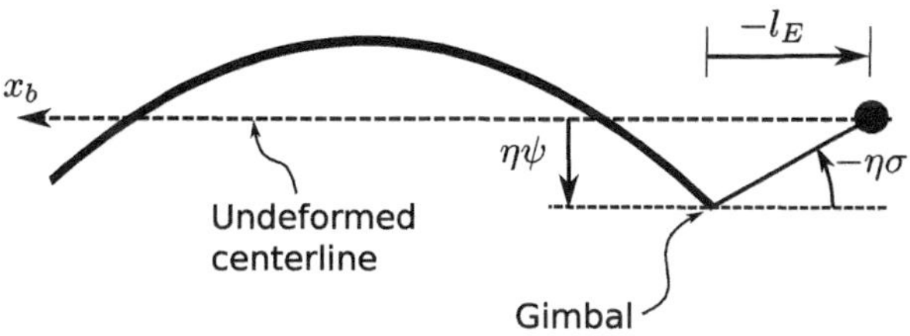

Figure 6.5 Beam (representing a rocket) connected to a point-mass nozzle, with zero stiffness at the gimbal point.

thus

$$S_E = l_E m_E \tag{6.2.23}$$

$$I_{EG} = l_E^2 m_E \tag{6.2.24}$$

with l_E defined as shown in Fig. 6.5. Inserting these into (6.2.4), we have

$$\Gamma_1 = m_E l_E \left(l_E \sigma - \psi \right) \tag{6.2.25}$$

However, if the actuator stiffness is actually zero, then the nozzle mass does not move even when the beam in Fig. 6.5 is undergoing oscillations, i.e., it will stay on the undeformed centerline. From the geometry of Fig. 6.5, since we define $\eta\sigma$ as the modal rotation for the outboard gimbal node, we have

$$\eta\psi = l_E \eta\sigma \tag{6.2.26}$$

Eliminating η from both sides and inserting the result into (6.2.25), we obtain

$$\Gamma_1 = 0 \tag{6.2.27}$$

and the system of Eqs. (6.2.15) becomes decoupled. Thus there is not any difficulty with the stability of the equations in this limiting case.

If the engine is not a point mass, but a real body with a finite moment of inertia about its own center of mass, this analysis becomes slightly more complex, but there will still be some point on the engine that remains on the undeformed centerline, so the basic result remains intact.

Case 2 – infinite actuator stiffness

All three cases examined herein deal with the homogeneous problem, i.e., the commanded gimbal angle is zero. For Case 2, since the actuator stiffness is infinite, $\beta = 0$. This corresponds to a locked actuator. The bending equation feels no effect from any nozzle motion relative to the deformed centerline. In this limiting case, one pair of eigenvalues and eigenvectors coming from (6.2.15) will be unchanged from what is obtained from the modal analysis alone. Another pair of infinite eigenvalues would have to be eliminated from the equations in order to proceed. This can be done by simply deleting the nozzle equation. Indeed, there would be no need for such an equation. A "dog-wags-tail" (DWT) analysis is only required to determine the effect of rocket motion (in the present case just the flex motion) on the gimbal angle β. If the actuator stiffness is infinite, there is no such effect. It is important to realize that this case still allows the actuator attach point compliance to be finite. This compliance will show up as the difference $\Delta\sigma$ between the inboard and outboard modal rotations at the gimbal point.

Case 3 – finite actuator stiffness

Assume that a FEM has been created that includes the attach point compliance $1/k_C$, but not the compliance of the actuator $1/k_A$. This case can be analyzed using the stability analysis techniques which will be discussed in more detail in Chapter 9.

If the outboard node at the gimbal point is used to define the parameter Γ_1 in (6.2.4), then the effect of k_C is incorporated into Γ_1. Eq. (6.2.15) can be written using the Laplace operator as

$$\begin{bmatrix} s^2 I_{Eb} + k_A & s^2\Gamma_1 \\ s^2\Gamma_1 & s^2 m_{B1} + k_{B1} \end{bmatrix} \begin{bmatrix} \beta \\ \eta_1 \end{bmatrix} = \mathbf{0} \tag{6.2.28}$$

This system has the characteristic polynomial

$$\left(s^2 I_{EG} + k_A\right)\left(s^2 m_{B1} + k_{B1}\right) - s^4\Gamma_1^2 = P(s) \tag{6.2.29}$$

which is stable if and only if the real parts of the roots of the equation $P(s) = 0$ are negative. If the roots occur in complex pairs with zero real parts, the solutions of (6.2.28) are periodic. The *root locus* is the trajectory of the roots of $P(s)$ in the complex plane as a function of a chosen parameter.

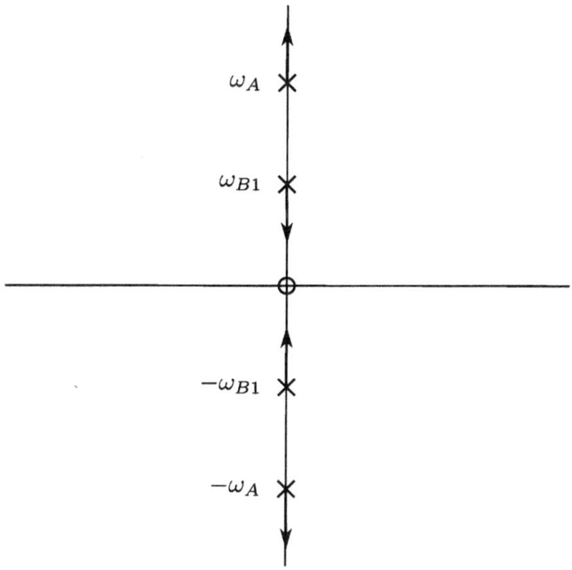

Figure 6.6 Root locus for finite actuator stiffness.

Dividing by $m_{B1}I_{EG}$ gives

$$\left(s^2+\omega_A^2\right)\left(s^2+\omega_{B1}^2\right)-s^4\Gamma_1^2/m_{B1}I_{Eb}=0 \tag{6.2.30}$$

One can create a root locus of this equation as a function of the quantity $\Gamma_1^2/m_{B1}I_{Eb}$, which plays the role of root locus gain. This would appear as in Fig. 6.6. The poles represent the roots when there is no coupling. The upper and lower loci reach plus and minus infinity, respectively, when the gain is equal to one. For any higher gain there is a positive real root.

Our first observation is that the sum of all the terms multiplying s^4 in the characteristic polynomial (6.2.29) must be positive for the system to be stable. This is the same as stating that the determinant D in (6.2.16) must be greater than zero. This illustrates how easy it is to confuse the requirement for system stability with the requirement that the mass matrix be positive definite. In this passive system we know there is no possibility of an actual instability. However, as soon as thrust is added, the system becomes active, i.e., has an external source of energy, and requires a control loop for stability. One can easily imagine that if such a control loop is being analyzed, and many modes are included, it may not be obvious that apparent instabilities in the control response have nothing to do with the actual stability of the control design and everything to do with the condition of the mass matrix. Another lesson from Case 3 is that both the actuator frequency and the bending frequency will shift due to the coupling. In one mode β and $\sigma\eta$ have the same sign, and in the other mode they have the opposite sign. In other words, in one mode the sign of β is such that engine swings in relative to what occurs from the flex motion, and in the other mode it swings out, as shown in Fig. 6.7. The term "pendulum mode" brings to mind a nozzle mounted on a gimbal that is free to rotate but fixed in translation. When the gimbal is attached to one end of a free-free beam, there is some translation of the gimbal point, but the motion still resembles that of a pendulum. In this situation, it is logical to label the modes of the system as shown in Fig. 6.7.

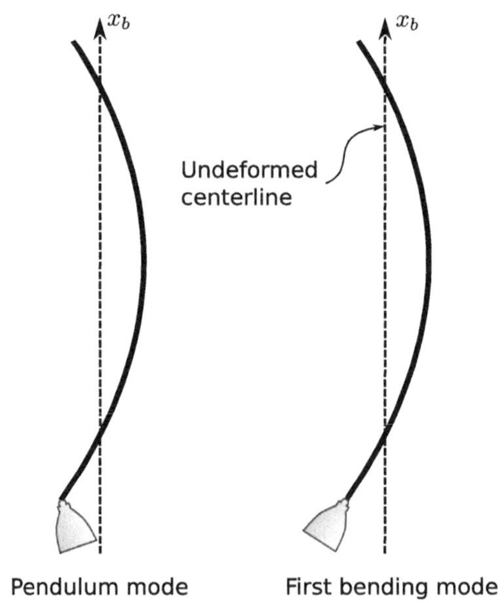

Figure 6.7 Modes obtained from the nozzle/bending system of equations.

When a large number of modes from the FEM are included, it may be possible to identify one of them as the pendulum mode due to the attach point compliance. The above analysis gives rise to the notion of deleting the pendulum mode from the finite element model. The idea is that this mode is first taken out but an additional mode is created when the bending equations are combined with the nozzle equation. This newly created mode should correspond to what was deleted, although it will be at a different frequency. This may be appropriate for initial analyses, but the next section discusses a better way that avoids such dubious meddling when the time comes for final verification of the control design.

Thrust vector servoelasticity (TVSE)

The stability problems that arise when thrust is included in the analysis are of two types. The most obvious issue is heading stability, which requires active control of the nozzle angle. For large rockets, lack of active control results in a slow divergence from the intended path. This is not of interest to the present discussion. The other type of problem is that the thrust can add energy to an oscillatory engine-flex interaction. An initial examination of this type of problem can make the assumption that the commanded nozzle angle is zero. To facilitate this discussion, we will continue with the above notion that the engine plus actuator can be represented as a mass-spring system with a natural frequency ω_A, even though a more sophisticated TVC model would be required for most purposes. More complex TVC models are discussed in Chapter 9.

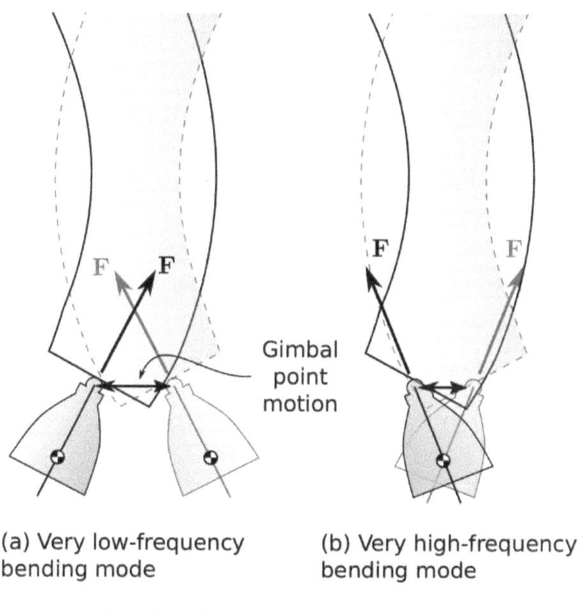

Figure 6.8 Motion with bending frequencies well below, or well above, the actuator bandwidth.

Fig. 6.8(a) shows the situation for a low-frequency bending mode, such that $\omega_B \ll \omega_A$. In this case, the actuator has no difficulty keeping the nozzle angle close to its commanded value of zero, and very little energy is added to the motion as the result of thrust. Fig. 6.8(b) shows the situation for $\omega_B \gg \omega_A$. In this case, the actuator is not at all successful at resisting the high-frequency motion, and the center of mass of the nozzle remains stationary. The thrust adds energy to the motion during 90 degrees of one phase of a cycle, and subtracts an equal amount of energy in the next 90 degrees. In contrast, when ω_B is in the same range as ω_A, we have the situation shown in Fig. 6.9. Because of TVC lag, thrust adds energy to the flexible mode during both outward motion and return motion.

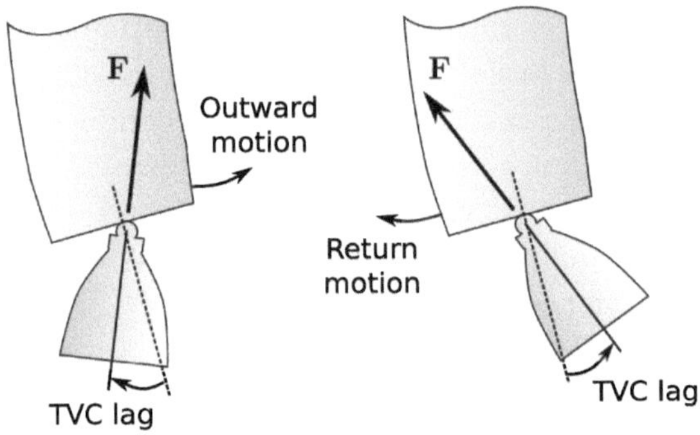

Figure 6.9 Motions with bending frequencies comparable to the actuator frequency.

This situation is an analog of a phenomenon in flexible aircraft known as aeroservoelasticity (ASE), where energy is extracted from the aerodynamic forces by a control surface. For rockets, the energy source is thrust, and the phenomenon is called thrust vector servoelasticity (TVSE).

The effects of thrust coupling to bending can be shown with a numerical example. The engine equation with damping is

$$\left(I_{Eb}s^2 + C_A s + k_A\right)\beta + \Gamma s^2 \eta = 0 \tag{6.2.31}$$

Illustrative numbers for a large solid rocket booster would be

$$\begin{aligned}
m_E &= \text{engine mass} = 620 \text{ slugs} \\
l_E &= \text{engine CM from gimbal point} = -2 \text{ ft} \\
I_{Eb} &= \text{engine inertia} = 20,000 \text{ slug} \cdot \text{ft}^2 \\
\psi_\beta &= \text{mode shape at gimbal point} = 0.009 \text{ slug}^{-1/2} \\
\sigma_\beta &= \text{mode slope at gimbal point} = 2.25 \times 10^{-4} \text{ ft}^{-1} \text{ slug}^{-1/2}
\end{aligned}$$

The bending equation with damping and thrust is

$$\Gamma s^2\beta + \left(s^2 + 2\zeta_B\omega_B s + \omega_B^2\right)\eta = -\psi_\beta F\beta \tag{6.2.32}$$

where

$$\begin{aligned} F &= \text{thrust} = 3 \times 10^6 \text{lb} \\ \zeta_B &= \text{flex damping ratio} = 0.005 \\ \omega_B &= 23 \text{ rad/sec} \end{aligned}$$

Typical characteristics for a large hydraulic TVC actuator in the vicinity of ω_B, that is, approximating the actuator as a spring, are

$$\begin{aligned} k_A &= 3.7 \times 10^7 \text{ ft} \cdot \text{lb per radian} \\ C_A &= \text{actuator damping coefficient} = 5.5 \times 10^5 \text{ ft} \cdot \text{lb} \cdot \text{sec per radian} \end{aligned}$$

The first term on the LHS is the excitation of the flexible response due to engine motion, i.e., the TWD effect. The engine and bending equations in matrix form become

$$\begin{bmatrix} I_{EG}s^2 + C_A s + k_A & \Gamma s^2 \\ \Gamma s^2 + \psi_\beta F & s^2 + 2\zeta_B\omega_B s + \omega_B^2 \end{bmatrix} \begin{bmatrix} \beta \\ \eta \end{bmatrix} = \begin{bmatrix} 0 \\ 0 \end{bmatrix} \tag{6.2.33}$$

Applying Kramer's rule to this equation yields the following characteristic polynomial

$$c_1 s^4 + c_2 s^3 + c_3 s^2 + c_4 s + c_5 = P(s) \tag{6.2.34}$$

where

$$c_1 = I_{EG} - \Gamma^2 \tag{6.2.35}$$

$$c_2 = 2I_{EG}\zeta_B\omega_B + C_A \tag{6.2.36}$$

$$c_3 = I_{EG}\omega_B^2 + 2C_A\zeta_B\omega_B + k_A - \Gamma\psi_\beta F \tag{6.2.37}$$

$$c_4 = C_A\omega_B^2 + 2k_A\zeta_B\omega_B \tag{6.2.38}$$

$$c_5 = k_A\omega_B^2 \tag{6.2.39}$$

Solving the characteristic equation $P(s) = 0$ yields two pairs of complex roots. With no coupling ($\Gamma = 0$) we have

$$\begin{aligned} \lambda_{1,2} &= -13.7500 \pm j40.7546 \\ \lambda_{3,4} &= -0.1150 \pm j22.9997 \end{aligned}$$

The first pair is heavily damped and represents the engine equation. The second pair represents the flex equation. With Γ set to its actual value we obtain

$$\lambda_{1,2} = -13.8871 \pm j40.6931$$
$$\lambda_{3,4} = -0.0737 \pm j23.0868$$

These roots differ only slightly from the values obtained when there is no coupling. The effect of coupling on the engine roots does not hold any particular interest. On the other hand, there is a keen interest in the bending roots. Fig. 6.10 shows how these roots shift when coupling is turned on. This figure is slightly unconventional in that the real and imaginary axes are plotted on different scales. The real axis is restricted to a narrow range so that the engine roots do not show up. If the axes were plotted on the same scale, the bending roots would be very close to the imaginary axis.

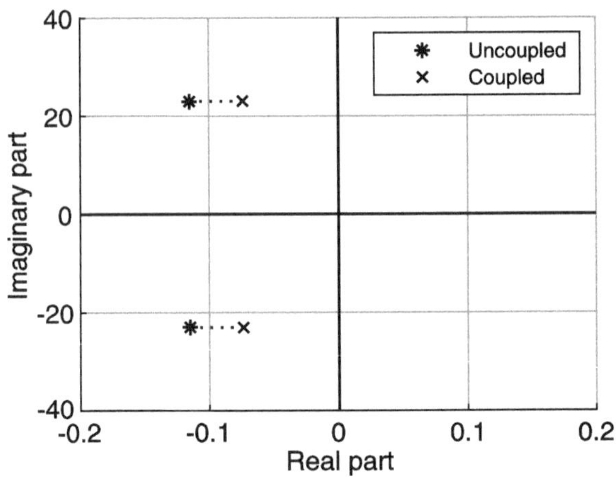

Figure 6.10 Bending roots for the uncoupled and coupled equations.

The uncoupled bending roots have the specified damping ratio of $\zeta_B = 0.005$. It can be seen that coupling has the effect of moving the real parts of these roots toward the imaginary axis, giving the coupled roots an effective damping of $\zeta \approx 0.003$. A 40% decrease in the damping of a structural mode is quite significant, especially for control design. It is not difficult to see that reasonable variations in the specified parameters could result in the real parts becoming positive, i.e., the system becoming unstable.

The phenomenon of reduced bending damping only occurs over a distinct range of frequencies. The upper limit for a given flexible mode is the *TWD-flex* frequency, which is the flexible body version of the rigid-body TWD frequency derived in Section 6.1. Consider again the mass-normalized form of the elastic equation (6.2.3) in the pitch plane, with thrust and a sinusoidal engine motion as in (6.1.1).

$$\ddot{\eta}_i + 2\zeta_{Bi}\Omega_{Bi}\dot{\eta}_i + \Omega_{Bi}^2\eta_{Bi} = -\psi_{z\beta i}F\beta_{Eby} - \left(I_{Ebyy}\sigma_{y\beta i} - m_E l_E \psi_{z\beta i}\right)\ddot{\beta}_{Eby}. \quad (6.2.40)$$

As in the analysis in the preceding section, we let $\ddot{\beta}_{Eby} = -\omega_A^2\beta_{Eby}$ where ω_A is the applied sinusoidal excitation frequency. Considering the instant when $\dot{\eta}_i = \eta_i = 0$,

$$\ddot{\eta}_i = -\psi_{z\beta i}F\beta_{Eby} + \left(I_{Ebyy}\sigma_{y\beta i} - m_E l_E \psi_{z\beta i}\right)\omega_A^2\beta_{Eby}. \quad (6.2.41)$$

It follows that the elastic generalized acceleration for mode i is zero if the terms on the RHS of Eq. (6.2.41) sum to zero. This occurs when

$$\omega_A^2 = \omega_{\mathrm{TWDflex}}^2 = \frac{\psi_{z\beta i}F}{I_{Ebyy}\sigma_{y\beta i} - m_E l_E \psi_{z\beta i}}. \quad (6.2.42)$$

Similarly to the rigid body case, $\omega_{\mathrm{TWDflex}}$ is the frequency at which inertial loads from sinusoidal engine motion will exactly cancel thrust forces on the rocket's flexible structure. This frequency is unique for each mode i, and it should be noted that this frequency does not depend directly on the bending frequencies. Below this frequency, a decrease in elastic damping will tend to increase coupling of the engine and structure, decreasing the stability of the bending roots as shown in Figure (6.10). Above the TWD-flex zero frequency, the coupling of the engine and structure is again stable, since the engine motion is dominated by inertial effects. Importantly, bending modes whose modal frequencies Ω_{Bi} are just above $\omega_{\mathrm{TWDflex}}$ may rapidly destabilize if shifting below the critical frequency due to parameter uncertainty or in-flight variations. Likewise, the response of the structure to autopilot inputs is shifted by 180 degrees above $\omega_{\mathrm{TWDflex}}$, which presents a challenge for flight control design.

6.3 Defining the finite element model

During the development of a rocket, finite element analyses will be called on to serve many purposes. Three examples in flight dynamics are (1) identify favorable locations for sensors, (2) provide mode shapes and frequencies for a detailed nonlinear simulation, (3) provide mode shapes and frequencies for a linearized state-space model that can be used for flight control design. In the support of other disciplines, additional models will be defined to assess loads, vibroacoustics, and so on. Since the mass of the rocket is continually changing as the propellant is depleted, a single FEM is rarely adequate – it is necessary to develop a set of FEMs representing different times of flight. For most rockets, several FEMs per stage would be required, so that the time interval and therefore mass difference between FEMs is never too large. Thus there is a strong desire to define the FEMs in such a way that a single set of models can be common to flight dynamics analyses, decreasing workload.

The phrase "component model" is used herein to mean either a FEM or a TVC model. These two components are combined into an "assembled model." In general, it cannot be assumed that the modes and frequencies coming out of a FEM will match those of the assembled model. That is, the FEM contains a set of frequencies and mode shapes, all of which will be shifted when the FEM is combined into the assembled model. One observation is that the number of degrees of freedom of the assembled model should be equal to the sum of the degrees of freedom of the two components. For the simple cases examined above, this became obvious. Each case examined has only two degrees of freedom, one from the nozzle equation and one from the bending equation. However, if both the FEM and the TVC model have many degrees of freedom, care must be taken to ensure that the assembled model does not represent the same degree of freedom twice.

In the analysis used above, it is assumed that the FEM includes all the masses of the rocket, including the engine. This is the integrated body approach. A major advantage of this approach is that only small changes to mode shapes and frequencies occur when the component models are assembled. This is the traditional approach for rocket analysis. It is important to interpret the word "assembled" in its mathematical sense – equations are assembled, not pieces of the rocket. The integrated body approach suffers the disadvantage that the mass matrix will not be positive definite if a sufficiently large number of modes is included in the dynamic model.

The alternative is the reduced body approach, i.e., the rocket mass in the FEM is reduced by the engine mass. This results in a more complicated mass matrix, and in many cases the added complexity may not be warranted. The number of modes that must be included increases with the size of the rocket and with its length/diameter ratio, unless a large investment in testing is undertaken to isolate those modes which are important. To this point, it is worth noting that the control design for the Apollo Saturn rockets

[4] was carried out using the integrated body approach, in spite of their large size, albeit while modeling only four lateral bending modes for the pitch and yaw dynamics. A large number of modes may also be required if the configuration is complex, as with the Space Shuttle. If a reduced body approach is properly defined, there is no limit to the number of modes that can be included while still retaining a positive definite mass matrix.

Both the FEM and the TVC model can be considered collections of masses and springs. Defining the FEM requires deciding which masses and which springs go into which component. Assuming the integrated body approach is taken, all the masses go into the FEM, and the question reduces to allocating the springs, as depicted in Fig. 6.11. A guiding principle is that each mass and each spring is counted only once. An exception to this principle occurs if the engine motion is prescribed, in which case the engine mass properties are included in both the FEM and the TVC model.

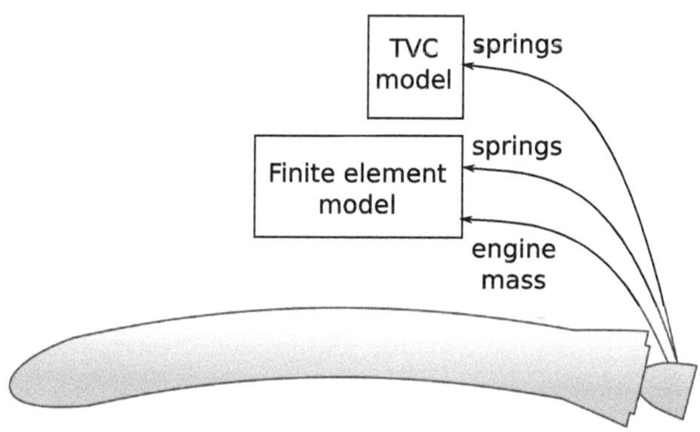

Figure 6.11 Allocation of masses and springs for the integrated body approach.

If there is a desire for the FEM to produce modes and frequencies that match those of the assembled model, then *all* of the compliance in Fig. 6.3, including both k_A and k_C, should be included in the FEM. Then the only purpose of the TVC model is to insert the effect of the commanded gimbal motion. This would seem to require a very simple TVC model, since it may be difficult to define a detailed model in such a way that the effect of the actuator compliance on just the commanded motion is included, but the effect of actuator compliance on nozzle motion due to externally applied torques is not included. Such an approach would not be suitable for capturing DWT effects. It might, however, be suitable for initial analyses using prescribed gimbal motion. For any given mode, for example the first bending mode, there may be a desire to know the location where the rotation due to bending is zero, since the relative distance between this and the angular rate sensor is of great interest. A FEM produced using this approach would be good for finding this location.

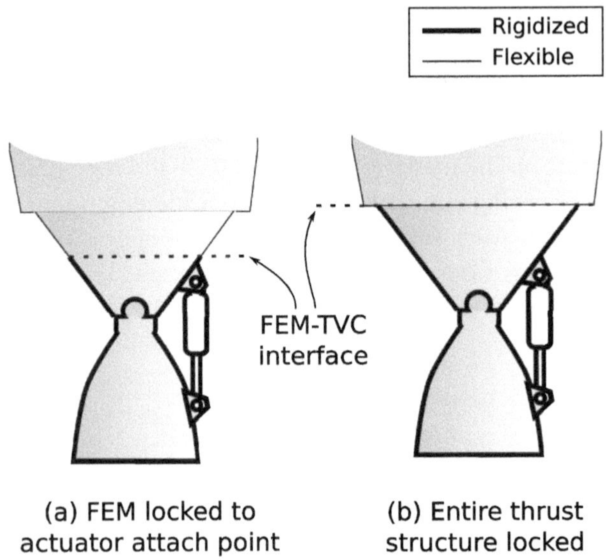

Figure 6.12 Locking the FEM.

A FEM used for final verification should take exactly the opposite approach, and embed all of the attach point compliances in the TVC model. In the FEM, the engine should be locked to the rocket body. In the days of simple structural dynamic models, before finite element techniques became mature, this was relatively easy. Typically the rocket was represented by a string of masses connected by torsional springs, and locking the engine simply meant that the spring connecting the last mass to the engine was made infinitely stiff. With the current state of the art yielding FEMs with millions of degrees of freedom, the following question arises: How far into the structure should things be locked? By "locked," we mean flexible elements are replaced with infinitely stiff elements and the associated flexible degrees of freedom are removed. A partial answer to this question is that it should be locked at least as far as the actuator attach point, as shown in Fig. 6.12. In the FEM, all of the elements shown as heavy lines, including the thrust cone or thrust structure, are locked. All of the light lines continue to be represented by the finite element mesh. It can be seen that with all the heavy lines locked, the engine and the lower part of the thrust structure become effectively one rigid body.

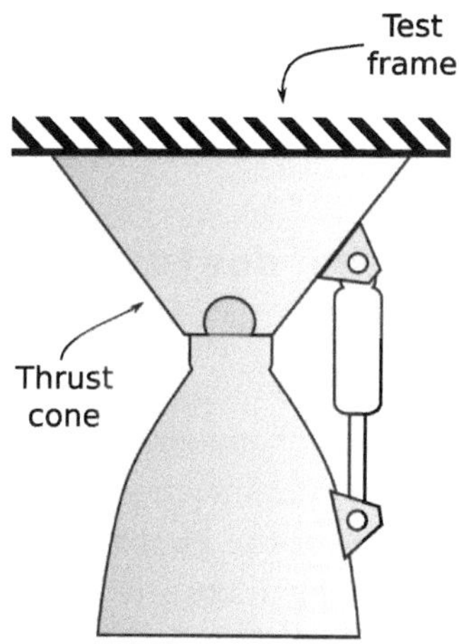

Figure 6.13 Notional test configuration for a TVC system.

Another approach to specifying how many elements to lock up is to anticipate that some kind of experimental verification of the attach point compliance and actuator dynamics will be required. A possible experimental test setup is shown in Fig. 6.13. For the purposes of model validation, hardware that is representative of the actual thrust cone is used. Thus a load can be applied to the nozzle, and the resulting motion will reflect the combined compliance of the actuator, the nozzle attach point, the attach point on the thrust cone, and any compliance of the thrust cone itself. The important point is that the design of this experiment and the design of the FEM should be coordinated. All the aforementioned compliances can be removed from (locked out of) the FEM. The test frame defines an unequivocal boundary that defines which springs go into the TVC and which go into the FEM. Thus the experiment provides a result that is directly usable in the TVC model.

The TVC model includes all of the compliances that are locked in the FEM. Usually, this model represents these compliances with simple springs. The stiffness of these springs might be derived from a separate FEM, from experimental results, or from some combination. If the integrated body approach is being used, all of the mass of the thrust cone and the engine is included as one locked mass in the FEM. There is no reason the interface has to be a flat plane, as depicted in the figures. The only requirement is that there be a clearly defined boundary between what is locked and what is not. The cleanest solution to the engine interface problem is to generate a reduced body FEM for the vehicle in which the engine is not included, and all of the elements on the gimbal side of the boundary have been locked. In this case, the mass of the thrust cone (but not the engine) is still included in the FEM.

6.4 Bending frequency shift due to thrust

Fig. 6.14 shows a flexible rocket with a thrust vector aligned with the local centerline. The term "force following" has been coined to convey the simple notion that in the absence of any gimbal angle from the Thrust Vector Control (TVC) mechanism, the direction of the thrust will follow the direction of the local centerline at the gimbal point and not the direction of the undeformed centerline (the x axis). It is important to include force following in the rotation equation. However, the effect of force following on the bending equation is not obvious. Fig. 6.14 introduces the concept of the neutral line. The angle this line makes with the x axis, γ, defines the boundary between having the thrust increase or decrease the deflection. This is not a fixed angle – it is proportional to the flex deflection. A force at an angle less than γ will act to increase the deflection, and a force at an angle greater than that of the neutral line will act to decrease the deflection. It remains to be established whether γ is greater than that of the deflection angle, as shown, or less.

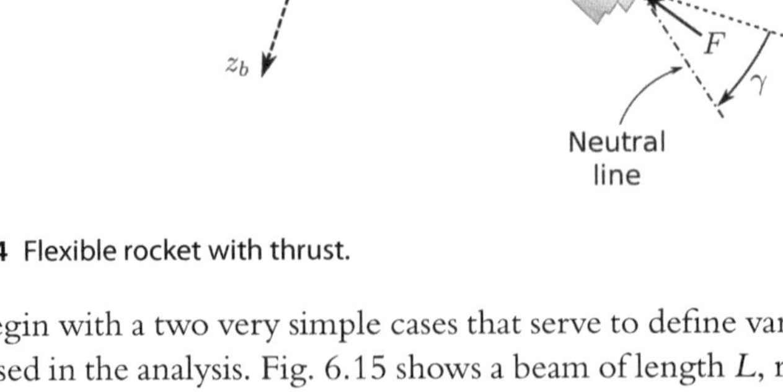

Figure 6.14 Flexible rocket with thrust.

We begin with a two very simple cases that serve to define various terms that are used in the analysis. Fig. 6.15 shows a beam of length L, represented

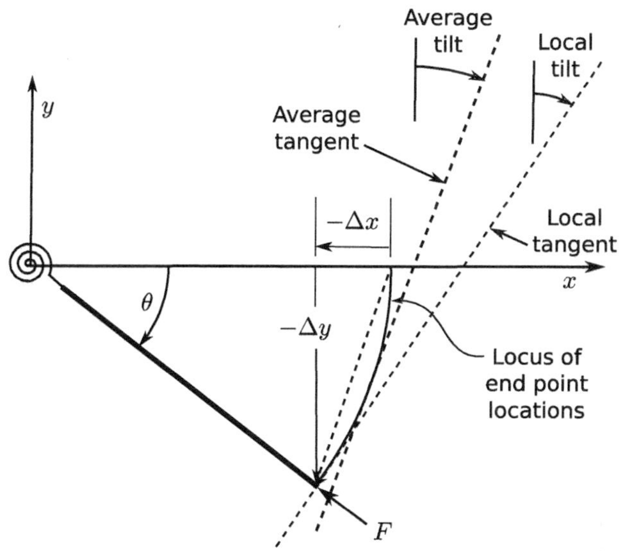

Figure 6.15 Beam pinned to the origin.

by the heavy line, that is pinned to the origin of an x-y coordinate frame. There is a torsional spring at the origin, which supplies a torque $-k\theta$ acting to align the beam to the x axis. There is a force F aligned with the beam. If the beam rotates about the origin, the locus of end points traces out a circular arc, as shown. For every position, there is a line that is tangent to the arc. The local tangent is the tangent that passes through the final end point. The average tangent is the line the passes through the end point corresponding to the average deflection $\theta/2$. For a final deflection θ, the end point coordinates are deflected by Δx and Δy.

$$\Delta x = -L(1-\cos\theta) \cong -\frac{L\theta^2}{2} \tag{6.4.1}$$

$$\Delta y = -L\sin\theta \cong -L\theta \tag{6.4.2}$$

As shown in Fig. 6.15, the average tangent is parallel to a line connecting the initial and final end points. We use the word "tilt" to describe an angle measured from the y axis, and "slope" to describe an angle measured from the x axis. The tilt of the average tangent is

$$\text{average tilt} = \tan^{-1}\left(\frac{\Delta x}{\Delta y}\right) \cong \frac{\Delta x}{\Delta y} \tag{6.4.3}$$

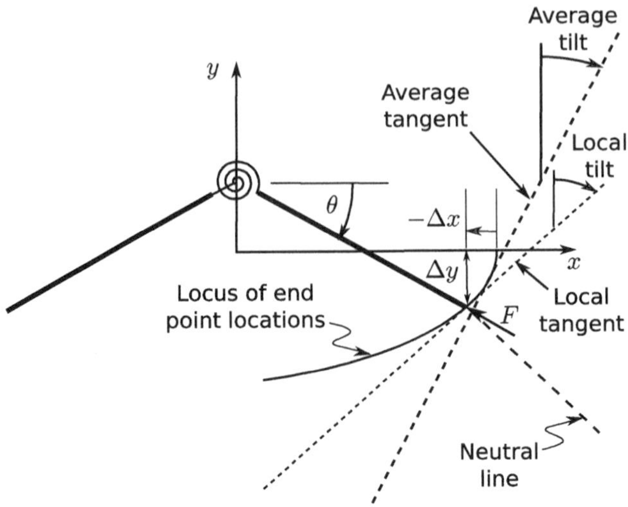

Figure 6.16 Two beams connected by a torsional spring.

Substituting from (6.4.1) and (6.4.2)

$$\text{average tilt} \cong \frac{\theta}{2} \tag{6.4.4}$$

For this simple system, the local tilt is the same as the beam deflection θ. Thus

$$\text{local tilt} = 2 \times \text{average tilt} \tag{6.4.5}$$

This relationship can be extended to any case in which the locus of end point locations is a circular arc.

One feature of this example is since F is aligned with the beam, it does not act to either increase or decrease the angle θ. Using language that is used to describe the principle of virtual work, since F is always normal to the local tangent, it does not do any work as θ changes, and thus does not cause any variation of θ.

Fig. 6.16 shows what might be considered the simplest possible example of a free free structure – two rigid beams connected by a torsional spring. It has exactly one flexible mode and three rigid-body modes, assuming the analysis is confined to a single plane. The flexible mode can be decoupled from the rigid-body modes by defining a body frame in which the x axis is aligned with the undeformed structure. For the flexible mode considered in isolation, the center of mass remains at the origin of this body frame.

Thus, the point at the center of each beam remains on the x axis. The torsional spring starts out at the origin when $\theta = 0$, and moves upward as θ increases, eventually reaching a position $x = 0$, $y = L/2$ when $\theta = \pi/2$ and the two beams are vertical, where L is the length of each beam. The locus of end point locations traces out a figure that resembles an elliptical arc. In a region of sufficiently small deflections θ, this locus is a circular arc, and we can apply (6.4.3) and (6.4.5). For this example, Δx is still given by (6.4.1), but we have

$$\Delta y = -\frac{L}{2}\sin\theta \tag{6.4.6}$$

Thus

$$\text{average tilt} \cong \frac{\Delta x}{\Delta y} = \theta \tag{6.4.7}$$

and the local tilt is 2θ. Thus the average tilt and the local tilt are both twice what they were in the previous example. More important, the force F is no longer normal to the local tangent. The direction of this normal is defined as the neutral line in Fig. 6.16. A positive deflection θ corresponds to a negative slope, with the result that the neutral line has a slope equal to the negative of the local tilt. A force acting along the neutral line, being normal to the local tangent, does no work on the structure. The principle of virtual work tells us that if no work is done on the structure, there is no effect on θ. All the energy from such a force goes into rigid-body motion. In contrast, since F is at slope that is smaller in magnitude than that of the neutral line, it will act to create a torque that increases θ. A force whose slope is steeper than that of the neutral line will act to flatten out the structure.

Now consider another simple example, a uniform free-free beam. We follow the usual practice of using separation of variables to represent the deflection y transverse to the beam. For a structure with M modes,

$$y(x,t) = \sum_{i=1}^{M} \psi_i(x)\,\eta_i(t) \tag{6.4.8}$$

The origin is placed at the center of the beam. We again let L be the length of the beam, but there is now only one beam, so the ends of the structure are at $\pm L/2$. The solution to this classic problem has been derived in many texts, for example [1] or [27], so the results will simply be presented. Mode shapes for this beam are divided into symmetric and asymmetric modes.

For the symmetric modes

$$\psi_i = A_i\left(\cosh\left(\frac{2\beta_i x}{L}\right) + \frac{\cosh\beta_i}{\cos\beta_i}\cos\left(\frac{2\beta_i x}{L}\right)\right) \quad i = 1, 3, 5, \cdots \tag{6.4.9}$$

For the asymmetric modes

$$\psi_i = A_i\left(\sinh\left(\frac{2\beta_i x}{L}\right) + \frac{\cosh\beta_i}{\cos\beta_i}\sin\left(\frac{2\beta_i x}{L}\right)\right) \quad i = 2, 4, 6, \cdots \tag{6.4.10}$$

where

$$\beta_1 = 2.365$$
$$\beta_2 = 3.926$$
$$\beta_i \cong \frac{(2i+1)\pi}{4}, \quad i > 2$$

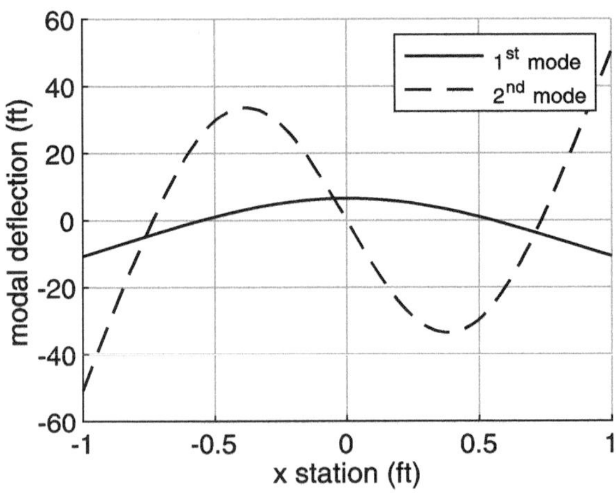

Figure 6.17 First two modes of a uniform free-free beam of length $L = 2$.

The amplitudes A_i of the mode shapes may be chosen arbitrarily. For our present purposes, it is convenient to choose $A_i = L/2$ for all modes. That means ψ_i has units of length and η is non-dimensional. No matter how the A_i are chosen, the local tilt is non-dimensional. As a further simplification, we choose $L = 2$. The first two mode shapes for this beam are shown in Fig. 6.17. This figure shows the length unit as feet, although any other unit could be used as long as compatible units are used throughout. The numbers shown in this figure are larger than what one would obtain using mass normalization, which simply means that with the chosen scheme for selecting A_i we can expect very small values of η_i. The resulting physical elastic deflections will come out the same regardless of how the A_i are chosen.

In a linear model of a beam, the x coordinate of the end remains at $L/2$ no matter how large the deflection, as in Fig. 6.17, which creates a fictitious elongation of the beam as the deflection increases. However, it is a straightforward matter to compute how much the beam has elongated in the model, which allows us to compute a correction. The length of each half of the beam in the model is given by

$$l = \int_0^{L/2} \left[1 + \left(\frac{\eta \, \mathrm{d}\psi}{\mathrm{d}x} \right)^2 \right]^{1/2} \mathrm{d}x \tag{6.4.11}$$

Assuming $\eta \ll 1$, this can be approximated by

$$l \approx \int_0^{L/2} \left[1 + \left(\frac{\eta \, \mathrm{d}\psi}{\mathrm{d}x} \right)^2 \right]^{1/2} \mathrm{d}x \approx \frac{L}{2} + \frac{\eta^2}{2} \int_0^{L/2} \left(\frac{\mathrm{d}\psi}{\mathrm{d}x} \right)^2 \mathrm{d}x \tag{6.4.12}$$

where we have used the approximation $(1+\epsilon)^{1/2} \approx 1+\epsilon/2$. Thus, each half of the beam is too long by the amount of the second term above. In terms of our previous variables,

$$\Delta x = -\frac{\eta^2}{2} \int_0^{L/2} \left(\frac{\mathrm{d}\psi}{\mathrm{d}x} \right)^2 \mathrm{d}x \tag{6.4.13}$$

Each end of the beam must be moved toward the center by this amount in order to maintain a constant length. Now define

$$\gamma_i \equiv \text{slope of the neutral line of mode } i \tag{6.4.14}$$

This can also be defined as

$$\gamma_i \equiv \text{- (local tilt of mode } i \text{ at the end point)} \tag{6.4.15}$$

This is the tilt of the locus of end point locations, not the tilt related to the modal rotation. The minus sign in (6.4.15) is necessary because a positive tilt in Fig. 6.15 or 6.16 results in a negative slope of the neutral line. See the discussion after (6.4.7). Using (6.4.3), (6.4.5), and (6.4.15),

$$\gamma_i = -\frac{2\Delta x}{\Delta y} = \frac{\eta_i^2 \int_0^{L/2} \left(\frac{\mathrm{d}\psi_i}{\mathrm{d}x} \right)^2 \mathrm{d}x}{\eta_i \psi_i (L/2)} \tag{6.4.16}$$

Thus the local tilt ends up being proportional to η_i. Let us further define

$$\bar{\gamma}_i = \frac{\gamma_i}{\eta_i} = \frac{\int_0^{L/2} \left(\frac{\mathrm{d}\psi_i}{\mathrm{d}x}\right)^2 \mathrm{d}x}{\psi_i(L/2)} \tag{6.4.17}$$

This is a constant, i.e. it does not vary with time. It can be considered a modal parameter, like $\psi_i(L/2)$. It also assumes the same sign as $\psi_i(L/2)$. A more general formula is

$$\bar{\gamma}_i = \frac{\int_{cm}^{gimbal} \left(\frac{\mathrm{d}\psi_i}{\mathrm{d}x}\right)^2 \mathrm{d}x}{\psi_{gimbal,i}} \tag{6.4.18}$$

For the first mode ($i = 1$) of the present problem we obtain

$$\bar{\gamma}_1 = -33.7881 \tag{6.4.19}$$

The numerical value for the derivative at the end of the beam is

$$\left(\frac{\mathrm{d}\psi_1}{\mathrm{d}x}\right)_{x=L/2} = -24.9515 \tag{6.4.20}$$

It is noteworthy that these numerical values are so close to each other. This means that for the first mode of a uniform beam, the thrust acts in a direction that is not very different from the neutral line. The numerical values in (6.4.19) and (6.4.20) will vary with the particular choice of beam length and normalization scheme, but their ratio is a number that is valid for any uniform beam

$$\frac{\bar{\gamma}_1}{(\mathrm{d}\psi_1/\mathrm{d}x)_{x=L/2}} = 1.35 \tag{6.4.21}$$

Thus the neutral line is at a slope that is only about one third greater than the slope at the end of the beam. This same ratio is derived in Reference [28].

For a beam structure, the following relationship is approximately correct:

$$\psi' \equiv \frac{\mathrm{d}\psi}{\mathrm{d}x} = \pm\sigma \tag{6.4.22}$$

The choice of sign depends on whether the x axis points forward or aft, and whether bending is about the pitch axis or the yaw axis. A noteworthy fact that may help clarify sign issues is that for the uniform beam, ψ and ψ' always have the same sign at the end corresponding to the positive x direction, and this can be expected to be true for the first few modes of a real rocket structure.

It is typical that during the creation of the finite element model, the coordinate frame is set up as in Fig. 6.2, with the positive x axis pointing aft, whereas the vehicle dynamic analysis invariably has the positive x axis pointing forward. Assuming that all the finite element variables have been converted to this latter setup, the above sign relationship between ψ and ψ' reverses twice. First, thrust is applied at the end corresponding to the negative x direction. Second, the transverse axis points down in Fig. 6.14, as opposed to up in Fig. 6.15. The net result is that ψ and ψ' will typically have the same sign at the point where the thrust is applied. The bending equation (subscript B) can be obtained from (2.2.24). If mass normalization has not been applied, as in the above uniform beam example, it is necessary to include m_{Bi}. For a rocket with the gimbal angle fixed at zero we obtain

$$\left(\ddot{\eta}_i + 2\zeta_{Bi}\Omega_{Bi}\dot{\eta}_i + \Omega_{Bi}^2\eta_i\right) m_{Bi} = F\psi_{\beta i}\left(\eta_i\bar{\gamma}_i - \eta_i\psi'_{\beta i}\right) \tag{6.4.23}$$

This equation is an approximation that only shows the effect of a single mode. Because of the quadratic nature of (6.4.13), it is not possible to produce an equation that has the modal effects decoupled into individual terms. If there is more than one mode, both terms within the parentheses on the right side of (6.4.23) will require some kind of summation. Since this is a small effect, it seems reasonable to confine our attention to the lowest frequency mode and ignore this nonlinear intermodal coupling. It should normally be the case that $\bar{\gamma}_i$ and $\psi'_{\beta 1}$ have the same sign, resulting in these terms subtracting on the right side of the above equation.

The most important point of this section is that including the force following term $\eta_i\psi'_{\beta i}$ by itself, without also including the $\eta_i\bar{\gamma}_i$ term, will result in the RHS of (6.4.23) applying a force of the wrong sign. One can justify leaving out both terms, if they are small compared to $m_{Bi}\Omega_{Bi}^2\eta_i$, but including one without the other is less accurate than having neither term.

If the mode shapes have been normalized such that $m_B = 1$, the above equation can be written as

$$\ddot{\eta}_i + 2\zeta_{Bi}\Omega_{Bi}\dot{\eta}_i + \left[\Omega_{Bi}^2 - F\psi_{\beta i}\left(\bar{\gamma}_i - \psi'_{\beta i}\right)\right]\eta_i = 0 \tag{6.4.24}$$

The natural frequency of the mode becomes shifted, as follows

$$\Omega_{i,\ \mathrm{Thrust}} = \sqrt{\Omega_{i,\ \mathrm{NoThrust}}^2 - F\psi_{\beta i}\left(\bar{\gamma}_i - \psi'_{\beta i}\right)} \tag{6.4.25}$$

where $\Omega_{i,\text{ NoThrust}} = \Omega_{Bi}$. If the thrust is strong enough, the frequency will shift down to zero and the rocket structure will buckle. Obviously, Ω_{Bi} has to be large enough to stay far away from this catastrophe. We can also see that for higher-frequency modes in which Ω_{Bi} is larger, the thrust effects will be less significant.

Thrust correction in vector form

The following vector equation shows the sign relationships for the y and z slopes at any point along the beam:

$$\begin{bmatrix} 0 \\ \frac{\partial \psi_y}{\partial x} \\ \frac{\partial \psi_z}{\partial x} \end{bmatrix} = -\mathbf{u}_e^\times \boldsymbol{\sigma} \tag{6.4.26}$$

where $\mathbf{u}_e$ is a unit vector in the x direction:

$$\mathbf{u}_e = \begin{bmatrix} 1 & 0 & 0 \end{bmatrix}^T \tag{6.4.27}$$

and $\boldsymbol{\sigma}$ is a vector of modal rotations

$$\boldsymbol{\sigma} = \begin{bmatrix} \sigma_x & \sigma_y & \sigma_z \end{bmatrix}^T$$

In Eq. (6.4.26), we can attach subscripts for any mode i and for the gimbal location β. With this vector approach, a more universal version of (6.4.23) can be written as

$$\left(\ddot{\eta}_i + 2\zeta_{Bi}\Omega_{Bi}\dot{\eta}_i + \Omega_{Bi}^2\eta_i\right) m_{Bi} = F\eta_i \boldsymbol{\psi}_{\beta i}^T \left(\bar{\boldsymbol{\gamma}}_i + \mathbf{u}_e^\times \boldsymbol{\sigma}_{\beta i}\right) \tag{6.4.28}$$

where

$$\bar{\boldsymbol{\gamma}}_i = \begin{bmatrix} 0 & \bar{\gamma}_{yi} & \bar{\gamma}_{zi} \end{bmatrix}^T \tag{6.4.29}$$

$$\bar{\gamma}_{yi} = \frac{\int_{cm}^{gimbal} \sigma_{zi}^2 \, dx}{\psi_{y\beta i}} \tag{6.4.30}$$

and

$$\bar{\gamma}_{zi} = \frac{\int_{cm}^{gimbal} \sigma_{yi}^2 \, dx}{\psi_{z\beta i}} \tag{6.4.31}$$

In Eq. (6.4.28), the products on the right multiply out to a scalar. This equation can be rearranged with everything on the left as in (6.4.24), and the frequency shift for any bending mode can be computed, even for a mode that is partially in the pitch plane and partially in the yaw plane.

Application to Ares I

The Constellation program was initiated by NASA as a follow-on to the Space Shuttle, with a goal of transporting astronauts to low earth orbit using a launch vehicle design derived from Space Shuttle components, the Ares I. Although the program was eventually canceled, the Ares I launch vehicle can be used to provide a good example of the frequency shift that occurs under thrusting conditions. This rocket had an unusually high length to diameter ratio: the vehicle was more than 315 feet tall but only 18 feet in diameter at its widest segment.

The following numerical values are representative of the first bending mode of the first stage of the Ares I:

$$\psi_{\beta 1} = \frac{.02}{\sqrt{\text{slinch}}} = \frac{.0058}{\sqrt{\text{slug}}}$$

$$\psi'_{\beta 1} = \frac{3.3 \times 10^{-4}\ \text{rad}}{\text{in}\sqrt{\text{slinch}}} = \frac{9.5 \times 10^{-5}\ \text{rad}}{\text{ft}\sqrt{\text{slug}}}$$

$$\Omega_{B1} = .898 \times 2\pi = 5.64\ \text{sec}^{-1}$$

The thrust for this rocket, provided by a massive 5-segment, 12-foot diameter solid rocket motor, is

$$F = 2,928,700\ \text{lbs}$$

The units are discussed in Appendix B. Assuming (6.4.21) is approximately correct for this rocket, it follows that

$$\bar{\gamma}_1 = 1.35\psi'_{\beta 1} = \frac{1.3 \times 10^{-4}\ \text{rad}}{\text{ft}\sqrt{\text{slug}}}$$

and hence

$$F\psi_{\beta 1}\left(\bar{\gamma}_1 - \psi'_{\beta 1}\right) = 0.56\ \text{sec}^{-2}$$

This is not negligible compared to $\Omega_{B1}^2 = 31.8\ \text{sec}^{-2}$. The ratio is

$$\frac{F\psi_{\beta 1}\left(\bar{\gamma}_1 - \psi'_{\beta 1}\right)}{\Omega_{B1}^2} = 0.018$$

The ratio of the frequency with thrust to the frequency without thrust would be

$$\frac{\Omega_{\text{Thrust}}}{\Omega_{\text{No Thrust}}} = \sqrt{1 - \frac{F\psi_{\beta 1}\left(\bar{\gamma}_1 - \psi'_{\beta 1}\right)}{\Omega^2_{B1}}} = .99$$

Thus the thrust reduces the bending frequency by about 1 percent.

Important points of Chapter 6

Section 6.2 shows why equations written for a totally passive system can give the appearance of instability. If an unusually large number of modes from the FEM must be retained in order to verify the control design, there is no guarantee that the mass matrix will remain positive definite. In such cases, it may be necessary to generate a FEM for the rocket body without the nozzle (the reduced body) and to use the reduced body approach.

Section 6.4 offers one insight that remains important even if the frequency shift due to thrust can be neglected. The two terms on the RHS of (6.4.23) must either be applied together or both left out. Leaving both terms out may be a reasonable option. If this option is selected, it is still necessary that the variation in thrust direction due to flex (force following) be included in the system translation and rotation equations, even though it is left out of the bending equations.

CHAPTER 7

Linearization

This chapter is concerned with the short-term dynamics of a rocket. "Short-term" means that changes in the mass properties as time progresses are ignored. In other words, a snapshot of the mass properties is taken at a certain point in the trajectory, and the stability of the system is analyzed based on the assumption that the mass properties are fixed. These simplified equations are linear, and can be used to analyze the dynamic stability or feedback control characteristics of a rocket at a specific point in the trajectory. The point at which the linear equations are constructed is called the *flight condition*. By assuring the behavior of the system is acceptable at several flight conditions spaced closely together, the analyst gains confidence that the overall rocket trajectory will meet design requirements. Final verification is performed in a time-varying nonlinear simulation.

In linear form, the equations can be cast into the augmented or *descriptor* state-space form

$$\mathbf{E}\dot{\mathbf{x}} = \mathbf{A}\mathbf{x} + \mathbf{B}\mathbf{u} \tag{7.0.1}$$

$$\mathbf{y} = \mathbf{C}\mathbf{x} + \mathbf{D}\mathbf{u} \tag{7.0.2}$$

and the first equation can be solved for an explicit system of linear equations,

$$\dot{\mathbf{x}} = \underbrace{\mathbf{E}^{-1}\mathbf{A}}_{\tilde{\mathbf{A}}}\mathbf{x} + \underbrace{\mathbf{E}^{-1}\mathbf{B}}_{\tilde{\mathbf{B}}}\mathbf{u} \tag{7.0.3}$$

where the descriptor matrix $\mathbf{E}$ contains the system mass matrix. The matrices $\tilde{\mathbf{A}}$, $\tilde{\mathbf{B}}$, $\mathbf{C}$, and $\mathbf{D}$ can be used in control system design and stability analysis, as discussed in Chapter 9.

One conceptual issue with linearizing the equations of motion is the rocket x velocity. Greensite [1] considers an accelerating reference frame, in which the x velocity is zero, and the only concern is with the small velocities that develop in the y and z directions. In essence, the x velocity of the frame is continually being subtracted from that of the rocket. This is essentially the same as the reference trajectory frame of Garner [3], which is the inertial frame used in this chapter. The present analysis is based on the results of Chapter 2, which at its core is a body frame analysis. As described

Dynamics and Simulation of Flexible Rockets
https://doi.org/10.1016/B978-0-12-819994-7.00012-1

in Chapter 10, the need to actually compute velocities in the body frame is avoided by converting the accelerations to the inertial frame. In this frame, the accelerations can be integrated to get velocities. It is not necessary to compute $\mathbf{v}$, the velocity in body coordinates, although it is not difficult to do this. This is simply the velocity in inertial coordinates transformed to the body frame, using either (2.1.2) or (2.1.4).

7.1 Scalar equations of motion

In this analysis, the linear equations are developed for a system with a single gimbaled engine and no RCS thrusters. The extensions to include multiple engines and RCS thrusters are straightforward. In order to further aid in the interpretation of the system equations, the following assumptions are used:

1. The curvature of the rocket trajectory balances the lateral component of gravity – a *gravity turn*. (An orbit can be considered a special case of a gravity turn in which the velocity equals the orbit velocity.)
2. The propellant tanks and the engine gimbal are on the centerline.
3. All angular motion is a small perturbation from a reference trajectory frame, which may be treated as an inertial frame (subscript I).
4. The nonlinear terms in the equations can be neglected.
5. The origin is located at the x position of the center of mass.
6. The moment of inertia matrix is diagonal.
7. There is no aerodynamic roll moment, and no roll acceleration ($\dot{\omega}_x = 0$).

Any of these assumptions may be relaxed with appropriate modifications, but these conditions are appropriate for many practical problems in the initial phases of analysis.

The total angular velocity is divided into a steady turn rate plus a small perturbation:

$$\boldsymbol{\omega}_{\text{tot}} = \boldsymbol{\omega}_0 + \boldsymbol{\omega} \tag{7.1.1}$$

where $\boldsymbol{\omega}_0 = \begin{bmatrix} 0 & \omega_{0y} & 0 \end{bmatrix}^T$, representing a slowly varying pitch rate. This amounts to a redefinition of the symbol $\boldsymbol{\omega}$. In this chapter, it represents the perturbation from $\boldsymbol{\omega}_0$. Typical values for $\boldsymbol{\omega}_0$ during ascent would be less than one degree per second. The first assumption above, that the rocket is in a gravity turn, may be written

$$g\mathbf{u}_v = \boldsymbol{\omega}_0^{\times}\mathbf{v} \tag{7.1.2}$$

where $\mathbf{u}_v$ is a unit vector in the downward vertical direction. This equation is only valid for the body y and z coordinates. There is a mismatch along the body x axis which does not have any consequences for the equations given below for the y and z axis motion. A more complete discussion of a gravity turn is given in Chapter 9.

It becomes convenient to redefine the first three elements of the state vector. In each of Eqs. (2.5.1) through (2.5.5), the following substitution using $\dot{\mathbf{v}}$ from (2.1.5) is made;

$$\mathbf{a}_b = \dot{\mathbf{v}} + \boldsymbol{\omega}_0^\times \mathbf{v} = \dot{\mathbf{v}} + g\mathbf{u}_v \tag{7.1.3}$$

This allows us to eliminate all effects of gravity, including the gravity torque about the origin and the gravity torque on the engine. In each case, the $g\mathbf{u}_v$ term on the left of (2.5.1) through (2.5.5) cancels the gravity term on the right hand side excitation vector, computed in Sections 5.2 and 5.3. The equations below use (2.1.2) to transform $\dot{\mathbf{v}}$ in the body frame to $\dot{\mathbf{v}}_I$, thus converting the first three elements of the state vector to the coordinates of what Garner [3] and Frosch and Vallely [4] call the reference trajectory frame. This is not a true inertial frame, since it is in fact accelerating relative to a frame that might be appropriate for navigation, such as a frame whose origin is the center of the earth. The reference trajectory frame should more accurately be considered a "quasi-inertial" frame. For the short-term dynamics, the distinction is not significant. Note that there is a subtle distinction between the above method for dealing with gravity and the discussion at the end of Section 5.3. The present approach is very specific and applies only to a gravity turn. Section 5.3 provides an approach for using sensed acceleration rather than total acceleration that is generally applicable but not as convenient as (7.1.3).

With a single engine on the centerline, the location of the gimbal point can be written

$$\mathbf{r}_G = \begin{bmatrix} X_G & 0 & 0 \end{bmatrix}^T \tag{7.1.4}$$

X_G is shown in Fig. 7.1. The x locations are positive if forward of the origin and negative if aft of the origin, thus X_G is a negative quantity. The first moment of inertia of the entire rocket about the origin is given by:

$$\mathbf{s}_T = \begin{bmatrix} 0 & S_{Ty} & S_{Tz} \end{bmatrix}^T \tag{7.1.5}$$

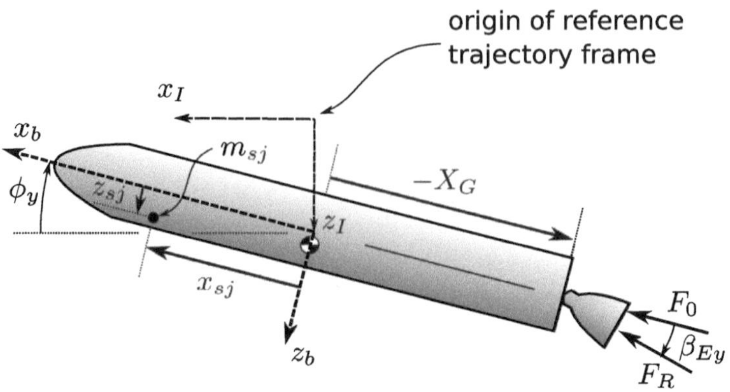

Figure 7.1 Parameters in the pitch plane.

The engine first moment of inertia has no y or z components:

$$\mathbf{s}_E = \begin{bmatrix} S_{Ex} & 0 & 0 \end{bmatrix}^T \tag{7.1.6}$$

This is related to the notation of Frosch and Vallely [4] by $S_{Ex} = -S_E$. It is useful to assume that the engine is rotationally symmetric. Thus its inertia matrix takes the form

$$\mathbf{I}_{EG} = \begin{bmatrix} I_{Ebxx} & 0 & 0 \\ 0 & I_{Eb} & 0 \\ 0 & 0 & I_{Eb} \end{bmatrix} \tag{7.1.7}$$

The engine angular acceleration is given by

$$\dot{\boldsymbol{\omega}}_E = \begin{bmatrix} 0 & \ddot{\beta}_{Ey} & \ddot{\beta}_{Ez} \end{bmatrix}^T \tag{7.1.8}$$

The body acceleration and slosh acceleration vectors are written in terms of their components

$$\dot{\mathbf{v}} \equiv \begin{bmatrix} \ddot{x}_b & \ddot{y}_b & \ddot{z}_b \end{bmatrix}^T \tag{7.1.9}$$

$$\ddot{\boldsymbol{\delta}}_{sj} = \begin{bmatrix} 0 & \ddot{y}_{sj} & \ddot{z}_{sj} \end{bmatrix}^T \tag{7.1.10}$$

To avoid confusion with the use of β to define the gimbal angle, the following substitutions are made for the aerodynamic quantities in Chapter 5;

$$\alpha_y \equiv \alpha \tag{7.1.11}$$

$$\alpha_z \equiv \beta. \tag{7.1.12}$$

Also consistent with the definitions in Chapter 5, the fixed component of the thrust magnitude is given by F_0, and the gimbaled component is F_R. The total thrust magnitude is

$$F = F_0 + F_R. \tag{7.1.13}$$

Substituting all the above simplifications into (2.5.1) yields three scalar equations:

$$m_T\ddot{x}_b = -S_{Tz}\dot{\omega}_y + S_{Ty}\dot{\omega}_z + F - D \tag{7.1.14}$$

$$m_T\ddot{y}_b = C_{Y\beta}\bar{q}S_{ref}\alpha_z + F_R\beta_{Ez} + F\sum_i \sigma_{z\beta i}\eta_i - S_{Ex}\ddot{\beta}_{Ez} - \sum_j m_{sj}\ddot{y}_{sj} \tag{7.1.15}$$

$$m_T\ddot{z}_b = -C_{N\alpha}\bar{q}S_{ref}\alpha_y - F_R\beta_{Ey} - F\sum_i \sigma_{y\beta i}\eta_i + S_{Ex}\ddot{\beta}_{Ey} - \sum_j m_{sj}\ddot{z}_{sj} \tag{7.1.16}$$

These three equations give the sensed acceleration. For the absolute acceleration, the effect of gravity would have to be added back. The summations over i represent the flexible modes, and the summations over j represent sloshing propellant masses. To convert to inertial coordinates, we consider the sensed acceleration of the center of mass along the inertial x axis. If ϕ_y and ϕ_z are both small angles this is

$$\ddot{x}_I = \frac{F - D}{m_T} = \bar{g} \tag{7.1.17}$$

where $D = C_A\bar{q}S_{ref}$ is the drag, and the subscript I is used to denote accelerations expressed in the quasi-inertial coordinates of the reference trajectory frame. The x_I axis is shown in Fig. 7.1. For a linearized analysis, Eq. (2.1.2) can be used to convert any vector from inertial coordinates to body coordinates, not just the velocity vector. For the acceleration vector the conversion is

$$\ddot{y}_b = \ddot{y}_I - \ddot{x}_I\phi_z + \ddot{z}_I\phi_x \tag{7.1.18}$$

$$\ddot{z}_b = \ddot{z}_I + \ddot{x}_I\phi_y - \ddot{y}_I\phi_x \tag{7.1.19}$$

In both of these expressions, the last term is a product of two small quantities and can be neglected. Substituting from (7.1.17),

$$m_T\ddot{y}_b = m_T\ddot{y}_I - (F - D)\,\phi_z \tag{7.1.20}$$

$$m_T \ddot{z}_b = m_T \ddot{z}_I + (F - D)\,\phi_y \tag{7.1.21}$$

Thus (7.1.15) and (7.1.16) become

$$m_T \ddot{y}_I = C_{Y\beta} \bar{q} S_{ref} \alpha_z + F_R \beta_{Ez} + F \sum \sigma_{z\beta i} \eta_i + (F - D)\,\phi_z - S_{Ex} \ddot{\beta}_{Ez} - \sum m_{sj} \ddot{y}_{sj} \tag{7.1.22}$$

$$m_T \ddot{z}_I = -C_{N\alpha} \bar{q} S_{ref} \alpha_y - F_R \beta_{Ey} - F \sum \sigma_{y\beta i} \eta_i - (F - D)\,\phi_y + S_{Ex} \ddot{\beta}_{Ey} - \sum m_{sj} \ddot{z}_{sj} \tag{7.1.23}$$

The main advantage of inertial (or quasi-inertial) coordinates for analyzing the short-term dynamics of an ascending rocket is that it simplifies the aerodynamics. $\ddot{z}_I$ can be integrated once to obtain an approximate value of the inertial velocity, whereas integrating $\ddot{z}_b$ does not produce a useful quantity if there is a non-zero body rotation rate.

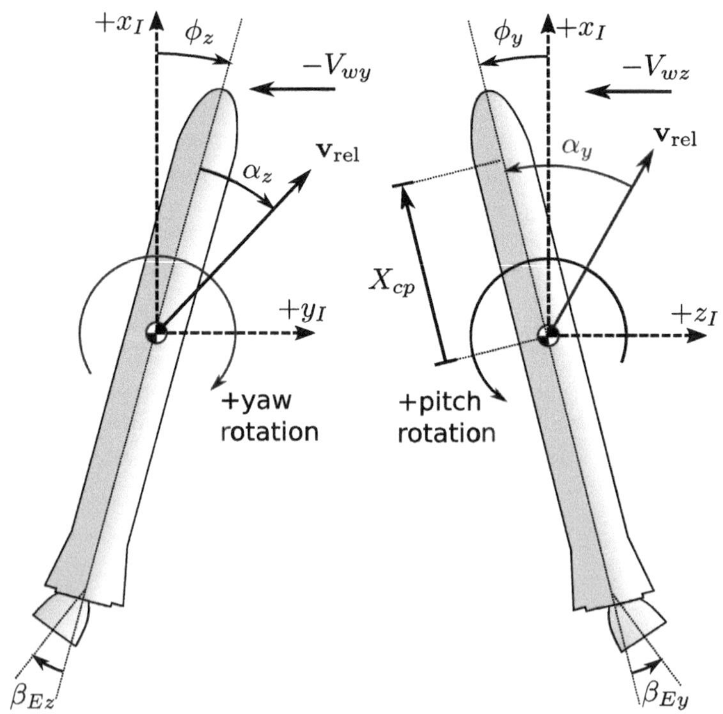

Figure 7.2 Sign conventions for pitch and yaw rotation in the trajectory frame. Positive rotations indicated. Wind velocity shown is negative.

Sign conventions for the trajectory frame in the pitch and yaw planes are shown in Fig. 7.2. Consistent with the above definitions, α_y is what is commonly called the angle of attack α and α_z is commonly called the sideslip angle β. It is important to recognize that α_y and α_z are defined differently from other angles such as β_{Ey}. They are indeed angles, but it can be misleading to think of them as rotation angles. They are really non-dimensional representations of the y and z components of velocity. Consider first the case in which there is no wind, and let $V = |\mathbf{v}|$. If $\dot{y}_b$ and $\dot{z}_b$ are small compared to V, we have $\alpha_y \approx \dot{z}_b/V$ and $\alpha_z \approx \dot{y}_b/V$. Note that the sign is positive for both the y axis and the z axis, and there is no need to invoke either a right hand or left hand rule. In contrast, the ϕ_y term in (7.1.19) above has the opposite sign of the ϕ_z term in (7.1.18).

It was mentioned in Chapter 5 that standard practice is to define the sideslip angle β using a left-handed rotation about the z axis, as in Fig. 5.4. Fig. 7.2 takes a slightly different approach that accomplishes the same net result. This figure shows $\alpha_z = \beta$ as the angle from the body x to the velocity vector, as opposed to Fig. 5.4 which has β as the angle from the velocity vector to the body x. This change allows all the angles in Fig. 7.2 to obey the right hand rule. Nielsen [29] uses an alternate coordinate system but there is a similar difference in the way α and β are defined. The present definitions are consistent with those of Greensite [1]. If there is wind, and the reference trajectory frame is being used, the pitch and yaw angles of attack can be computed from

$$\alpha_y = -\frac{V_{wz}}{V} + \phi_y + \frac{\dot{z}_I}{V} \text{ (angle of attack)} \tag{7.1.24}$$

$$\alpha_z = -\frac{V_{wy}}{V} - \phi_z + \frac{\dot{y}_I}{V} \text{ (angle of sideslip)} \tag{7.1.25}$$

V_{wy} and V_{wz} are the y and z velocities of the wind.

Turning now to the rotation equation, we have

$$\mathbf{g} = \begin{bmatrix} g_x & g_y & g_z \end{bmatrix}^T \tag{7.1.26}$$

The applied moments are the aerodynamic moment, the moments that arise from the lateral component of thrust, and the fuel offset moment from Eqs. (5.2.3) and (5.2.4). The moment due to the engine mass being offset from the centerline when $\boldsymbol{\beta}_E$ is nonzero is neglected. The components of the torque are

$$g_x = 0 \tag{7.1.27}$$

$$g_y = C_{Y\beta}\bar{q}S_{ref}X_{cp}\alpha_z + X_G\left(F_R\beta_{Ey} + F\sum\sigma_{y\beta\,i}\eta_i\right) + F\sum\psi_{z\beta\,i}\eta_i - \frac{F-D}{m_T}\sum_j m_{sj}\,z_{sj} \quad (7.1.28)$$

$$g_z = C_{N\alpha}\bar{q}S_{ref}X_{cp}\alpha_y + X_G\left(F_R\beta_{Ez} + F\sum\sigma_{z\beta\,i}\eta_i\right) - F\sum\psi_{y\beta\,i}\eta_i + \frac{F-D}{m_T}\sum_j m_{sj}\,y_{sj} \quad (7.1.29)$$

where X_{cp} is the distance of the center of pressure forward of the origin, and X_G is defined in Fig. 7.1. Note that this is normally a negative quantity. Starting with (2.5.2), again using $\bar{g} = (F-D)/m_T$, and going through a process that parallels that leading up to (7.1.22) and (7.1.23), we obtain

$$\dot{\omega}_y I_{yy} = C_{N\alpha}\bar{q}S_{ref}X_{cp}\alpha_y + X_G\left(F_R\beta_{Ey} + F\sum\sigma_{\beta\,i}\eta_i\right) + F\sum\psi_{z\beta i}\eta_i + (X_G S_E - I_{EG})\ddot{\beta}_{Ey} + \sum_j m_{sj}\left(X_{sj}\ddot{z}_{sj} - \bar{g}z_{sj}\right) - S_{Tz}\left(S_{Ty}\ddot{\phi}_z - S_{Tz}\ddot{\phi}_y + F - D\right)/m_T \quad (7.1.30)$$

$$\dot{\omega}_z I_{zz} = C_{Y\beta}\bar{q}S_{ref}X_{cp}\alpha_z + X_G\left(F_R\beta_{Ez} + F\sum\sigma_{z\beta\,i}\eta_i\right) - F\sum\psi_{y\beta i}\eta_i + (X_G S_E - I_{EG})\ddot{\beta}_{Ez} - \sum_j m_{sj}\left(X_{sj}\ddot{y}_{sj} - \bar{g}y_{sj}\right) + S_{Ty}\left(S_{Ty}\ddot{\phi}_z - S_{Tz}\ddot{\phi}_y + F - D\right)/m_T \quad (7.1.31)$$

The rotation equations are not changed by conversion to inertial coordinates.

Slosh equations

From the second assumption at the beginning of this section, the equilibrium positions of the slosh masses lie on the centerline;

$$\mathbf{r}_{sj} = \begin{bmatrix} X_{sj} & 0 & 0 \end{bmatrix}^T. \quad (7.1.32)$$

For the present linearized analysis, it is adequate to use the equilibrium position of the slosh mass. For a full nonlinear model, $\mathbf{r}_{sj}$ would acquire time-varying components in y and z as the slosh mass moves around, i.e., the vector $\boldsymbol{\delta}_{sj}$ in Fig. 2.1 must be taken into account. The y and z components of the slosh mode shape are

$$\psi_{yji} \equiv \psi_{yi}\left(X_{sj}\right) \quad (7.1.33)$$

$$\psi_{zji} \equiv \psi_{zi}\left(X_{sj}\right) \tag{7.1.34}$$

The definition (7.1.33) can be interpreted as "the constant ψ_{yji} is the function ψ_{yi} evaluated at the position X_{sj}." Integrating both sides of (7.1.10) gives

$$\delta_{sj} = \begin{bmatrix} 0 & y_{sj} & z_{sj} \end{bmatrix}^T \tag{7.1.35}$$

From (2.5.4) and (5.3.10), the linearized yaw and pitch plane slosh equations become, after using (7.1.3), (7.1.20) and (7.1.21)

$$\ddot{y}_{sj} + 2\zeta_{sj}\Omega_{sj}\dot{y}_{sj} + \Omega_{sj}^2 y_{sj} = -\ddot{y}_I + \bar{g}\phi_z - X_{sj}\ddot{\phi}_z - \sum_i \psi_{yji}\ddot{\eta}_i \tag{7.1.36}$$

$$\ddot{z}_{sj} + 2\zeta_{sj}\Omega_{sj}\dot{z}_{sj} + \Omega_{sj}^2 z_{sj} = -\ddot{z}_I - \bar{g}\phi_y + X_{sj}\ddot{\phi}_y - \sum_i \psi_{zji}\ddot{\eta}_i \tag{7.1.37}$$

Bending equations

From (2.5.5), (2.5.10), and (5.3.30), the bending equations can be derived;

$$\begin{aligned}\left(\ddot{\eta}_i + 2\zeta_{Bi}\Omega_{Bi}\dot{\eta}_i + \Omega_{Bi}^2\eta_i\right) m_{Bi} &= -\sum_j m_{sj}\left(\psi_{yji}\ddot{y}_{sj} + \psi_{zji}\ddot{z}_{sj}\right) \\ &\quad + \left(\psi_{z\beta i}\ddot{\beta}_{Ey} - \psi_{y\beta i}\ddot{\beta}_{Ez}\right) S_{Ex} - \left(\sigma_{y\beta i}\ddot{\beta}_{Ey} + \sigma_{z\beta i}\ddot{\beta}_{Ez}\right) I_{EG} \\ &\quad - F_R\left(\psi_{z\beta i}\beta_{Ey} - \psi_{y\beta i}\beta_{Ez}\right) + \bar{q}S_{ref}\left(C_{Y\eta i}\alpha_z - C_{N\eta i}\alpha_y\right)\end{aligned} \tag{7.1.38}$$

where the aerodynamic bending coefficients $C_{Y\eta i}$, $C_{N\eta i}$ have been included. These terms represent, to first order, the effect of nonzero angle of attack on the distributed aerodynamic bending loads of the vehicle structure. A more detailed model for including these and the rigid-body effects of distributed aerodynamics is discussed in Section 7.3.

Output equations

The output equations are used to describe the relative rotations or accelerations at a given point on the rocket body where a physical sensor is located. Here, the term "relative" means relative to the trajectory frame, noting that in this chapter the definitions of $\boldsymbol{\omega}$ and $\dot{\mathbf{v}}_I$ are used to indicate a perturbation from the accelerating quasi-inertial reference frame of the gravity-turn trajectory. Three types of sensors are considered: an angular displacement sensor (such as the attitude output of an IMU), an angular rate sensor (such as a body-mounted rate gyro), and a body-mounted accelerometer.

In modern practice, the IMU angle is the output of the navigation equations, which process the discrete, nonholonomic body angle increments to produce an attitude (such as a quaternion) that relates the vehicle attitude to a chosen navigation frame. In the present analysis, the details of this process are neglected, which is common for initial stability studies. At a later point in the analysis, it may become necessary to include the processing details of the IMU angles insofar as they affect the control system, for example, filters, latency, and sensor dynamics. When referring to a sensed angle, we really mean the sensed difference between the vehicle attitude and the commanded trajectory frame attitude, noting that the quasi-constant angular rate $\boldsymbol{\omega}_0$ is being continuously integrated by the navigation system to produce the commanded trajectory frame attitude. This matter is discussed more fully in Chapter 9.

For the angular sensors, the output expressions are simply the sum of the rigid and elastic components of the angle and angular rate. Let $\boldsymbol{\sigma}_{cm}$ be the mode slope at the measurement location $\mathbf{r}_m$; that is,

$$\boldsymbol{\sigma}_{cmi} \equiv \boldsymbol{\sigma}_i(\mathbf{r}_m). \tag{7.1.39}$$

The subscript c is attached to this mode slope in order to associate it with the matrix $\mathbf{C}$. Recall that in previous chapters $\boldsymbol{\psi}_{ji}$ with two index subscripts represents a mode shape at slosh location j. The subscript c indicates that the mode shape or slope is taken at a measurement location rather than a slosh location, even when m and i are given their integer values. The mth sensor location may, in general, include any combination of attitude, attitude rate, and acceleration outputs. The sensed attitude variables in body coordinates are given by

$$\hat{\boldsymbol{\phi}}_m = \boldsymbol{\phi} + \sum_i \boldsymbol{\sigma}_{cmi}\eta_i \tag{7.1.40}$$

and the sensed rates are given by

$$\hat{\boldsymbol{\omega}}_m = \boldsymbol{\omega} + \sum_i \boldsymbol{\sigma}_{cmi}\dot{\eta}_i. \tag{7.1.41}$$

The sensed quantity is denoted by a hat symbol. The sensed angular rate equation holds for both linear and nonlinear simulations, whether $\boldsymbol{\omega}$ represents a perturbation or the total angular rate.

The sensed acceleration is slightly more challenging. In the angle and angular rate equations, the elastic deformation of the sensor input axes has been neglected as it is a second-order effect. As will be shown, the

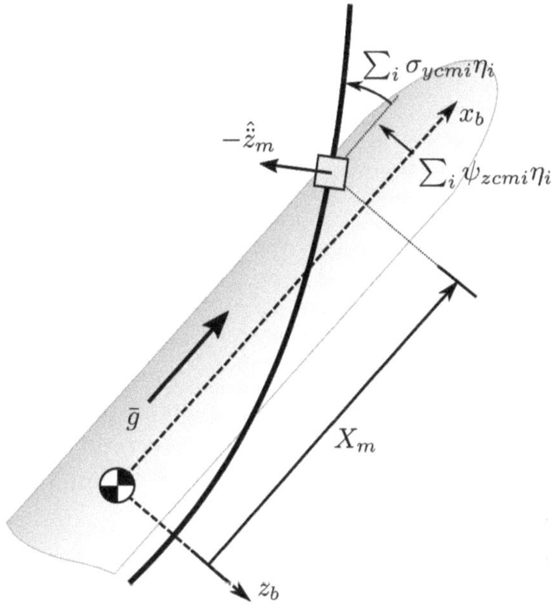

Figure 7.3 Accelerometer on a flexible rocket body.

accelerometer output contains a significant contribution from this deformation. The linearized accelerometer equation contains four major components:

1. Rigid-body ("truth") translational acceleration of the body frame;
2. Translational acceleration due to elastic deformation of the structure to which the sensor is attached;
3. Translational acceleration due to body angular acceleration measured at the accelerometer station;
4. Measurement error due to dynamic misalignment of the sensor with respect to the thrust axis caused by elastic rotation of the sensor.

In a full nonlinear simulation, the true inertial position of an accelerometer located at $\mathbf{r}_m$ in body coordinates is given by

$$\mathbf{r}_{mI} = \mathbf{r}_I + \mathbf{C}_{Ib}\left(\mathbf{r}_m + \sum_i \boldsymbol{\psi}_{cmi}\eta_i\right) \tag{7.1.42}$$

where the mode shape at the sensor location is

$$\boldsymbol{\psi}_{cmi} \equiv \boldsymbol{\psi}_i(\mathbf{r}_m). \tag{7.1.43}$$

The relationships involving the sensed acceleration are depicted in Fig. 7.3. The acceleration at the sensor can be derived by differentiating Eq. (7.1.42) twice and expressing the result in the body frame. This can be performed with help of Eq. (2.1.6), noting that we must use the quantity $\boldsymbol{\omega}_{\text{tot}}$ to represent the total angular rate. It follows that the velocity at the sensor is

$$\dot{\mathbf{r}}_{mI} = \mathbf{v}_I + \mathbf{C}_{Ib}\left(\dot{\mathbf{r}}_m + \sum_i \boldsymbol{\psi}_{cmi}\dot{\eta}_i\right) + \mathbf{C}_{Ib}\boldsymbol{\omega}_{\text{tot}}^{\times}\left(\mathbf{r}_m + \sum_i \boldsymbol{\psi}_{cmi}\eta_i\right). \quad (7.1.44)$$

Since the origin of the coordinate system is coincident with the axial position of the center of mass, the value $\dot{\mathbf{r}}_m$ includes the effect of a moving center of mass due to burning propellant. However, this effect is very small and can be neglected, so it is assumed that $\dot{\mathbf{r}}_m = 0$. Using (7.1.44),

$$\begin{aligned}\ddot{\mathbf{r}}_{mI} = \dot{\mathbf{v}}_I + \mathbf{C}_{Ib}\sum_i \boldsymbol{\psi}_{cmi}\ddot{\eta}_i + 2\mathbf{C}_{Ib}\boldsymbol{\omega}_{\text{tot}}^{\times}\sum_i \boldsymbol{\psi}_{cmi}\dot{\eta}_i \\ + \mathbf{C}_{Ib}\dot{\boldsymbol{\omega}}^{\times}\left(\mathbf{r}_m + \sum_i \boldsymbol{\psi}_{cmi}\eta_i\right) \\ + \mathbf{C}_{Ib}\boldsymbol{\omega}_{\text{tot}}^{\times}\boldsymbol{\omega}_{\text{tot}}^{\times}\left(\mathbf{r}_m + \sum_i \boldsymbol{\psi}_{cmi}\eta_i\right). \quad (7.1.45)\end{aligned}$$

We use Eq. (2.1.5) to express the inertial acceleration $\dot{\mathbf{v}}_I$ in terms of the total body frame acceleration $\mathbf{a}_b$. Premultiplying Eq. (7.1.45) by $\mathbf{C}_{bI}$, it follows that the acceleration at the sensor location is

$$\begin{aligned}\ddot{\mathbf{r}}_m = \mathbf{a}_b + \sum_i \boldsymbol{\psi}_{cmi}\ddot{\eta}_i \\ + 2\boldsymbol{\omega}_{\text{tot}}^{\times}\sum_i \boldsymbol{\psi}_{cmi}\dot{\eta}_i + \dot{\boldsymbol{\omega}}_{\text{tot}}^{\times}\left(\mathbf{r}_m + \sum_i \boldsymbol{\psi}_{cmi}\eta_i\right) \\ + \boldsymbol{\omega}_{\text{tot}}^{\times}\boldsymbol{\omega}_{\text{tot}}^{\times}\left(\mathbf{r}_m + \sum_i \boldsymbol{\psi}_{cmi}\eta_i\right). \quad (7.1.46)\end{aligned}$$

Finally, the relationship $\mathbf{C}_{bm}$ between the measurement frame and the body frame can be expressed as a small rotation, where

$$\hat{\ddot{\mathbf{r}}}_m = \mathbf{C}_{bm}\ddot{\mathbf{r}}_m = \left\{\mathbf{1} - \left(\sum_i \boldsymbol{\sigma}_{cmi}\eta_i\right)^{\times}\right\}\ddot{\mathbf{r}}_m. \quad (7.1.47)$$

The value $\hat{\ddot{\mathbf{r}}}_m$ denotes the output of the sensor, and $\ddot{\mathbf{r}}_m$ is the true acceleration at the sensor location. The measurement axes are perturbed by local rotation according to Eq. (7.1.47) where $\boldsymbol{\sigma}_{cmi}$, the mode slope at the accelerometer location, is the same as in (7.1.39). This term is particularly important for rockets since large quasi-steady accelerations are projected onto the sensor axes as the sensor rotates elastically with respect to the body. Substituting Eq. (7.1.46) into (7.1.47) yields a complete, albeit lengthy, accelerometer equation with the familiar Coriolis and centripetal accelerations that arise due to flexible motions. Using Eq. (3.2.32), ignoring terms of order two and higher, and assuming that $\dot{\boldsymbol{\omega}}_0 \approx \mathbf{0}$, this expression simplifies to

$$\hat{\ddot{\mathbf{r}}}_m \approx \mathbf{a}_b' + \mathbf{a}_{b0}^\times \sum_i \boldsymbol{\sigma}_{cmi}\eta_i + \sum_i \boldsymbol{\psi}_{cmi}\ddot{\eta}_i - \mathbf{r}_m^\times \dot{\boldsymbol{\omega}}$$
$$+ \boldsymbol{\omega}_0^\times \boldsymbol{\omega}_0^\times \mathbf{r}_m + \boldsymbol{\omega}_0^\times \left(2\sum_i \boldsymbol{\psi}_{cmi}\dot{\eta}_i + \boldsymbol{\omega}_0^\times \sum_i \boldsymbol{\psi}_{cmi}\eta_i \right) \tag{7.1.48}$$

where on the gravity turn,

$$\mathbf{a}_b' = \begin{bmatrix} 0 & \ddot{y}_b & \ddot{z}_b \end{bmatrix}^T \tag{7.1.49}$$

and the steady acceleration $\mathbf{a}_{b0}$ is simply

$$\mathbf{a}_{b0} = \begin{bmatrix} \bar{g} & 0 & 0 \end{bmatrix}^T. \tag{7.1.50}$$

Only the first four terms in Eq. (7.1.48) are typically used in a linear analysis. The fifth term is usually small, except when the sensor is not located on the rocket centerline and the rocket is rolling. The remaining terms, while first order, can usually be neglected unless the ascent trajectory turning rate is very high.

For the present analysis, the three sensor expressions can be expanded into their components in the y and z directions, letting the location of any sensor be described by

$$\mathbf{r}_m = \begin{bmatrix} X_m & Y_m & Z_m \end{bmatrix}^T. \tag{7.1.51}$$

The angle and angular rate sensor expressions are

$$\hat{\phi}_{ym} = \phi_y + \sum_i \sigma_{ycmi}\eta_i \tag{7.1.52}$$

$$\hat{\phi}_{zm} = \phi_z + \sum_i \sigma_{zcmi}\eta_i \tag{7.1.53}$$

$$\hat{\omega}_{ym} = \omega_y + \sum_i \sigma_{ycmi}\dot{\eta}_i \tag{7.1.54}$$

$$\hat{\omega}_{zm} = \omega_z + \sum_i \sigma_{zcmi}\dot{\eta}_i. \tag{7.1.55}$$

The accelerometer equation can be expanded by using (7.1.20) and (7.1.21) to rewrite the body frame components of $\mathbf{a}_b$ in terms of their trajectory-frame quantities. The first-order Coriolis and centripetal terms are neglected.

$$\hat{\ddot{y}}_m = \ddot{y}_I - \bar{g}\left(\phi_z + \sum_i \sigma_{zcmi}\eta_i\right) + \sum_i \psi_{ycmi}\ddot{\eta}_i + X_m\dot{\omega}_z \tag{7.1.56}$$

$$\hat{\ddot{z}}_m = \ddot{z}_I + \bar{g}\left(\phi_y + \sum_i \sigma_{ycmi}\eta_i\right) + \sum_i \psi_{zcmi}\ddot{\eta}_i - X_m\dot{\omega}_y \tag{7.1.57}$$

This completes the required set of equations for the simplified case.

7.2 State-space model

For control design purposes, it is customary to represent the coupled system equations in the form given in Eqs. (7.0.1) and (7.0.2), and reduce the problem to either the pitch axis or yaw axis. By "reduce," we mean that all motion in the other plane is zero. These linear equations can be cast into this form as a set of block diagonal matrices, which are scalable to any number of propellant tanks or flexible modes [30]. For the purposes of the present discussion, the pitch axis will be considered for the case where the origin of the quasi-inertial frame is colocated with the center of mass, or $\mathbf{s}_T = \mathbf{0}$.

Let the system state vector be comprised of the vehicle rigid-body states, the sloshing propellant states, and the elastic generalized coordinates, such that

$$\mathbf{x} = \begin{bmatrix} \mathbf{x}_r \\ \mathbf{x}_s \\ \mathbf{x}_f \end{bmatrix} \tag{7.2.1}$$

where $\mathbf{x}_r$ is the rigid-body state vector, $\mathbf{x}_s$ is the sloshing propellant state vector, and $\mathbf{x}_f$ is the flex state vector. These state vectors contain the components (for the pitch axis)

$$\mathbf{x}_r = \begin{bmatrix} \phi_y & \omega_y & \dot{z}_I \end{bmatrix}^T \tag{7.2.2}$$

$$\mathbf{x}_s = \begin{bmatrix} z_{s1} & \dots & z_{sN} & \dot{z}_{s1} & \dots & \dot{z}_{sN} \end{bmatrix}^T = \begin{bmatrix} \mathbf{z}_s^T & \dot{\mathbf{z}}_s^T \end{bmatrix}^T \tag{7.2.3}$$

$$\mathbf{x}_f = \begin{bmatrix} \eta_1 & \dots & \eta_M & \dot{\eta}_1 & \dots & \dot{\eta}_M \end{bmatrix}^T = \begin{bmatrix} \boldsymbol{\eta}^T & \dot{\boldsymbol{\eta}}^T \end{bmatrix}^T. \tag{7.2.4}$$

The inputs consist of the gimbal angle β_{Ey} and the gimbal acceleration $\ddot{\beta}_{Ey}$, as well as the lateral component of the wind velocity V_{wz}; thus,

$$\mathbf{u} = \begin{bmatrix} \beta_{Ey} & \ddot{\beta}_{Ey} & V_{wz} \end{bmatrix}^T. \tag{7.2.5}$$

The integrated model (7.0.1) can be written in block matrix form as

$$\begin{bmatrix} \mathbf{E}_r \\ \mathbf{E}_s \\ \mathbf{E}_f \end{bmatrix} \begin{bmatrix} \dot{\mathbf{x}}_r \\ \dot{\mathbf{x}}_s \\ \dot{\mathbf{x}}_f \end{bmatrix} = \begin{bmatrix} \mathbf{A}_r & \mathbf{A}_{rs} & \mathbf{A}_{rf} \\ \mathbf{A}_{sr} & \mathbf{A}_s & \mathbf{A}_{sf} \\ \mathbf{A}_{fr} & \mathbf{A}_{fs} & \mathbf{A}_f \end{bmatrix} \begin{bmatrix} \mathbf{x}_r \\ \mathbf{x}_s \\ \mathbf{x}_f \end{bmatrix} + \begin{bmatrix} \mathbf{B}_r \\ \mathbf{B}_s \\ \mathbf{B}_f \end{bmatrix} \mathbf{u}. \tag{7.2.6}$$

In this form, the blocks of the descriptor matrix $\mathbf{E}$ are rectangular matrices having the same number of rows as the rigid, slosh, and flex states, respectively, and a number of columns equal to the total number of states in the model. Likewise, each diagonal block of the $\mathbf{A}$ matrix is a square matrix that represents the coupling of the rigid, slosh, or flex states, respectively, to themselves, while each off-diagonal block represents coupling between one set of dynamic motions (e.g., slosh) and another (e.g., flex). Note that in this integrated state model form, the state vector contains both velocities and displacements, so the descriptor matrix $\mathbf{E}$ and the system matrix $\mathbf{A}$ are not symmetric. As long as the mass matrix is positive definite, the descriptor matrix should likewise be positive definite. The system matrix $\mathbf{A}$ may have eigenvalues with positive real parts. This is expected for most large rockets, since the airframe is statically unstable without closed-loop control.

In order to facilitate the conversion to a matrix form, the scalar pitch axis equations from Section 7.1 are summarized below with dynamic terms grouped first, followed by inputs. Accelerations are on the LHS of the equations, followed by velocities and displacements on the RHS, followed finally by external inputs. Eq. (7.1.24) has been used to replace the angle of attack α_y with quantities that depend on the state variables, and since only pitch motion is considered, gimbal motions about the z axis appearing in the bending equations are set to zero. In addition, the quantity $(F - D)/m_T$ has been replaced by the acceleration $\bar{g}$ as in Eq. (7.1.17), and we assume mass-normalized bending modes $m_{Bi} = 1$.

Rotation

$$I_{yy}\dot{\omega}_y - \sum_j m_{sj}X_{sj}\ddot{z}_{sj} = -\sum_j m_{sj}\bar{g}z_{sj} + C_{N\alpha}\bar{q}S_{ref}X_{cp}\left(\phi_y + \frac{\dot{z}_I}{V}\right)$$
$$+ X_G\left(F_R\beta_{Ey} + F\sum \sigma_{y\beta\,i}\eta_i\right) + F\sum \psi_{z\beta i}\eta_i$$
$$+ (X_G S_E - I_{EG})\ddot{\beta}_{Ey} - X_{cp}\frac{C_{N\alpha}\bar{q}S_{ref}}{V}V_{wz} \quad (7.2.7)$$

Translation

$$m_T\ddot{z}_I + \sum m_{sj}\ddot{z}_{sj} = -m_T\bar{g}\phi_y - C_{N\alpha}\bar{q}S_{ref}\left(\phi_y + \frac{\dot{z}_I}{V}\right)$$
$$- F\sum \sigma_{y\beta\,i}\eta_i - F_R\beta_{Ey} + S_{Ex}\ddot{\beta}_{Ey} + \frac{C_{N\alpha}\bar{q}S_{ref}}{V}V_{wz} \quad (7.2.8)$$

Slosh

$$\ddot{z}_{sj} - X_{sj}\ddot{\phi}_y + \ddot{z}_I + \sum_i \psi_{zji}\ddot{\eta}_i = -\Omega_{sj}^2 z_{sj} - 2\zeta_{sj}\Omega_{sj}\dot{z}_{sj} - \bar{g}\phi_y \quad (7.2.9)$$

Flex

$$\ddot{\eta}_i + \sum_j m_{sj}\psi_{zji}\ddot{z}_{sj} = -\bar{q}S_{ref}C_{N\eta i}\left(\phi_y + \frac{\dot{z}_I}{V}\right)$$
$$- \Omega_{Bi}^2\eta_i - 2\zeta_{Bi}\Omega_{Bi}\dot{\eta}_i - F_R\psi_{z\beta\,i}\beta_{Ey}$$
$$+ \left(\psi_{z\beta i}S_{Ex} - \sigma_{y\beta i}I_{EG}\right)\ddot{\beta}_{Ey} + \frac{\bar{q}S_{ref}C_{N\eta i}}{V}V_{wz} \quad (7.2.10)$$

Sensor

$$\hat{\phi}_{ym} = \phi_y + \sum_i \sigma_{ycmi}\eta_i \quad (7.2.11)$$

$$\hat{\omega}_{ym} = \omega_y + \sum_i \sigma_{ycmi}\dot{\eta}_i \quad (7.2.12)$$

$$\hat{\ddot{z}}_m = \ddot{z}_I + \bar{g}\left(\phi_y + \sum_i \sigma_{ycmi}\right) + \sum_i \psi_{zcmi}\ddot{\eta}_i - X_m\dot{\omega}_y \quad (7.2.13)$$

The translation and attitude dynamics of Eqs. (7.2.7) and (7.2.8) can be grouped in the order given by the state vector (2.5.17), noting that for small perturbations $\dot{\phi}_y = \omega_y$. The rigid-body system matrix is

$$\mathbf{A}_r = \begin{bmatrix} 0 & 1 & 0 \\ C_{N\alpha}\bar{q}S_{ref}X_{cp} & 0 & \frac{C_{N\alpha}\bar{q}S_{ref}}{V}X_{cp} \\ -m_T\bar{g} - C_{N\alpha}\bar{q}S_{ref} & 0 & \frac{-C_{N\alpha}\bar{q}S_{ref}}{V} \end{bmatrix} \quad (7.2.14)$$

A refined version of this expression that contains aerodynamic damping terms is given in the section below entitled Distributed Aerodynamics. The rigid-body input matrix is

$$\mathbf{B}_r = \begin{bmatrix} 0 & 0 & 0 \\ X_G F_R & (X_G S_E - I_{EG}) & \frac{-C_{N\alpha}\bar{q}S_{ref}}{V}X_{cp} \\ -F_R & S_{Ex} & \frac{C_{N\alpha}\bar{q}S_{ref}}{V} \end{bmatrix}. \quad (7.2.15)$$

The system matrices for the N slosh dynamics equations are constructed from Eq. (7.2.9) as

$$\mathbf{A}_s = \begin{bmatrix} \mathbf{O}_{N\times N} & \mathbf{I}_{N\times N} \\ -\boldsymbol{\Omega}_s^2 & -2\boldsymbol{\zeta}_s\boldsymbol{\Omega}_s^2 \end{bmatrix} \quad (7.2.16)$$

with a null input matrix $\mathbf{B}_s = [\mathbf{0}_{2N\times 2}]$. Note that since the sloshing equations do not depend directly on the input, the only excitation of the slosh dynamics is through coupling with the rigid-body dynamics and flexible dynamics.

The flex system matrix $\mathbf{A}_f$ can be computed using the M equations in (7.2.10) such that

$$\mathbf{A}_f = \begin{bmatrix} \mathbf{O}_{M\times M} & \mathbf{I}_{M\times M} \\ -\boldsymbol{\Omega}_B^2 & -2\zeta_B\boldsymbol{\Omega}_B \end{bmatrix}. \quad (7.2.17)$$

Direct input excitation of flex results from the last three terms in (7.2.10). These are used to form the flex input matrix

$$\mathbf{B}_f = \left[\begin{array}{c|c|c} & \mathbf{O}_{M\times 3} & \\ -F_R\psi_{z\beta 1} & \psi_{z\beta 1}S_{Ex} - \sigma_{y\beta 1}I_{EG} & \frac{\bar{q}S_{ref}}{V}C_{N\eta 1} \\ \vdots & \vdots & \vdots \\ -F_R\psi_{z\beta M} & \psi_{z\beta M}S_{Ex} - \sigma_{y\beta M}I_{EG} & \frac{\bar{q}S_{ref}}{V}C_{N\eta M} \end{array}\right], \quad (7.2.18)$$

which uses the wind bending coefficients defined by (5.3.26). In comparing the slosh and flex matrices, note that it is typical to use a different value of the slosh damping for each tank and flight condition. Thus, the values of

ζ_{sj} can be used to construct a diagonal matrix $\boldsymbol{\zeta}_s$ as needed. For flex, the assumed damping ratio for all bending modes is usually the same, so it is possible to use a scalar $\zeta_{Bi} = \zeta_B$.

State coupling matrices

The coupling between the rigid body, flex, and slosh dynamics can be represented by the off-diagonal coupling matrices in Eq. (7.2.6). The contributions of slosh and flex to the rigid-body accelerations are represented by the matrices $\mathbf{A}_{rs}$ and $\mathbf{A}_{rf}$, respectively, where

$$\mathbf{A}_{rs} = \left[\begin{array}{cccc|c} & \mathbf{0}_{1\times N} & & & \\ -\bar{g}m_{s1} & -\bar{g}m_{s2} & \dots & -\bar{g}m_{sN} & \mathbf{O}_{3\times N} \\ & \mathbf{0}_{1\times N} & & & \end{array}\right] \tag{7.2.19}$$

and

$$\mathbf{A}_{rf} = \left[\begin{array}{ccc|c} & \mathbf{0}_{1\times M} & & \\ F(X_G\sigma_{y\beta 1}+\psi_{z\beta 1}) & \dots & F(X_G\sigma_{y\beta M}+\psi_{z\beta M}) & \mathbf{O}_{3\times M} \\ -F\sigma_{y\beta 1} & \dots & -F\sigma_{y\beta M} & \end{array}\right]. \tag{7.2.20}$$

If $\mathbf{s}_T = \mathbf{0}$, and under the present assumption that all of the mass is included in the finite element model, there is no direct coupling of rigid-body motion to the flexible modes in Eq. (7.2.10). However, since the angle of attack affects the excitation of flex through the wind bending coefficients defined by (5.3.26), there is a coupling term that depends on both the rigid-body trajectory-relative pitch angle and the lateral velocity.

$$\mathbf{A}_{fr} = \begin{bmatrix} & \mathbf{O}_{M\times 3} & \\ -\bar{q}S_{ref}C_{N\eta 1} & 0 & \frac{-\bar{q}S_{ref}}{V}C_{N\eta 1} \\ \vdots & \vdots & \vdots \\ -\bar{q}S_{ref}C_{N\eta M} & 0 & \frac{-\bar{q}S_{ref}}{V}C_{N\eta M} \end{bmatrix} \tag{7.2.21}$$

There is also no coupling of the sloshing velocity or displacement states into the bending equations or vice-versa. It follows that

$$\mathbf{A}_{fs} = \mathbf{A}_{sf}^T = \mathbf{O}_{2M\times 2N}. \tag{7.2.22}$$

However, due to the relationship of the perturbation angle ϕ_y and the body acceleration $\ddot{z}_b$ by Eq. (7.1.19), the slosh motion is coupled to the rigid-body pitch angle by

$$\mathbf{A}_{sr} = \left[\begin{array}{c|c} \mathbf{0}_{N\times 1} & \\ -\bar{g}\mathbf{1}_{N\times 1} & \mathbf{O}_{2N\times 2} \end{array} \right]. \tag{7.2.23}$$

Descriptor matrix

The descriptor matrix $\mathbf{E}$ is similar to a mass matrix. The three components on the left side of (7.2.6) can be constructed by inspection of the pitch equations to yield the expressions in Eqs. (7.2.24), (7.2.25), and (7.2.26) below.

Output matrices

It is common for large rockets to have more than one rate gyro or accelerometer, so in the model the sensor output equations are repeated to reflect as many sensors as are installed. It is important to note that the sensor location vector and modal data will be different if the physical locations of the sensors are different. Conversely, an instrument package such as an Inertial Measurement Unit (IMU) can produce multiple outputs from the same location. Suppose the rocket contains an angle sensor, an attitude rate sensor, and an accelerometer, all located at measurement location $m = 1$, plus a rate sensor at a second location $m = 2$. The output vector becomes

$$\mathbf{y} = \left[\begin{array}{cccc} \hat{\phi}_{y1} & \hat{\omega}_{y1} & \hat{\omega}_{y2} & \hat{\ddot{z}}_1 \end{array} \right]^T \tag{7.2.27}$$

This can be implemented using Eqs. (7.2.11) through (7.2.13). As will be shown, the outputs depend on both the state variables $\mathbf{x}$ and the inputs $\mathbf{u}$.

The components of the $\mathbf{C}$ and $\mathbf{D}$ matrices in (7.0.2) for the sensed outputs are constructed from row vectors for each sensor. For example, the $\mathbf{C}$ and $\mathbf{D}$ matrices for the case described in Eq. (7.2.27) are

$$\mathbf{C} = \begin{bmatrix} \mathbf{c}_\phi^T \\ \mathbf{c}_{\omega 1}^T \\ \mathbf{c}_{\omega 2}^T \\ \mathbf{c}_{\ddot{z}}^T \end{bmatrix} \quad \mathbf{D} = \begin{bmatrix} \mathbf{d}_\phi^T \\ \mathbf{d}_{\omega 1}^T \\ \mathbf{d}_{\omega 2}^T \\ \mathbf{d}_{\ddot{z}}^T \end{bmatrix}. \tag{7.2.28}$$

Each row of $\mathbf{C}$ corresponding to an angle sensor can be written using Eq. (7.2.11) as

$$\mathbf{c}_\phi^T = \left[\begin{array}{ccc|ccccc} 1 & 0 & 0 & \mathbf{0}_{1\times 2N} & \sigma_{ycm1} & \ldots & \sigma_{ycmM} & \mathbf{0}_{1\times M} \end{array} \right] \tag{7.2.29}$$

$$\mathbf{E}_r = \left[\begin{array}{ccc|c|cccc|c} 1 & 0 & 0 & & 0 & 0 & \cdots & 0 & \\ 0 & I_{yy} & 0 & \mathbf{O}_{3\times N} & -m_{s1}X_{s1} & -m_{s2}X_{s2} & \cdots & -m_{sN}X_{sN} & \mathbf{O}_{3\times 2M} \\ 0 & 0 & m_T & & m_{s1} & m_{s2} & \cdots & m_{sN} & \end{array}\right] \tag{7.2.24}$$

$$\mathbf{E}_s = \left[\begin{array}{ccc|c|c|cccc} & \mathbf{O}_{N\times 3} & & & & & \mathbf{O}_{N\times M} & & \\ 0 & -X_{s1} & 1 & & & \psi_{z11} & \psi_{z12} & \cdots & \psi_{z1M} \\ 0 & -X_{s2} & 1 & \mathbf{I}_{2N\times 2N} & \mathbf{O}_{2N\times M} & \psi_{z21} & \psi_{z22} & \cdots & \psi_{z2M} \\ \vdots & \vdots & \vdots & & & \vdots & \vdots & \ddots & \vdots \\ 0 & -X_{sN} & 1 & & & \psi_{zN1} & \psi_{zN2} & \cdots & \psi_{zNM} \end{array}\right] \tag{7.2.25}$$

$$\mathbf{E}_f = \left[\begin{array}{c|cccc|c} & & \mathbf{O}_{N\times M} & & & \\ & m_{s1}\psi_{z11} & m_{s2}\psi_{z21} & \cdots & m_{sN}\psi_{zN1} & \\ \mathbf{O}_{2M\times 3} & m_{s1}\psi_{z12} & m_{s2}\psi_{z22} & \cdots & \psi_{zN2} & \mathbf{I}_{2M\times 2M} \\ & \vdots & \vdots & \ddots & \vdots & \\ & m_{s1}\psi_{z1M} & m_{s2}\psi_{z2M} & \cdots & \psi_{zNM} & \end{array}\right] \tag{7.2.26}$$

For the simple example proposed above, there would be only one such row, with $m = 1$. On the other hand, there are two rate sensors, each with their own row

$$\mathbf{c}_{\omega 1}^{T} = \left[\begin{array}{ccc|ccccc} 1 & 0 & 0 & \mathbf{0}_{1\times 2N} & \mathbf{0}_{1\times M} & \sigma_{yc11} & \cdots & \sigma_{yc1M} \end{array} \right]. \tag{7.2.30}$$

$$\mathbf{c}_{\omega 2}^{T} = \left[\begin{array}{ccc|ccccc} 1 & 0 & 0 & \mathbf{0}_{1\times 2N} & \mathbf{0}_{1\times M} & \sigma_{yc21} & \cdots & \sigma_{yc2M} \end{array} \right] \tag{7.2.31}$$

No sensor to input coupling is present for the angle and angular rate sensors, so

$$\mathbf{d}_{\phi}^{T} = \mathbf{d}_{\omega}^{T} = \left[\begin{array}{ccc} 0 & 0 & 0 \end{array} \right]. \tag{7.2.32}$$

In order to construct the accelerometer output matrices, the descriptor matrix in Eq. (7.0.1) must first be inverted to provide an explicit solution for the state derivatives. The resulting matrices are given in Eq. (7.0.3). In practice, the values of $\tilde{\mathbf{A}}$ and $\tilde{\mathbf{B}}$ can only be computed numerically, so matrix indexing notation is used to refer to the elements of these matrices. For example, the value $\tilde{\mathbf{A}}_{[3,\ldots]}$ is the 3rd row and all columns of the matrix $\tilde{\mathbf{A}}$.

From Eq. (7.2.13), a row of $\mathbf{C}$ corresponding to an accelerometer is

$$\begin{aligned} \mathbf{c}_{\ddot{z}}^{T} &= \tilde{\mathbf{A}}_{[3,\ldots]} - X_m \tilde{\mathbf{A}}_{[2,\ldots]} \\ &\quad + \begin{bmatrix} \psi_{czm1} \\ \vdots \\ \psi_{czmM} \end{bmatrix}^{T} \tilde{\mathbf{A}}_{[(3+2N+M+1)\cdots(3+2N+2M),\ldots]} \\ &\quad + \bar{g} \left[\begin{array}{ccc|ccccc} 1 & 0 & 0 & \mathbf{0}_{1\times 2N} & \sigma_{ycm1} & \cdots & \sigma_{ycmM} & \mathbf{0}_{1\times M} \end{array} \right]. \end{aligned} \tag{7.2.33}$$

Finally, the corresponding row of $\mathbf{D}$ is

$$\begin{aligned} \mathbf{d}_{\ddot{z}}^{T} &= \tilde{\mathbf{B}}_{[3,\ldots]} - X_m \tilde{\mathbf{B}}_{[2,\ldots]} \\ &\quad + \begin{bmatrix} \psi_{zcm1} \\ \vdots \\ \psi_{zcmM} \end{bmatrix}^{T} \tilde{\mathbf{B}}_{[(3+2N+M+1)\cdots(3+2N+2M),\ldots]}. \end{aligned} \tag{7.2.34}$$

7.3 Distributed aerodynamics

In most rocket stability analysis models, aerodynamic effects are incorporated using the steady rigid body force and moment coefficients presented in Chapter 5. These coefficients are reduced to an equivalent center of pressure and normal force slope, which can then be used in the linearized pitching and yawing equations. The effect of wind loads is approximated by the line load distribution introduced in Eqs. (5.3.25) and (5.3.26), which yield the wind bending coefficients that can be incorporated into the linear bending equations (7.2.10).

The aerodynamic force and moment coefficients used to compute the center of pressure X_{cp} are usually taken from a set of static wind tunnel test conditions or CFD runs. Each flight condition contains data at multiple incidence angles. Once the center of pressure has been computed, the coefficient slopes $C_{N\alpha}$ at each flight condition are derived via curve fitting or finite differencing of the data.

The bending line load data $\partial C_{N\alpha}/\partial x$ or $\partial C_{Y\beta}/\partial x$ can be computed from the pressure distribution on the body. These so-called "distributed forces" are a function of local incidence angle. An equivalent normal or side force coefficient can be calculated by integrating the distributed forces over the body.

The use of the rigid-body $C_{N\alpha}$, $C_{Y\beta}$ combined with the wind bending coefficients assumes that the angles of attack and sideslip are the same everywhere on the rocket body. If the rocket is bending, rotating, or penetrating a spatially varying wind field, this is not the case. If the line load data is available, an enhancement to the model can be incorporated which approximates all of these effects simultaneously and eliminates the need for two sources of the aerodynamic coefficients. This model is called a *distributed aerodynamics* or *static aeroelastic model.*

The term "static aerodynamics" is used herein to denote the case of a rigid body at a steady angle of attack. In the present approach, static *data* is extended to a dynamic *analysis*. The force on each slice of the vehicle is assumed to be the same as what comes from static aerodynamics, using the local angle of attack in conjunction with the line load data. This might be called a "quasi-static" approach, in which the local angle of attack is varied according to the pitching and flexing motion. In reality, the flow condition at each station affects the flow at downstream stations. This is especially true at supersonic speeds for rockets with asymmetric protuberances. Thus the quasi-static approach cannot always be trusted. It is worth noting that this approach is *nonconservative* in that it always produces a damping force. In reality, downstream effects may create the possibility of negative damping for structural dynamic motions. Because of this possibility, the quasi-static approach should not be used to compute enhanced structural damping unless there are some independent spot checks of the results. Although slow and resource-intensive, coupled structural-aerodynamic computational aeroelastic (CAE) methods have become increasingly accurate during the past few decades and are the logical means of such checking [31].

Pitch damping data could in theory be obtained from wind tunnel tests in which the test article is oscillated in pitch. In practice, this can be quite problematic. Only the largest facilities have a test section that allows a long slender model to be moved in this fashion without introducing significant uncertainties in the data.

Bear in mind that the basic purpose of a state space model is to serve as a tool for the design of a control system. There is always a tradeoff to be made between ease of use and accuracy, and for a state space model the former is an important requirement. In the later stages of rocket development, the focus turns to verification of the design, which requires a nonlinear simulation – a different tool that places a higher premium on accuracy. For the state space model, one approach that should be considered is the use of correction factors to bring the quasi-static approach into acceptable agreement with results from a more sophisticated aerodynamic analysis at fewer flight conditions. This is a good way to align the results from a state space model with those from a nonlinear simulation.

The development in this section will describe the linearized distributed aerodynamic equations for the pitch plane using the method presented in Reference [32]. The development for the yaw plane is similar. The wind velocity is assumed to be the same everywhere on the rocket body, although an extension that allows the wind velocity to vary along the x direction is straightforward.

Consider a flexible rocket as shown in Fig. 7.4. Each station location on the vehicle is denoted by the subscript h, related to the origin by the position X_h (positive forward of the origin). Employing Eq. (7.1.24), the local angle of attack at station X_h is given by

$$\alpha_{yh} = \phi_{yh} + \frac{\dot{z}_{Ih}}{V} + \alpha_w \tag{7.3.1}$$

where

$$\alpha_w = -\frac{V_{wz}}{V} \tag{7.3.2}$$

is the angle of attack due to wind. In numerical calculations, the subscript h becomes an index with values ranging from 1 to H, where H is the number of centerline nodes. Each station X_h has a different velocity relative to the

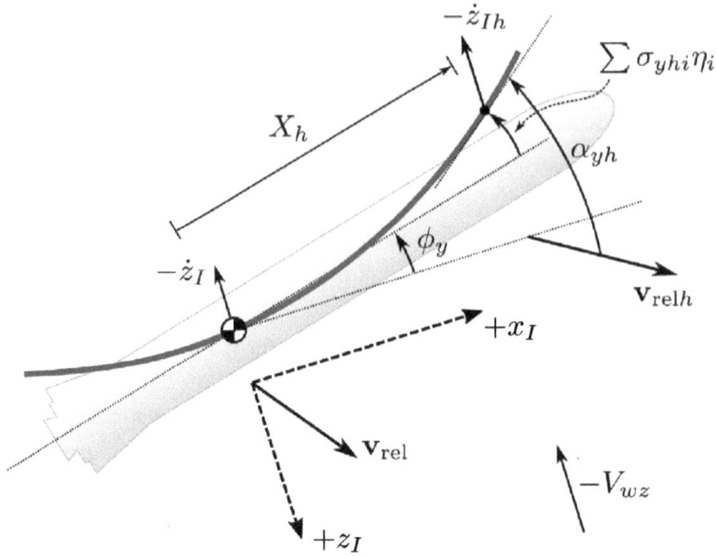

Figure 7.4 Flexible rocket with local angle of attack.

surrounding air mass, and the local velocity consists of the sum of the rigid, elastic, and wind velocities. The total local angle of attack α_{yh} is the angle between the local velocity vector $\mathbf{v}_{\mathrm{rel}h}$ and the deformed centerline.

Since the entire body is rotating, the expression for the lateral velocity of point X_h relative to the x_I axis also contains rigid rotation and elastic translation components;

$$\dot{z}_{Ih} = \dot{z}_I + \sum_i \psi_{zhi}\dot{\eta}_i - X_h\dot{\phi}_y \tag{7.3.3}$$

The local rotation is given by

$$\phi_{yh} = \phi_y + \sum \sigma_{yhi}\eta_i \tag{7.3.4}$$

where ψ_{zhi}, σ_{yhi} are the mode shape and slope at station X_h. These values are typically taken from a special set of finite element nodes that are representative of the global bending of the structure, and are sometimes called *centerline load path* nodes. The important aspect of choosing these nodes is that global, low-frequency motions are represented. As with the aerodynamic line loads, the choice of FEM nodes should be coordinated with their intended use. Higher-frequency modes that involve small-scale local deformations of the vehicle structure should not be used in computing the distributed aerodynamics.

Combining Eqs. (7.3.2) through (7.3.4) using Eq. (7.3.1) and noting that $\omega_y = \dot{\phi}_y$ in the linearized frame, the local angle of attack becomes

$$\alpha_{yh} = \phi_y + \frac{\dot{z}_I}{V} - \frac{X_h}{V}\omega_y - \frac{V_{wz}}{V} + \sum_i \left(\sigma_{yhi}\eta_i + \frac{1}{V}\psi_{zhi}\dot{\eta}_i\right). \tag{7.3.5}$$

The finite element model is used to model motion at distinct physical points. In theory, the line load distribution $\partial C_{N\alpha}/\partial x$ is a continuous quantity. To use the line load in the model, it is necessary to convert the data into discrete section loads so that it can be matched with centerline nodes in the FEM. A notional example of this process for a launch vehicle is shown in Fig. 7.5. This simply requires computing the values $C_{N\alpha h}$ that satisfy

$$C_{N\alpha} = \int_0^L \frac{\partial C_{N\alpha}}{\partial x} dx \approx \sum_{h=1}^{H} C_{N\alpha h} \tag{7.3.6}$$

Note that spacing of the centerline nodes in the FEM will dictate the width of the rectangles in Fig. 7.5. The coefficients $C_{N\alpha h}$ can be computed via any suitable quadrature scheme, such as the trapezoidal rule. The aerodynamic line load data may also contain a scaling factor, such as a reference diameter, but this factor has been omitted from Eq. (7.3.6).

The aerodynamic force and moment at each station X_h are computed in the same way as their rigid-body counterparts. The total normal force is

$$f_{z\,\mathrm{aero}} = -\bar{q}S_{ref} \sum_h C_{N\alpha h}\alpha_{yh} \tag{7.3.7}$$

and the moment is

$$g_{y\,\mathrm{aero}} = \bar{q}S_{ref} \sum_h X_h C_{N\alpha h}\alpha_{yh}. \tag{7.3.8}$$

Note that in this expression, for each element h the quantity X_h takes the place of the center of pressure X_{CP}. The elastic generalized aerodynamic force is likewise

$$Q_{\eta i\,\mathrm{aero}} = \bar{q}S_{ref} \sum_h \psi_{zhi} C_{N\alpha h} \left\{ \left(\phi_y + \frac{\dot{z}_I}{V} - \frac{X_h}{V}\omega_y - \frac{V_{wz}}{V}\right) + \left(\sum_\rho \sigma_{yh\rho}\eta_\rho + \sum_\rho \frac{1}{V}\psi_{zh\rho}\dot{\eta}_\rho\right)\right\} \tag{7.3.9}$$

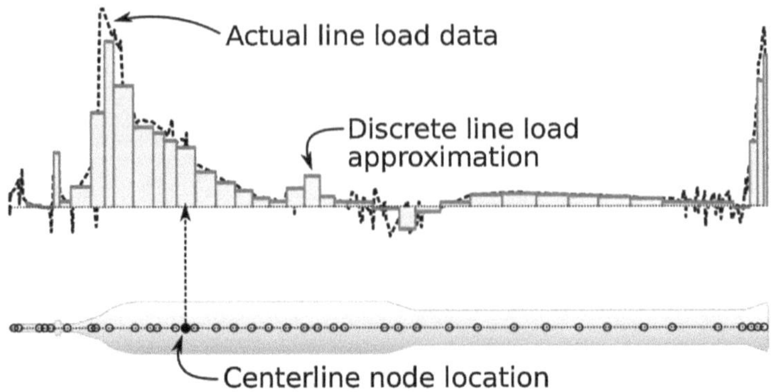

Figure 7.5 Matching the continuous line load to the FEM.

where the local angle of attack expression has been expanded to highlight the fact that inner summations appear inside the main summation. In the second line of (7.3.9), the index ρ plays the same role as the index i in (7.3.5). Thus the excitation of each mode i involves a summation over all the modes (including i), i.e., a summation over all ρ. These seemingly complex expressions can be implemented using matrix computations, as shown below. It is important to notice that distributed aerodynamics represents a coupling of elastic motion into itself through external generalized forces. Thus it is possible for an instability to develop via aeroelasticity alone, without any contribution from another variable such as slosh or gimbal motion.

Let the discrete load coefficient vector be given by

$$\mathbf{c}_{N\alpha} = \begin{bmatrix} C_{N\alpha 1} & C_{N\alpha 2} & \dots & C_{N\alpha H} \end{bmatrix}^T \tag{7.3.10}$$

and using Eq. (7.3.6), let the usual rigid-body normal force coefficient, without a subscript, be

$$C_{N\alpha} \equiv \sum_{h=1}^{H} C_{N\alpha h}. \tag{7.3.11}$$

It will also be convenient to define the station location vector and its element-wise square as

$$\mathbf{l}_a = \begin{bmatrix} X_1 & X_2 & \dots & X_H \end{bmatrix}^T \tag{7.3.12}$$

$$\mathbf{l}_a^2 = \begin{bmatrix} X_1^2 & X_2^2 & \dots & X_H^2 \end{bmatrix}^T. \tag{7.3.13}$$

In this section, the subscript "a" will be used to denote quantities involving the aeroelastic model.

Using (7.2.7) and (7.2.8) with the distributed aerodynamic forces (7.3.7) and moments (7.3.8), the rigid-body dynamics matrix $\mathbf{A}_r$ can be modified as

$$\mathbf{A}_r = \begin{bmatrix} 0 & 1 & 0 \\ \bar{q}S_{ref}\mathbf{l}_a^T\mathbf{c}_{N\alpha} & -\frac{\bar{q}S_{ref}}{V}\left[\mathbf{l}_a^2\right]^T\mathbf{c}_{N\alpha} & \frac{\bar{q}S_{ref}}{V}\mathbf{l}_a^T\mathbf{c}_{N\alpha} \\ -m_T\bar{g}-\bar{q}S_{ref}C_{N\alpha} & \frac{\bar{q}S_{ref}}{V}\mathbf{l}_a^T\mathbf{c}_{N\alpha} & -\frac{\bar{q}S_{ref}}{V}C_{N\alpha} \end{bmatrix}. \tag{7.3.14}$$

This should be compared to (7.2.14). If the line loads have been correctly computed, the first and third columns should be identical. Note the newly introduced elements in the middle column of Eq. (7.3.14). These elements introduce pitch damping plus a normal force due to pitch rate. The matrix (7.3.14) accounts for the first three terms in the local angle of attack expression (7.3.5) that depend on the rigid-body states.

It is helpful to define the binary operator

$$\mathbf{a} \circ \mathbf{b} = \begin{bmatrix} a_1b_1 & a_2b_2 & \dots & a_nb_n \end{bmatrix}^T \tag{7.3.15}$$

which is known as the *Hadamard product* and represents element-wise multiplication of two vectors or matrices of the same size. This is equivalent to the ".*" operation appearing in numerical programming languages such as *MATLAB*® and *Octave*. If the mode shape and slope at each centerline node location are given by ψ_{zhi}, σ_{yhi}, define the aeroelastic coupling matrices

$$\boldsymbol{\Psi}_{za} \equiv \begin{bmatrix} \psi_{z11} & \psi_{z21} & \dots & \psi_{zH1} \\ \psi_{z12} & \psi_{z22} & \dots & \psi_{zH2} \\ \vdots & \vdots & \ddots & \vdots \\ \psi_{z1M} & \psi_{z2M} & \dots & \psi_{zHM} \end{bmatrix} \tag{7.3.16}$$

$$\bar{\boldsymbol{\sigma}}_{ya} \equiv \begin{bmatrix} \sigma_{y11} & \sigma_{y21} & \dots & \sigma_{yH1} \\ \sigma_{y12} & \sigma_{y22} & \dots & \sigma_{yH2} \\ \vdots & \vdots & \ddots & \vdots \\ \sigma_{y1M} & \sigma_{y2M} & \dots & \sigma_{yHM} \end{bmatrix} \tag{7.3.17}$$

which are each of dimension $M \times H$ over M modes and H station locations. These matrices have the subscript a to indicate that they are aeroelastic coupling matrices. The first subscript in each of the above matrices is either y or z throughout, and the second and third subscripts can be considered the first and second indices. These indices have the first index as the column and the second as the row – the reverse of the usual convention. The first index refers to the centerline node location as in Fig. 7.5, not to be confused with either a slosh location or a measurement location.

Excitation of the rigid body dynamics from flex already occurs due to the force follower terms (functions of η_i) given in Eqs. (7.2.7) and (7.2.8) and expressed in matrix form in (7.2.20). It will be necessary to modify this matrix to include the effects of distributed aerodynamics. Let us redefine

$$\mathbf{A}_{rf} = \mathbf{A}_{rf1} + \mathbf{A}_{rf2} \tag{7.3.18}$$

where $\mathbf{A}_{rf1}$ is the same expression as in Eq. (7.2.20) and $\mathbf{A}_{rf2}$ is a new distributed aerodynamics coupling matrix to be defined. We can expand the elastic components of Eq. (7.3.5) in (7.3.7) and (7.3.8) in order to write

$$\bar{q}S_{ref}\sum_h X_h C_{N\alpha h}\sum_i \sigma_{yhi}\eta_i = \bar{q}S_{ref}\begin{bmatrix} \sum_h X_h C_{N\alpha h}\sigma_{yh1} & \dots & \sum_h X_h C_{N\alpha h}\sigma_{yhM} \end{bmatrix}\boldsymbol{\eta} \tag{7.3.19}$$

where the vector form of the generalized displacement

$$\boldsymbol{\eta} = \begin{bmatrix} \eta_1 & \eta_2 & \dots & \eta_M \end{bmatrix}^T$$

has been used. Using the Hadamard product (7.3.15) and the aeroelastic coupling matrix (7.3.17), the row vector in Eq. (7.3.19) can be rewritten such that

$$\bar{q}S_{ref}\sum_h X_h C_{N\alpha h}\sum_i \sigma_{yhi}\eta_i = \bar{q}S_{ref}\left[\bar{\boldsymbol{\sigma}}_{ya}\left(\mathbf{l}_a \circ \mathbf{c}_{N\alpha}\right)\right]^T \boldsymbol{\eta}. \tag{7.3.20}$$

Similar results follow for the remaining terms in (7.3.7) and (7.3.8). These are

$$\bar{q}S_{ref}\sum_h X_h C_{N\alpha h}\sum_i \frac{1}{V}\psi_{zhi}\dot{\eta}_i = \frac{\bar{q}S_{ref}}{V}\left[\boldsymbol{\Psi}_{za}\left(\mathbf{l}_a \circ \mathbf{c}_{N\alpha}\right)\right]^T \dot{\boldsymbol{\eta}} \tag{7.3.21}$$

$$\bar{q}S_{ref}\sum_h C_{N\alpha h}\sum_i \sigma_{yhi}\eta_i = \bar{q}S_{ref}\left[\bar{\boldsymbol{\sigma}}_{ya}\mathbf{c}_{N\alpha}\right]^T \boldsymbol{\eta} \tag{7.3.22}$$

$$\bar{q}S_{ref}\sum_h C_{N\alpha h}\sum_i \frac{1}{V}\psi_{zhi}\dot{\eta}_i = \frac{\bar{q}S_{ref}}{V}\left[\boldsymbol{\Psi}_{za}\mathbf{c}_{N\alpha}\right]^T \dot{\boldsymbol{\eta}}. \tag{7.3.23}$$

It follows that the distributed aerodynamics flex-to-rigid coupling matrix can be compactly expressed as

$$\mathbf{A}_{rf2} = \begin{bmatrix} \mathbf{0}_{1\times M} & \mathbf{0} \\ \bar{q}S_{ref}\left[\bar{\boldsymbol{\sigma}}_{ya}\left(\mathbf{l}_a \circ \mathbf{c}_{N\alpha}\right)\right]^T & \frac{\bar{q}S_{ref}}{V}\left[\boldsymbol{\Psi}_{za}\left(\mathbf{l}_a \circ \mathbf{c}_{N\alpha}\right)\right]^T \\ -\bar{q}S_{ref}\left[\bar{\boldsymbol{\sigma}}_{ya}\mathbf{c}_{N\alpha}\right]^T & -\frac{\bar{q}S_{ref}}{V}\left[\boldsymbol{\Psi}_{za}\mathbf{c}_{N\alpha}\right]^T \end{bmatrix}. \tag{7.3.24}$$

Inspecting Eq. (7.3.9) and using the same approach, the block matrix $\mathbf{A}_{fr}$ (which is usually zero in the absence of distributed aerodynamics) may be replaced with

$$\mathbf{A}_{fr} = \begin{bmatrix} \mathbf{0}_{M\times 1} & \mathbf{0}_{M\times 1} & \mathbf{0}_{M\times 1} \\ \bar{q}S_{ref}\boldsymbol{\Psi}_{za}\mathbf{c}_{N\alpha} & -\frac{\bar{q}S_{ref}}{V}\boldsymbol{\Psi}_{za}\left(\mathbf{l}_a \circ \mathbf{c}_{N\alpha}\right) & \frac{\bar{q}S_{ref}}{V}\boldsymbol{\Psi}_{za}\mathbf{c}_{N\alpha} \end{bmatrix}. \tag{7.3.25}$$

The last step in assembling the distributed aerodynamics system matrices is to compute the flex coupling effects in the flex system matrix (7.2.17). Let this quantity be redefined as

$$\mathbf{A}_f = \begin{bmatrix} \mathbf{O}_{M\times M} & \mathbf{I}_{M\times M} \\ -\boldsymbol{\Omega}_B^2 - \mathbf{K}_a & -2\zeta_B\boldsymbol{\Omega}_B^2 - \mathbf{D}_a \end{bmatrix} \tag{7.3.26}$$

where $\mathbf{K}_a$ and $\mathbf{D}_a$ are the aeroelastic "stiffness" and "damping" matrices, respectively.

Via expansion and reordering of summations, the first term involving double sums in Eq. (7.3.9) can be written for each i as

$$\bar{q}S_{ref}\sum_h \psi_{zhi}C_{N\alpha h}\sum_\rho \sigma_{yh\rho}\eta_\rho = \bar{q}S_{ref}\sum_h C_{N\alpha h}\begin{bmatrix} \psi_{zhi}\sigma_{yh1} & \dots & \psi_{zhi}\sigma_{yhM} \end{bmatrix}\boldsymbol{\eta} \tag{7.3.27}$$

and by replicating this expression over all $i = 1\dots M$ elastic equations, it can be written as the sum of H $M \times M$ square matrices; that is,

$$\begin{bmatrix} \bar{q}S_{ref}\sum_h \psi_{zh1}C_{N\alpha h}\sum_\rho \sigma_{yh\rho}\eta_\rho \\ \bar{q}S_{ref}\sum_h \psi_{zh2}C_{N\alpha h}\sum_\rho \sigma_{yh\rho}\eta_\rho \\ \vdots \\ \bar{q}S_{ref}\sum_h \psi_{zhM}C_{N\alpha h}\sum_\rho \sigma_{yh\rho}\eta_\rho \end{bmatrix} = \bar{q}S_{ref}\sum_h \boldsymbol{\Upsilon}'_h\boldsymbol{\eta} \tag{7.3.28}$$

where the aeroelastic coupling matrix

$$\boldsymbol{\Upsilon}'_h = C_{N\alpha h}\begin{bmatrix} \psi_{zh1} \\ \psi_{zh2} \\ \vdots \\ \psi_{zhM} \end{bmatrix}\begin{bmatrix} \sigma_{yh1} & \sigma_{yh2} & \dots & \sigma_{yhM} \end{bmatrix} \tag{7.3.29}$$

is formed from the outer product of the vectors containing the mode shapes and slopes at the locations X_h. Using the same procedure, the second term involving double sums in Eq. (7.3.9) can be expressed as

$$\begin{bmatrix} \bar{q}S_{ref}\sum_h \psi_{zh1} C_{N\alpha h}\sum_\rho \frac{1}{V}\psi_{zh\rho}\dot{\eta}_\rho \\ \bar{q}S_{ref}\sum_h \psi_{zh2} C_{N\alpha h}\sum_\rho \frac{1}{V}\psi_{zh\rho}\dot{\eta}_\rho \\ \vdots \\ \bar{q}S_{ref}\sum_h \psi_{zhM} C_{N\alpha h}\sum_\rho \frac{1}{V}\psi_{zh\rho}\dot{\eta}_\rho \end{bmatrix} = \frac{\bar{q}S_{ref}}{V}\sum_h \boldsymbol{\Upsilon}_h\dot{\boldsymbol{\eta}} \tag{7.3.30}$$

where the matrix

$$\boldsymbol{\Upsilon}_h = C_{N\alpha h}\begin{bmatrix} \psi_{zh1} \\ \psi_{zh2} \\ \vdots \\ \psi_{zhM} \end{bmatrix}\begin{bmatrix} \psi_{zh1} & \psi_{zh2} & \dots & \psi_{zhM} \end{bmatrix} \tag{7.3.31}$$

consists of the outer product of the mode shapes at locations X_h with itself. Comparing these expressions with Eq. (7.3.26), the aeroelastic stiffness and damping matrices are

$$\mathbf{K}_a = -\bar{q}S_{ref}\sum_h \boldsymbol{\Upsilon}'_h\boldsymbol{\eta} \tag{7.3.32}$$

$$\mathbf{D}_a = -\frac{\bar{q}S_{ref}}{V}\sum_h \boldsymbol{\Upsilon}_h\dot{\boldsymbol{\eta}}. \tag{7.3.33}$$

The nature of these matrices is important for the analysis of aeroelastic stability. Note that both are linear in the dynamic pressure, and the damping matrix is linear in the inverse of the velocity. When summed with the modal stiffness and damping in the flexibility equations (7.3.26), the resultant matrix $\mathbf{A}_f$ must be positive definite for stability of the uncoupled flexibility equations.

For axisymmetric slender vehicles, a general trend in this type of model is that $\mathbf{K}_a$ tends to decrease the bending frequencies, while $\mathbf{D}_a$ tends to slightly increase the bending damping. In the limiting case as $\mathbf{K}_a$ becomes large, one possibility is that a bending frequency drops to zero, and the associated complex roots become a real pair. Similar to analysis of the bending frequency shift presented Chapter 6, this represents buckling and vehicle structural failure. This phenomenon of *static aeroelastic divergence* is unlikely, but can be exacerbated in rocket designs that are slender, flexible, and have forward-located aerosurfaces such as canards.

As mentioned in the opening discussion of distributed aerodynamics, all this is based on quasi-static aerodynamics. The results are missing aerodynamic phenomena that can only be captured with a more sophisticated model. If the analyst does not have access to an unsteady CFD analysis that can be used for spot checks, one defensible approach is to accept the "bad news" coming from the matrix $\mathbf{K}_a$ but to reject the "good news" coming from $\mathbf{D}_a$.

Calculating the wind input using distributed coefficients in the input matrices (7.2.15) and (7.2.18) is straightforward and results in replacing the third column in both with the appropriate expressions from above. Using the same approach as used to develop (7.3.14) and (7.3.25), these input matrices become

$$\mathbf{B}_r = \begin{bmatrix} 0 & 0 & 0 \\ X_G F_R & (X_G S_E - I_{EG}) & -\frac{\bar{q} S_{ref}}{V} \mathbf{1}_a^T \mathbf{c}_{N\alpha} \\ -F_R & S_{Ex} & \frac{\bar{q} S_{ref}}{V} C_{N\alpha} \end{bmatrix} \tag{7.3.34}$$

$$\mathbf{B}_f = \left[\begin{array}{c|c|c} \multicolumn{3}{c}{\mathbf{O}_{M\times 3}} \\ -F_R \psi_{z\beta 1} & \psi_{z\beta 1} S_{Ex} - \sigma_{y\beta 1} I_{EG} & \\ \vdots & \vdots & -\frac{\bar{q} S_{ref}}{V} \boldsymbol{\Psi}_{za} \mathbf{c}_{N\alpha} \\ -F_R \psi_{z\beta M} & \psi_{z\beta M} S_{Ex} - \sigma_{y\beta M} I_{EG} & \end{array} \right] \tag{7.3.35}$$

As long as the coefficients $C_{N\alpha h}$ have been correctly computed, these should give the same results as (7.2.15) and (7.2.18). This completes the development of the distributed aerodynamics model.

CHAPTER 8

Simulation parameters

There are two main categories of parameter variation. The first category can be called expendables variation. This includes all the changes to the rocket that take place as the expendables (mainly propellant) are used up. The most important of these are changes to the mass properties and structural dynamic properties. One might be tempted to use the term "time variation" rather than "expendables variation", since for any given rocket flight the idea is to compute how the various properties change as the time of flight progresses. The latter term is used because it is more accurate to view the slosh parameters, the thrust, and the FEM parameters as functions of propellant fraction, rather than functions of time, where propellant fraction is the ratio of propellant mass to initial propellant mass. This is also known as burn fraction.

The second category is parametric uncertainty, commonly called dispersions. Virtually all large rocket development programs go through a stage of Monte Carlo analysis in which random dispersions are introduced to the values of various parameters and their effect on the flight dynamics is assessed. Since the costs of building and testing large rockets are so high, it is common to perform almost all of the design verification by simulation. The subsystem models, such as the FEM, TVC, and slosh models, are anchored to test data. A detailed statistical assessment of the design, based on Monte Carlo analysis, is then used to demonstrate that the rocket is safe to fly, and that the likelihood of a failure is acceptably low.

To take an example, one of the most important properties of a rocket is the mass. The initial mass of a rocket can be considered an initial condition, and the mass flow rate can be considered a parameter. A Monte Carlo simulation disperses both the initial conditions and the parameters. When properly conducted, such a simulation should be able to reveal how realistic variations of the mass might combine with other variations to produce unfavorable results. An excellent discourse on the subject of Monte-Carlo analysis has been provided by Hanson and Beard [33]; a brief summary is presented in Chapter 10.

A full nonlinear simulation will require several external analyses to create the necessary parameters. Three examples are the FEM, the aerodynamic coefficients, and the slosh model. These external analyses are used to pre-

Dynamics and Simulation of Flexible Rockets
https://doi.org/10.1016/B978-0-12-819994-7.00013-3

calculate values that can be entered into numerical look-up tables. The term "granularity" is used to describe the range of validity of these tables. Thus fine-grained FEM tables will have FEMs for many values of the propellant fraction. There should be some coordination between the granularity of these tables and the anticipated range of Monte Carlo dispersions. For instance, if a wide range of dispersions is used, it may not be as necessary to have fine-grained tables.

8.1 Thrust dispersions

Typical variations for the thrust of liquid rocket engines are 1 or 2 percent. In contrast, solid rocket engines may exhibit relatively large variations of their thrust profiles. A solid engine is said to "burn hot" if the thrust is higher than nominal, and "burn cold" if the thrust is less than nominal. These changes are mostly caused by variations in the initial temperature of the propellant. They do not normally cause a significant change in the total impulse. That is, a hot engine achieves its higher thrust from an increase in the mass flow rate, so that the propellant is used up more rapidly and the thrust does not last as long. Fig. 8.1 shows three notional thrust profiles for a solid engine with a nominal burn time of 60 seconds. For the hot profile, thrust is 110 percent of nominal but the burn time is only 90 percent of nominal. Each of the three profiles has the same area under the curve (same total impulse).

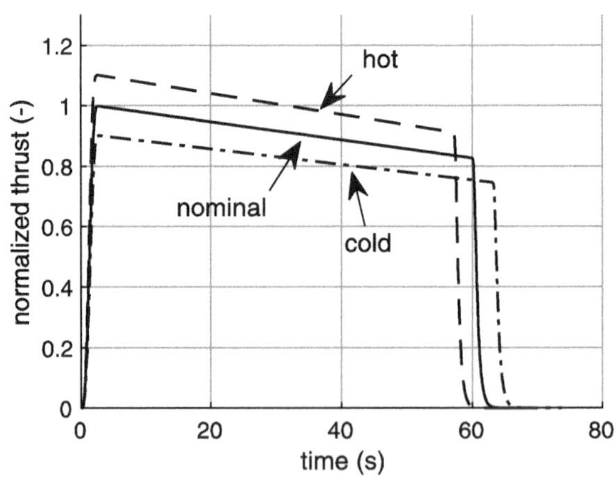

Figure 8.1 Typical thrust profile variations for a solid rocket engine.

8.2 Finite element parameters

Recall from Chapter 2 that the main outputs of a FEM are the mode frequencies plus the eigenvector matrix

$$\mathbf{\Phi} = \begin{bmatrix} | & \cdots & | \\ \boldsymbol{\phi}_1 & \cdots & \boldsymbol{\phi}_M \\ \downarrow & \cdots & \downarrow \end{bmatrix} \tag{8.2.1}$$

If the rocket is axisymmetric, the structural dynamics package that produces $\mathbf{\Phi}$ has no way to distinguish the y axis from the z axis. The bending modes will come in pairs, with one pair at the lowest bending mode frequency, one pair at the next frequency, etc. There may be torsional modes and axial modes interspersed with these bending modes. In all likelihood, the mode shapes (eigenvectors) from the lowest frequency mode pair will not align with the y and z axes. Thus, instead of having a "pitch bending mode" and a "yaw bending mode", we end up with a pair of modes at arbitrary angles relative to the y and z axes. Fig. 8.2 shows typical centerline node deflections of such a pair, both of which have shape components in both the y and z axes.

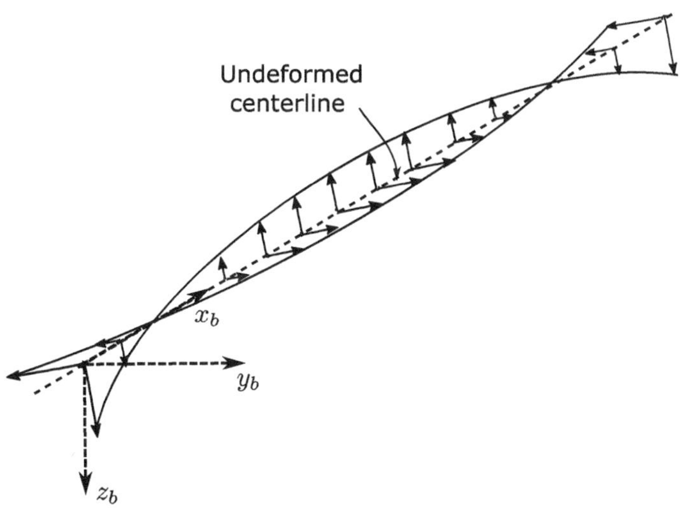

Figure 8.2 First bending mode shapes ψ_1, ψ_2 of a set of centerline nodes.

It is expected that the x components of these deflection vectors will be quite small. Occasionally, all of the deflections come out in one plane, but this is not essential. It is typical that the one member of a mode pair will lie in a plane at right angles to the plane of the other, as shown in Fig. 8.2.

If there is a need to use these modes in a frequency domain simulation representing motion in a single plane, it may be desirable to "clock" the mode pairs around the x axis, as shown in Fig. 8.3. The idea is that a mode pair can be replaced by an equivalent pair in which one mode is in a plane with the y axis (the yaw mode) and one mode is in a plane with the z axis (the pitch mode). In that way, one can perform a design and analysis of decoupled pitch or yaw motions.

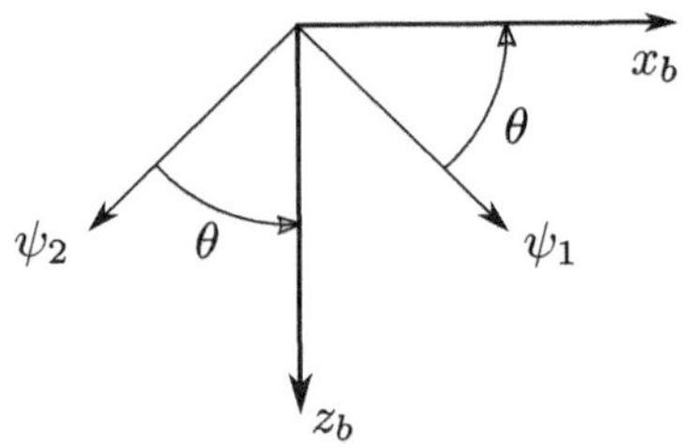

Figure 8.3 Clocking a mode pair onto the *y* and *z* axes.

Let us say that the first two modes come out as a pair, and the desire is to clock these as just described. In order to perform this operation, it is necessary to pick out one particular node from the first two eigenvectors. One choice is to look for the node with the largest amplitude and choose that. A better choice is the node at the gimbal point, since this point assumes such a prominent role in the excitation of bending. Using overbars to represent the clocked mode shapes, we can write for the gimbal node

$$\begin{bmatrix} \bar{\psi}_{y\beta 1} \\ \bar{\psi}_{z\beta 1} \end{bmatrix} = \begin{bmatrix} \cos\theta & \sin\theta \\ -\sin\theta & \cos\theta \end{bmatrix} \begin{bmatrix} \psi_{y\beta 1} \\ \psi_{z\beta 1} \end{bmatrix} \tag{8.2.2}$$

If we wish to align this to the y axis, we must choose θ to make $\bar{\psi}_{z\beta 1} = 0$

$$-\psi_{y\beta 1}\sin\theta + \psi_{z\beta 1}\cos\theta = 0$$

Thus

$$\theta = \tan^{-1}\left(\frac{\psi_{z\beta 1}}{\psi_{y\beta 1}}\right)$$

and

$$\bar{\psi}_{y\beta 1} = \psi_{y\beta 1}\cos\theta + \psi_{z\beta 1}\sin\theta$$

The vector for the clocked first mode shape at the gimbal point becomes

$$\bar{\boldsymbol{\psi}}_{\beta 1} = \begin{bmatrix} 0 & \bar{\psi}_{y\beta 1} & 0 \end{bmatrix}^T$$

If the original mode shape vectors all lie in a plane, the clocked vectors will all end up on the y axis. The same process is applied to the mode shape of the second mode, this time setting the y component to zero, giving

$$\bar{\boldsymbol{\psi}}_{\beta 2} = \begin{bmatrix} 0 & 0 & \bar{\psi}_{z\beta 2} \end{bmatrix}^T$$

Under normal circumstances, it should be possible to carry out both processes using the same angle θ. If different angles are required, an average angle can be taken. A significant difference between the required clocking angles indicates that the notion of clocking may not be appropriate for this mode pair. This frequently occurs if the structure is not axisymmetric, and there is a significant "twisting" component contained within the bending modes.

It is not necessary to employ clocking if the objective is a time domain simulation, or if the linearization is not reduced to a single plane. The dynamic equations in the previous chapters are formulated to accept eigenvectors in the coupled form that emerges from a finite element analysis.

As the rocket depletes its fuel, the eigenvalues and eigenvectors will change. A new FEM must be supplied for periodic intervals of flight time, typically every 10 seconds, or about every five percent variation in mass. It may appear to be advantageous to improve the accuracy of a simulation by interpolating between FEMs. Suppose, for example, that FEMs are available for flight times of 20 seconds and 30 seconds. One might think it is reasonable that the FEM parameters for 25 seconds could be obtained by interpolation, i.e., by averaging the values for 20 and 30 seconds. This concept runs into multiple problems, one of which is the clocking issue just described. There is no reason to assume that the orientation of a mode at 20 seconds will match that for 30 seconds. Thus, although it has just been stated that clocking is unnecessary for a time domain simulation, if interpolation is used it does become necessary.

In addition, any eigenvector can be negated and it is still a valid eigenvector. An eigenvector in the 30 second FEM may have a negative sign relative to the eigenvector in the 20 second FEM. This may not be too serious, since it is certainly possible to define a clocking scheme that also solves the sign problem.

The biggest challenge is mode identification. Suppose, for example, that the modes of two FEMs are arranged by frequency, and the results come out as in Table 8.1. In the 20 second FEM, the first torsional mode is the fifth mode, whereas in the 30 second FEM it is the third mode. For interpolation to work at all, it is necessary to make sure these modes are properly identified and matched up. Defining an automated process to do this creates another layer of complication. Doing this manually can be immensely tedious, especially for a complex rocket requiring a large number of modes.

Table 8.1 Hypothesized order of finite element modes.

Mode No.	20 second FEM	30 second FEM
1	1st bending	1st bending
2	1st bending	1st bending
3	2nd bending	1st torsion
4	2nd bending	2nd bending
5	1st torsion	2nd bending

Although interpolation can be used successfully for low-frequency bending modes if the foregoing issues are overcome, there is yet another problem with this scheme. Let us define "mode integrity" as the property of satisfying (2.2.29) and (2.2.30), plus having modes that are orthogonal. These conditions are repeated here. The modes $\boldsymbol{\psi}_i$ are chosen such that they conserve linear momentum and angular momentum about the center of mass;

$$\int \boldsymbol{\psi}_i(\mathbf{r})\,\mathrm{d}m = 0 \quad \forall i \tag{8.2.3}$$

$$\int \mathbf{r}^{\times}\boldsymbol{\psi}_i(\mathbf{r})\,\mathrm{d}m = 0 \quad \forall i. \tag{8.2.4}$$

Secondly, the modes are orthogonal:

$$\int \boldsymbol{\psi}_i^T \boldsymbol{\psi}_k \,\mathrm{d}m = 0,\ i \neq k \tag{8.2.5}$$

$$\int \boldsymbol{\psi}_i^T \boldsymbol{\psi}_k \,\mathrm{d}m = m_{Bi},\ i = k \tag{8.2.6}$$

Even if one starts with two perfectly valid sets of eigenvalues and eigenvectors, there is no guarantee that the result of interpolation will have mode integrity. If a FEM deviates too far from having mode integrity, numerical difficulties or nonphysical results will turn up in the solution of the dynamic equations.

As if the concept of interpolation didn't have enough problems, there is the issue that it adds significantly to the computational burden, particularly if new modal parameters are interpolated at each time step. The bottom line of all these considerations is that interpolating between FEMs is something to be avoided. It is far preferable to improve the accuracy of the solution by increasing the number of FEMs (and decreasing the amount mass variation between them). Selecting a set of FEMs should be done in conjunction with selecting the range of the Monte-Carlo dispersions. It may be acceptable to use fewer FEMs along with larger uncertainties on the FEM data in the Monte Carlo analysis.

Dispersions of the modal parameters

Eq. (2.2.6) provides the matrix equivalent of (8.2.5) and (8.2.6);

$$\boldsymbol{\Phi}^T \mathbf{M}_B \boldsymbol{\Phi} = \mathbf{m}_B \tag{8.2.7}$$

where

$$\mathbf{m}_B \equiv \mathrm{diag}\left(m_{B1} \cdots m_{BM}\right). \tag{8.2.8}$$

We also have, from (2.2.7)

$$\boldsymbol{\Phi}^T \mathbf{K}_B \boldsymbol{\Phi} = \mathbf{m}_B \boldsymbol{\Omega}_B^2 \tag{8.2.9}$$

Creating useful dispersions of the FEM parameters is not as straightforward as simply varying the values of the eigenvectors $\boldsymbol{\phi}_i$ or their components $\boldsymbol{\psi}_i$ and $\boldsymbol{\sigma}_i$. To illustrate why this is so, let us say that we start with a nominal eigenvector matrix $\boldsymbol{\Phi}$ that satisfies mass normalization, such that $\mathbf{m}_B = \mathbf{1}$, and the desire is to create one member of a set of dispersed $\boldsymbol{\Phi}$ matrices. We now use overbars to denote this member and all its associated variables. (These overbars are not to be confused with the overbars used in the above clocking analysis.) Just to create a simple example, suppose $\bar{\boldsymbol{\Phi}} = 2\boldsymbol{\Phi}$. That is, every eigenvector is multiplied by two. Mass normalization becomes invalidated, but this is not really a problem, since one can let $\bar{\mathbf{m}}_B$ become a diagonal matrix of 4's. The non-normalized version of the ith bending equation (2.2.24) is

$$\bar{m}_{Bi}\left(\ddot{\bar{\eta}}_i + 2\zeta_{Bi}\Omega_{Bi}\dot{\bar{\eta}}_i + \Omega_{Bi}^2\bar{\eta}_i\right) = \sum_n \left(\bar{\boldsymbol{\psi}}_{ni}^T \mathbf{f}_n + \bar{\boldsymbol{\sigma}}_{ni}^T \mathbf{g}_n\right) \tag{8.2.10}$$

Relative to the nominal situation, i.e., this equation without the overbars, all the parameters on the LHS are multiplied by 4 and all those on the RHS are multiplied by two, with the result that the modal amplitudes become half of what they are in the nominal situation;

$$\bar{\eta}_i = \eta_i/2. \tag{8.2.11}$$

Thus, for all i,

$$\bar{\boldsymbol{\psi}}_i\bar{\eta}_i = \boldsymbol{\psi}_i\eta_i \tag{8.2.12}$$

and there is no net change in the response. This should come as no surprise. Multiplying the eigenvectors by any scalar amounts to renormalizing, and nothing really changes.

The process of defining FEM dispersions should begin with the FEM mass and stiffness matrices $\mathbf{M}_B$ and $\mathbf{K}_B$. If we divide both $\mathbf{M}_B$ and $\mathbf{K}_B$ by two, we can leave the eigenvectors unchanged. The net result is

$$\bar{\mathbf{m}}_B = \text{diag}\left(\frac{1}{2}\cdots\frac{1}{2}\right) \tag{8.2.13}$$

$$\bar{\Omega}^2_{Bi} = \Omega^2_{Bi} \tag{8.2.14}$$

$$\bar{\eta}_i = 2\eta_i \tag{8.2.15}$$

$$\bar{\boldsymbol{\psi}}_i\bar{\eta}_i = 2\boldsymbol{\psi}_i\eta_i \tag{8.2.16}$$

Eq. (8.2.16) tells us that for a given excitation, each response is multiplied by two. There is no change in the frequency of any mode. The important thing to note about this case is that in (8.2.10) the only parameter that changes is m_{Bi}. Thus varying m_{Bi} would be a good basis for testing the robustness of the control system. The situation is slightly more complicated if we divide $\mathbf{M}_B$ by 2 and leave $\mathbf{K}_B$ unchanged. We can still use the same eigenvectors, and we still have $\bar{\mathbf{m}}_B = \text{diag}(0.5\cdots 0.5)$, but the frequencies will change:

$$\bar{\Omega}^2_{Bi} = 2\Omega^2_{Bi} \tag{8.2.17}$$

If the excitation on the RHS of (8.2.10) is a step function, then the initial $\ddot{\eta}$ will be twice what it is in the nominal case, just as it is for the previous case in which $\mathbf{M}_B$ and $\mathbf{K}_B$ are both divided by two. The response at subsequent times will reflect the change in natural frequencies.

If we divide $\mathbf{K}_B$ by two and leave $\mathbf{M}_B$ unchanged, then the only parameter that changes in (8.2.10) is the natural frequency Ω_{Bi}. The initial $\ddot{\eta}$ from a step function is the same as in the nominal case.

The above discussion suggests a simple strategy for dispersions. One can impose one set of dispersions on the frequencies (e.g., vary all the frequencies by ±10 percent), and another set on the m_{Bi}'s. The Monte Carlo process will generate combinations of both sets of dispersions, so this becomes the equivalent of two sets of scalar multiplications of the $\mathbf{M}_B$ and $\mathbf{K}_B$ matrices. As long as the discussion is confined to such scalar multiplication, there is little added value to applying dispersions to the eigenvectors.

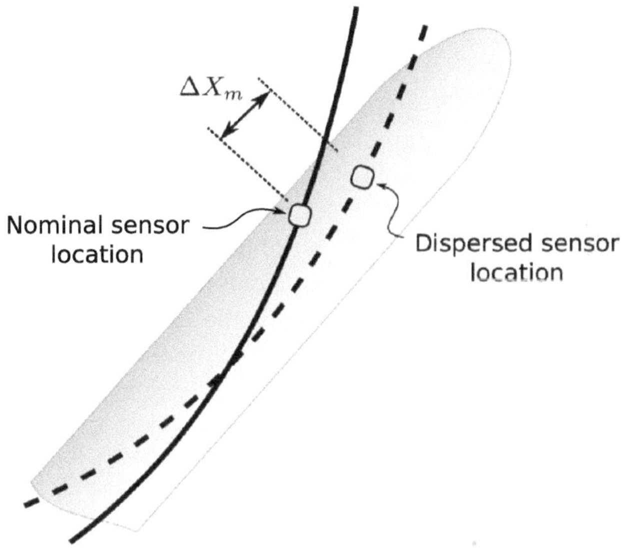

Figure 8.4 First bending mode shape with dispersed sensor location.

One limitation of scalar multiplication is that the mode shapes do not really change their *shape*. It would certainly be possible to vary individual elements of the $\mathbf{M}_B$ and $\mathbf{K}_B$ matrices (perhaps by multiplying by some matrix) to produce such shape alterations. However, a much simpler alternative is available. It turns out that applying dispersions to the locations of the sensors has essentially the same effect as making changes to the shape of a mode. The concept is shown in Fig. 8.4. Let us say that the solid curve represents the nominal mode shape that is used in the analysis, and the dashed curve is the "what if" curve; i.e., the analyst wants to know what would happen if the dashed curve is what actually occurs in the structure. The sensor measures angular rate, so it is responsive to the slope of the mode at the nominal sensor location. The nominal curve predicts an angular rate that is greater than the "what if" curve, since the slope of the nominal curve is greater than that of the "what if" curve at this location. Rather than generating a whole separate finite element analysis to produce an actual curve, it is easier to simply find a location on the nominal curve where the slope matches that of the "what if" curve, and temporarily pretend that is the actual sensor location. The stability margins predicted while using this dispersed location should be the same as for the modified mode shape. In practice, the analyst simply varies the location of the sensor (e.g., ΔX_m) by some portion of the length of the rocket, typically ± 10 percent.

Mode selection

A mode selection algorithm may be required to "down-select" a set of flexible modes for use in control design and stability analysis tools. To do this, we return to the problem discussed in Chapter 2, in which the only DOFs are those in the FEM itself. Motions such as slosh motion or engine motion, which are treated as DOFs in the full dynamics of the rocket, are treated as inputs to the FEM dynamics in the present section. It turns out to be convenient to analyze this subset of the larger problem by casting the equations in first-order form, and using the mathematical approach and notation that is commonly applied to control system analysis. For a system with M modes we define

$$\mathbf{x} = \begin{bmatrix} \eta_1 & \eta_2 & \cdots & \eta_M & \dot{\eta}_1 & \dot{\eta}_2 & \cdots & \dot{\eta}_M \end{bmatrix}^T \tag{8.2.18}$$

and write

$$\dot{\mathbf{x}} = \mathbf{A}\mathbf{x} + \mathbf{B}\mathbf{u} \tag{8.2.19}$$

$$\mathbf{y} = \mathbf{C}\mathbf{x} \tag{8.2.20}$$

The input vector is given by

$$\mathbf{u} = \begin{bmatrix} \mathbf{u}_1 \\ \mathbf{u}_2 \\ \vdots \\ \mathbf{u}_J \end{bmatrix}, \quad \mathbf{u}_j = \begin{bmatrix} f_{xj} & f_{yj} & f_{zj} & g_{xj} & g_{yj} & g_{zj} \end{bmatrix}^T \tag{8.2.21}$$

and $\mathbf{u}_j$ contains the components of the input force and torque at some "input" node. These nodes would typically be gimbal locations or the point of application of a slosh force. Each input node is associated with two indices – the j index from the above simple sequence and a p index from a larger indexing scheme that must provide unique numbers for all relevant nodes. The p indices might be the same as the node numbers (gridpoint numbers) from the original FEM. We will use the term "FEM index" for p, even though it may turn out to be convenient to define some intermediate and simpler indexing scheme. The mapping between indices is defined using a subscript notation. That is, p_1 is the FEM index of the first input node, and so on.

The output vector is

$$\mathbf{y} = \begin{bmatrix} \mathbf{y}_1 \\ \mathbf{y}_2 \\ \vdots \\ \mathbf{y}_K \end{bmatrix}, \quad \mathbf{y}_k = \begin{bmatrix} \dot{\delta}_{xk} & \dot{\delta}_{yk} & \dot{\delta}_{zk} & | & \dot{\varepsilon}_{xk} & \dot{\varepsilon}_{yk} & \dot{\varepsilon}_{zk} \end{bmatrix}^T \qquad (8.2.22)$$

The output nodes can be chosen as the sensor locations. The outputs $\dot{\delta}, \dot{\varepsilon}$ are the perturbations due to elasticity in terms of translational and angular rates, respectively. The notation for the angle components is chosen to reflect the assumption that the angles themselves (the integrals of the rates) are sufficiently small that they can be taken in any order, i.e., there is no need for concern about angular kinematics. As with the inputs, each output node is associated with two indices, the output index k and a FEM index q. To summarize,

$$\begin{aligned} i &= \text{mode index} \\ j &= \text{input gridpoint index} \\ k &= \text{output gridpoint index} \\ p_j &= \text{FEM index of the } j^{th} \text{ input gridpoint} \\ q_k &= \text{FEM index of the } k^{th} \text{ output gridpoint} \end{aligned}$$

Presuming that the flex dynamics have been diagonalized as described in Chapter 2, the homogeneous dynamics are given by

$$\mathbf{A} = \begin{bmatrix} & [\mathbf{O}_{M\times M}] & & & [\mathbf{1}_{M\times M}] & \\ -\Omega_{B1}^2 & \cdots & 0 & -2\zeta_1\Omega_{B1} & \cdots & 0 \\ \vdots & \ddots & \vdots & \vdots & \ddots & \vdots \\ 0 & \cdots & -\Omega_{BM}^2 & 0 & \cdots & -2\zeta_M\Omega_{BM} \end{bmatrix} \qquad (8.2.23)$$

The input and output matrices **B** and **C** consist of the eigenvectors associated with each input and output node p, q, respectively; that is,

$$\mathbf{B} = \begin{bmatrix} \mathbf{B}_1 & \mathbf{B}_2 & \cdots & \mathbf{B}_J \end{bmatrix} \tag{8.2.24}$$

where

$$\begin{aligned}\mathbf{B}_j &= \begin{bmatrix} & & [\mathbf{O}_{M\times 6}] & & & \\ \psi_{xp_j1} & \psi_{yp_j1} & \psi_{zp_j1} & \sigma_{xp_j1} & \sigma_{yp_j1} & \sigma_{zp_j1} \\ \psi_{xp_j2} & \psi_{yp_j2} & \psi_{zp_j2} & \sigma_{xp_j2} & \sigma_{yp_j2} & \sigma_{zp_j2} \\ \vdots & \vdots & \vdots & \vdots & \vdots & \vdots \\ \psi_{xp_jM} & \psi_{yp_jM} & \psi_{zp_jM} & \sigma_{xp_jM} & \sigma_{yp_jM} & \sigma_{zp_jM} \end{bmatrix} \\ &= \begin{bmatrix} [\mathbf{O}_{M\times 6}] \\ \mathbf{\Psi}_{p_j}^T \quad \bar{\boldsymbol{\sigma}}_{p_j}^T \end{bmatrix}\end{aligned} \tag{8.2.25}$$

and

$$\mathbf{C} = \begin{bmatrix} \mathbf{C}_1 \\ \mathbf{C}_2 \\ \vdots \\ \mathbf{C}_K \end{bmatrix} \tag{8.2.26}$$

where

$$\begin{aligned}\mathbf{C}_k &= \begin{bmatrix} & \psi_{xq_k1} & \psi_{xq_k2} & \cdots & \psi_{xq_kM} \\ [\mathbf{O}_{3\times M}] & \psi_{yq_k1} & \psi_{yq_k2} & \cdots & \psi_{yq_kM} \\ & \psi_{zq_k1} & \psi_{zq_k2} & \cdots & \psi_{zq_kM} \\ & \sigma_{xq_k1} & \sigma_{xq_k2} & \cdots & \sigma_{xq_kM} \\ [\mathbf{O}_{3\times M}] & \sigma_{yq_k1} & \sigma_{yq_k2} & \cdots & \sigma_{yq_kM} \\ & \sigma_{zq_k1} & \sigma_{zq_k2} & \cdots & \sigma_{zq_kM} \end{bmatrix} \\ &= \begin{bmatrix} [\mathbf{O}_{3\times M}] & \mathbf{\Psi}_{q_k} \\ [\mathbf{O}_{3\times M}] & \bar{\boldsymbol{\sigma}}_{q_k} \end{bmatrix}\end{aligned} \tag{8.2.27}$$

The symbol $\mathbf{\Psi}$ (upper case bold) represents a $3 \times M$ matrix composed of 3×1 vectors $\boldsymbol{\psi}$ (lower case bold), one for each mode. Likewise, the symbol $\bar{\boldsymbol{\sigma}}$ represents a $3 \times M$ matrix composed of one 3×1 vector $\boldsymbol{\sigma}$ for each mode. An overbar is used instead of upper case sigma, in order to avoid confusion with the summation sign. The system transfer function matrix can be found using the Laplace transform, which will be discussed further in Chapter 9.

The transfer function matrix from input to output is found from (8.2.19) and (8.2.20) as

$$\mathbf{H}(s) = \mathbf{C}[s\mathbf{1} - \mathbf{A}]^{-1}\mathbf{B} \tag{8.2.28}$$

therefore, the transfer function matrix for any input-output pair j, k is

$$\mathbf{H}_{jk}(s) = \mathbf{C}_k[s\mathbf{1} - \mathbf{A}]^{-1}\mathbf{B}_j \tag{8.2.29}$$

Define the *modal velocity gain matrix*

$$\mathbf{K}_{jk} = \mathbf{C}_k\mathbf{B}_j = \begin{bmatrix} \boldsymbol{\Psi}_{q_k}\boldsymbol{\Psi}_{p_j}^T & \boldsymbol{\Psi}_{q_k}\bar{\boldsymbol{\sigma}}_{p_j}^T \\ \bar{\boldsymbol{\sigma}}_{q_k}\boldsymbol{\Psi}_{p_j}^T & \bar{\boldsymbol{\sigma}}_{q_k}\bar{\boldsymbol{\sigma}}_{p_j}^T \end{bmatrix} \tag{8.2.30}$$

The selection algorithm consists of finding the modes i that contribute significantly to the modal velocity gain with respect to any relevant input-output pair j, k. Note that if j has the same value as k this does not necessarily mean they represent the same node. It is, of course, possible for a FEM input node index p_j to represent the same node as a FEM output node index q_k, where j and k have the same value. For instance, with a simplified FEM the same node at the aft end of the vehicle might represent both the gimbal point and the location of a rate gyro.

Consider the modal gain from a force input at node j to sensed angular rate at node k. This modal gain is a driving factor in the design of autopilot filters where j represents an input node (such as the rocket gimbal) and k represents a rate gyro. The total gain in the pitch plane, from z force input to y angular rate, is calculated from the appropriate row and column of the product in (8.2.30). The gain is simply

$$k_{jk}^{z\to\dot{\phi}_y} = \sum_{i=1}^{M} \sigma_{yq_ki}\psi_{zp_ji} \tag{8.2.31}$$

Likewise, from z force input to z velocity, the gain is

$$k_{jk}^{z\to\dot{z}} = \sum_{i=1}^{M} \psi_{zq_ki}\psi_{zp_ji} \tag{8.2.32}$$

Thus, for a single plane of motion, the problem of modal identification is reduced to finding the modes i that dominantly contribute to the sums (8.2.31), (8.2.32) over a set of input-output pairs (j, k). More generally, the modal velocity gain matrix has the form

$$\mathbf{K}_{jk} = \sum_{i=1}^{M} \begin{bmatrix} \boldsymbol{\psi}_{q_k i} \boldsymbol{\psi}_{p_j i}^T & \boldsymbol{\psi}_{q_k i} \boldsymbol{\sigma}_{p_j i}^T \\ \boldsymbol{\sigma}_{q_k i} \boldsymbol{\psi}_{p_j i}^T & \boldsymbol{\sigma}_{q_k i} \boldsymbol{\sigma}_{p_j i}^T \end{bmatrix} = \sum_{i=1}^{M} \mathbf{k}_{jki} \tag{8.2.33}$$

or the sum of the outer products of the eigenvectors $\boldsymbol{\psi}_{q_k i}$, $\boldsymbol{\psi}_{p_j i}$, $\boldsymbol{\sigma}_{q_k i}$, $\boldsymbol{\sigma}_{p_j i}$ associated with each mode i. Note that for each mode, the modal gain matrix $\mathbf{k}_{jk\,i}$ contains four 3×3 partitions:

- The $(1, 1)$ block is translation to translation modal gain.
- The $(1, 2)$ block is rotation to translation modal gain.
- The $(2, 1)$ block is translation to rotation modal gain.
- The $(2, 2)$ block is rotation to rotation modal gain.

Special treatment is required for the aerodynamic forces. These are distributed over the body rather than being applied at any particular node. Combining (5.3.20), (5.3.24), and (5.3.30) gives

$$\mathbf{c}_{EFi}^T \dot{\boldsymbol{\omega}}_E + \sum_j m_{sj} \boldsymbol{\psi}_{ji}^T \ddot{\mathbf{a}}_{sj} + m_{B\,i} \ddot{\eta}_i = -m_{Bi} \Omega_{Bi}^2 \eta_i + \bar{q} S_{ref} \left(C_{Y\eta\,i} \alpha_z - C_{N\eta\,i} \alpha_y \right) \tag{8.2.34}$$

where we have left out all the Q_η terms in (5.3.29) except the aerodynamic generalized forces. We can further assume that the model is mass-normalized so $m_{Bi} = 1$. Following the opening paragraph of this section, we delete the first two terms of this expression, since in the present analysis the engine and slosh motions are not DOFs of the state vector but instead their effects arise from force "inputs" at the gimbal and slosh nodes. To further simplify this expression, consider just the contribution from the pitch angle of attack. This leaves

$$\ddot{\eta}_i = -\Omega_{Bi}^2 \eta_i - \bar{q} S_{ref} C_{N\eta\,i} \alpha_y \tag{8.2.35}$$

This is almost in the form required by (8.2.19). The one remaining problem is that (8.2.19) demands an input $\mathbf{u}$ with dimensions of force or torque. To accomplish this, we write (8.2.35) as follows:

$$\ddot{\eta}_i = -\Omega_{Bi}^2 \eta_i + \lambda_{zj\,i} f_z \tag{8.2.36}$$

where

$$f_z = -\bar{q} S_{ref} C_{N\eta\,i} \alpha_y \tag{8.2.37}$$

and

$$\lambda_{zji} \equiv C_{N\eta i}/C_{N\alpha} \tag{8.2.38}$$

This is obtained by multiplying and dividing the last term in (8.2.35) by the lift curve slope $C_{N\alpha}$. A subscript j is added to λ to signify that this creates one additional force input channel, although one that is not associated with any particular gridpoint. Thus the total number of inputs J is increased by one. Note that the entire mode shape along the length of the rocket is involved in the definition (5.3.26) of $C_{N\eta i}$. This is what Frosch and Vallely [4] call the wind bending force coefficient. We can vectorize (8.2.38) by defining

$$\boldsymbol{\lambda}_{ji} = \begin{bmatrix} 0 & \frac{C_{Y\eta i}}{C_{Y\beta}} & \frac{C_{N\eta i}}{C_{N\alpha}} \end{bmatrix}^T = \begin{bmatrix} 0 & \lambda_{yji} & \lambda_{zji} \end{bmatrix}^T \tag{8.2.39}$$

The aerodynamic input matrix becomes

$$\mathbf{B}_j = \begin{bmatrix} & & [\mathbf{O}_{M\times 6}] & & & \\ 0 & \lambda_{yj1} & \lambda_{zj1} & 0 & 0 & 0 \\ 0 & \lambda_{yj2} & \lambda_{zj2} & 0 & 0 & 0 \\ \vdots & \vdots & \vdots & \vdots & \vdots & \vdots \\ 0 & \lambda_{yjM} & \lambda_{zjM} & 0 & 0 & 0 \end{bmatrix} = \begin{bmatrix} [\mathbf{O}_{M\times 6}] & \\ \boldsymbol{\Lambda}_j^T & \mathbf{O}_{M\times 3} \end{bmatrix} \tag{8.2.40}$$

The effect of a wind gust is to impose a distributed *force* along the length of the rocket. The torque acting on each longitudinal segment dx of the rocket (as used in Eq. (5.3.26)) is not normally computed. That is why the last three columns of this matrix are zero. The above formulation is not in conflict with the fact that the integrated effect of the forces on all the segments may produce a net torque about some reference point.

The aero gain matrix is

$$\mathbf{K}_{jk} = \sum_{i=1}^{M} \begin{bmatrix} \boldsymbol{\psi}_{q_k i}\boldsymbol{\lambda}_{ji}^T & \mathbf{O} \\ \boldsymbol{\sigma}_{q_k i}\boldsymbol{\lambda}_{ji}^T & \mathbf{O} \end{bmatrix} = \sum_{i=1}^{M} \mathbf{k}_{jki} \tag{8.2.41}$$

Eqs. (8.2.39) through (8.2.41) only apply to the particular value of j assigned to aerodynamics.

The above derivation produces a simple result from the outputs two relatively complex analyses – structural dynamics and aerodynamics. As is always the case, the result is only as good as the underlying inputs. It is therefore a good idea for the dynamicist to consult with his or her colleagues in these other disciplines to determine whether the assumptions of the present analysis are consistent with their understanding. The reader is also encouraged to review the Distributed Aerodynamics section of the previous chapter.

The modal selection algorithm must identify the modes that contribute significantly to (8.2.33) or (8.2.41) over all input-output pairs. The algorithm is as follows:

1. For each mode i compute the sum of the 2-norm

$$s_i = \sum_{j=1}^{J} \sum_{k=1}^{K} \|\mathbf{k}_{jki}\|_2 \tag{8.2.42}$$

2. Normalize the sum (8.2.42) and apply a frequency weighting α_i if desired, using

$$\tilde{s}_i = \alpha_i \frac{s_i}{\max(s_i)} \tag{8.2.43}$$

3. Determine those values of i for which $\tilde{s}_i > \varepsilon$, where ε is a selection threshold, and

$$0 < \varepsilon < 1 \tag{8.2.44}$$

A typical value for the selection threshold is $\varepsilon = 0.05$. With this value, any mode that contributes less than five percent of the maximum value of s_i is excluded.

An example of the above method, using typical modal data for a large launch vehicle, is shown in Fig. 8.5. The nodes used include the gimbal and the attitude rate sensor. A frequency weighting of $\alpha_i = \omega_i^{-1}$ is applied and data is evaluated in a frequency range of 0-10 Hz. As can be seen, the first few bending modes are heavily influential, as well as several modes at higher frequency. The remaining modes have little effect if only the gimbal and rate sensor grid points are considered. While these modes may be important for other reasons, the primary modes of interest for the control system (involving the gimbal and the sensor) are easily identified.

8.3 Transition between finite element models

At particular times of flight, or more accurately, particular values of the propellant fraction, it becomes necessary to switch from one FEM representing a heavier rocket state to the next FEM representing a lighter rocket state. When these times come, it is necessary to set the modal amplitudes

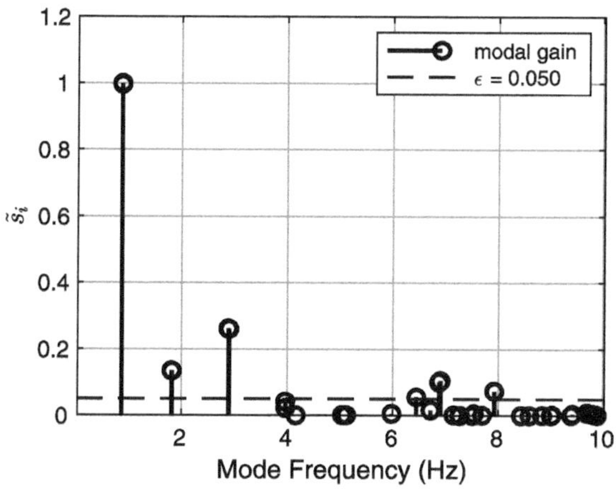

Figure 8.5 Modal gain matrix 2-norms for all modes below 10 Hz.

and velocities η, $\dot{\eta}$ for the new model. One possibility is to reset the states to zero, but this amounts to removing energy from the structure and is underconservative. The simplest idea is to ignore the initialization issue altogether and use the states from the previous model. If this is done, it is found that unrealistic changes may take place in the physical response. It is thus desirable to initialize the η, $\dot{\eta}$ in a way that comes as close as possible to reproducing the final state of the previous model.

Section 2.2 describes how the eigenvector matrix $\mathbf{\Phi}$ can be used as a similarity transformation to diagonalize the structural dynamic equations. The physical degrees of freedom are given by (2.2.14):

$$\mathbf{x}_{\text{full}} = \mathbf{\Phi}\eta_{\text{full}} \tag{8.3.1}$$

This is the same type of vector as (2.2.2), and is not related to the $\mathbf{x}$ of the previous section. The subscript "full" has been added to signify that all the modes, and all the DOFs, are included. In this case, $\mathbf{\Phi}$ is a very large, square matrix. For the following analysis, only a few modes, and only a subset of the DOFs, would be retained. We now define $\mathbf{x}$ as a vector containing only the physical DOFs of interest, and η as a vector containing the amplitudes of the retained modes. Thus

$$\mathbf{x} = \mathbf{\Psi}\eta \tag{8.3.2}$$

where the matrix

$$\boldsymbol{\Psi} = \begin{bmatrix} \phi_{11} & \cdots & \phi_{1M} \\ \vdots & \ddots & \vdots \\ \phi_{R1} & \cdots & \phi_{RM} \end{bmatrix} \tag{8.3.3}$$

has a row dimension R equal to the number of retained physical DOFs and a column dimension M equal to the number of retained modes.

The elements of $\mathbf{x}$ are picked out of $\mathbf{x}_{\text{full}}$ by judicious choice, and will typically represent gridpoints along the centerline. These in turn define which elements ϕ_{ri} of the matrix in (8.3.3) are picked out of $\boldsymbol{\Phi}$. Since R is not required to be the same as M, $\boldsymbol{\Psi}$ will not, in general, be a square matrix. It is necessary that $R \geq M$, and it turns out that an accurate solution is obtained by making R substantially greater than M. The contributions to the eigenvectors at each node can be divided into translational components $\boldsymbol{\psi}_{ni}$ and rotational components $\boldsymbol{\sigma}_{ni}$ as in (2.2.21). The notation $\boldsymbol{\Psi}$ is a reflection of the fact that the contributions from rotations of the finite elements can usually be neglected, so that the translational displacements are the only significant contributors to the energy of the system. Thus the matrix $\boldsymbol{\Psi}$ will normally be made up of displacement vectors $\boldsymbol{\psi}_{ni}$. However, if some rotational DOFs turn out to be important, there is no problem with including these in $\boldsymbol{\Psi}$.

The kinetic and potential energies of this dynamic system are invariant under coordinate transformations. In the full system of equations, the kinetic energy associated with flex is given by

$$T_{\text{flex, full}} = \frac{1}{2}\dot{\mathbf{x}}_{\text{full}}^T \mathbf{M}_B \dot{\mathbf{x}}_{\text{full}} \tag{8.3.4}$$

If the eigenvectors within $\boldsymbol{\Phi}$ are chosen such that the generalized mass matrix is mass-normalized, this becomes, via (2.2.11) and (2.2.14),

$$T_{\text{flex, full}} = \frac{1}{2}\dot{\boldsymbol{\eta}}_{\text{full}}^T \dot{\boldsymbol{\eta}}_{\text{full}} \tag{8.3.5}$$

Once a suitable set of modes has been chosen for the truncated model, we instead compute

$$T_{\text{flex, trunc}} \approx \frac{1}{2}\dot{\boldsymbol{\eta}}^T \dot{\boldsymbol{\eta}} \tag{8.3.6}$$

This can be compared to Eq. (2.3.11). The flex kinetic energy is the integrated body energy T_{IB} minus the energy associated with rigid-body translation and rotation. At this point we drop the subscript "flex" from the expressions for energy, since it can be understood that the present section is not concerned with any of the other contributions to energy that are discussed in Chapter 2.

The potential, or elastic strain energy, is given by

$$V_{\text{full}} = \frac{1}{2}\mathbf{x}_{\text{full}}^T \mathbf{K}_B \mathbf{x}_{\text{full}} \tag{8.3.7}$$

Using a process similar to that for the kinetic energy, the strain energy for the truncated system is given by

$$V_{\text{trunc}} \approx \frac{1}{2}\boldsymbol{\eta}^T \boldsymbol{\Omega}_B^2 \boldsymbol{\eta} \tag{8.3.8}$$

where $\boldsymbol{\Omega}_B^2$ is a diagonal matrix containing the eigenvalues of the selected modes. As with the kinetic energy, the strain energy of the truncated system is less than or equal to that of the full system. That is, $T_{\text{trunc}} \leq T_{\text{full}}$ and $V_{\text{trunc}} \leq V_{\text{full}}$.

Model transition using least squares

All of the following discussion applies only to the truncated models. We thus drop the subscript "trunc." Consider two finite element models at times t_0 and t_1, separated by a small time gap (typically the time step of a simulation). Given any solution $\boldsymbol{\eta}_0$ at time t_0, the physical displacements of the model are given by

$$\mathbf{x}_0 = \boldsymbol{\Psi}_0 \boldsymbol{\eta}_0 \tag{8.3.9}$$

The physical displacements at t_1 will be given by

$$\mathbf{x}_1 = \boldsymbol{\Psi}_1 \boldsymbol{\eta}_1 \tag{8.3.10}$$

Since $\boldsymbol{\Psi}_1$ has R rows and M columns, and we have presumably chosen $R > M$, this matrix equation supplies more individual equations than unknowns. We seek a new solution for $\boldsymbol{\eta}_1$ at time t_1 that closely approximates the physical displacement $\mathbf{x}_0$ such that $\left\| \boldsymbol{\Psi}_1 \boldsymbol{\eta}_1 - \mathbf{x}_0 \right\|_2$, the norm error of the approximation, is minimized. This is a standard estimation problem and a solution is readily found using the Moore-Penrose pseudoinverse given by

$$\boldsymbol{\eta}_1 = \boldsymbol{\Psi}_1^{\dagger} \mathbf{x}_0 \tag{8.3.11}$$

where

$$\boldsymbol{\Psi}_1^{\dagger} \equiv \left(\boldsymbol{\Psi}_1^T \boldsymbol{\Psi}_1\right)^{-1} \boldsymbol{\Psi}_1^T \tag{8.3.12}$$

If the same process is repeated with the velocities, rather than displacements, we find that

$$\dot{\boldsymbol{\eta}}_1 = \boldsymbol{\Psi}_1^{\dagger}\dot{\mathbf{x}}_0 \tag{8.3.13}$$

which uses exactly the same pseudoinverse.[1] Having obtained both $\boldsymbol{\eta}_1$ and $\dot{\boldsymbol{\eta}}_1$, we now have a complete specification of the new state.

Note that it is necessary that both models use the same set of DOFs, so they have the same dimension R, but it is not necessary that they both have the same number of modes M. If it were the case that both models, before and after transition, contained the full set of modes corresponding to all of the modeled mass in the original FEM, the matched displacement and velocity would provide approximately the same strain energy and kinetic energy, respectively, at time t_1. However, when the FEM is significantly truncated, the physical solution is an approximation. Both models will have a certain amount of energy truncation, but there is no guarantee that the same amount energy is truncated from each. Thus, energy continuity is not guaranteed. It is, of course, desirable to choose the number of modes so as to make the difference between the energies of the full system and the truncated system acceptably small, but it may turn out that significant differences remain for any practical number of modes. This is particularly true with complex rocket configurations.

From the standpoint of the t_0 model, the displacements at $t = t_0$ are the natural outcome of the applied forces. In contrast, from the standpoint of the t_1 model, the displacements to be matched are arbitrary. Since the set of basis vectors available to represent the physical space is substantially smaller than in the full model, those bases (mode shapes) must assume unrealistic combinations in order to match an arbitrary physical displacement. While the resultant displacement can be satisfied (in a least squares sense) with very little error, the resultant energy required may not be realistic. It thus turns out that the least squares solution requires more energy than that contained in the model of the previous time step. A general proof of this is not offered here. Suffice it to say that it would be somewhat surprising if the eigenvalues and eigenvectors for $t = t_1$ required *less* energy than that for the t_0 model. Let us focus on the strain energy, although similar arguments can be made for the kinetic energy. The physical displacements to be matched

[1] The matrix $\boldsymbol{\Psi}^T\boldsymbol{\Psi}$ can be very ill-conditioned; direct use of (8.3.12) may yield errors. More numerically robust methods for computation of the pseudoinverse are common: for example, see MATLAB's® "\" and *pinv* functions, which employ QR decomposition.

come from the t_0 model, and one normally expects the largest part of the displacement to be produced by the lowest frequency mode, with diminishing contributions from modes of higher and higher frequencies. For the t_1 model, the first mode is unlikely to do a better job of approximating the overall shape than the first mode of the previous model. Thus, getting the physical displacements to match requires more "correction" from the higher frequency modes. An unrealistic dose of energy is injected into the model at the time of transition, resulting in a "twang" in the physical response. Assuming the model has adequate damping, after a short time the twang will dissipate this energy and bring the solution to a realistic state, but there is a risk that the model will create the appearance of instability where no such instability exists.

Least squares with a quadratic inequality (LSQI) constraint

A least squares problem is said to be *regularized* when the model parameter estimate $\mathbf{y} \in \mathbb{R}^n$ that achieves

$$\min \left\| \mathbf{A}\mathbf{y} - \mathbf{b} \right\|_2 \tag{8.3.14}$$

is subject to lie within a hyperellipsoid in $\mathbb{R}^n$, which can be defined as the norm inequality

$$\left\| \mathbf{B}\left(\mathbf{y} - \mathbf{y}_0\right) \right\|_2 \leq \alpha \tag{8.3.15}$$

The regularized problem is useful in certain estimation applications, for example, where a nonlinear process is linearized at $\mathbf{y}_0$ and a model parameter estimate is refined via iteration. At each step, the new least-squares estimate for the parameters is constrained to lie within a trust region around the current linearization. There exist algorithms [34,35] that readily solve the regularized least squares problem in Eqs. (8.3.14) and (8.3.15), relying on a generalized singular value decomposition (GSVD) which diagonalizes the problem into a set of scalar objectives and constraints. The method requires the determination of a Lagrange multiplier via Newton-Rhapson iteration. Note the equivalence of (8.3.15) to the quadratic inequality constraint

$$\left(\mathbf{y} - \mathbf{y}_0\right)^T \mathbf{Q} \left(\mathbf{y} - \mathbf{y}_0\right) \leq \alpha^2 \tag{8.3.16}$$

where $\mathbf{Q} = \mathbf{B}^T\mathbf{B}$, a symmetric, positive-definite matrix. Given any symmetric, positive-definite (and therefore diagonalizable) $\mathbf{Q}$ we can always find a $\mathbf{B}$ and write the inequality in the form of (8.3.15). The quadratic form in (8.3.16) describes a hyperellipsoid, and can be used to express the potential and kinetic energies (8.3.6) and (8.3.8). In the previous non-regularized analysis, a least squares approximation for both the displacements and the velocities was obtained using the same pseudoinverse. Here, we formulate separate constrained least squares solutions for the displacement and velocity problems. The displacements are found from

$$\min \left\| \boldsymbol{\Psi}_1 \boldsymbol{\eta}_1 - \mathbf{x}_0 \right\|_2 \text{ subject to } \left\| \boldsymbol{\Omega}_{B1} \boldsymbol{\eta}_1 \right\|_2 \le \sqrt{2V_0} \tag{8.3.17}$$

That is, we substitute $\mathbf{A} = \boldsymbol{\Psi}_1$, $\mathbf{b} = \mathbf{x}_0$, $\mathbf{y} = \boldsymbol{\eta}_1$, $\mathbf{y}_0 = \mathbf{0}$, $\mathbf{Q} = \boldsymbol{\Omega}_{B1}^2$, $\mathbf{B} = \sqrt{\boldsymbol{\Omega}_{B1}^2}$ and $\alpha^2 = 2V_0$. This determines the generalized displacements such that the new potential energy $V_1 \le V_0$. For the velocities, we find

$$\min \left\| \boldsymbol{\Psi}_1 \dot{\boldsymbol{\eta}}_1 - \dot{\mathbf{x}}_0 \right\|_2 \text{ subject to } \left\| \dot{\boldsymbol{\eta}}_1 \right\|_2 \le \sqrt{2T_0} \tag{8.3.18}$$

such that the new kinetic energy $T_1 \le T_0$. Note that $V_0 = \frac{1}{2}\boldsymbol{\eta}_0^T \boldsymbol{\Omega}_{B0}^2 \boldsymbol{\eta}_0$ and $T_0 = \frac{1}{2}\dot{\boldsymbol{\eta}}_0^T \dot{\boldsymbol{\eta}}_0$.

There are only three possible cases with LSQI. The least-squares solution is either inside the hyperellipsoid, on the boundary, or outside the boundary. In the first two cases, the LSQI solution is the same as the least squares (LS) solution. In the last case, since the LS solution is outside the hyperellipsoid and this is the smallest norm error solution, it appears that the LSQI solution *must* lie on the boundary of the feasible set since it is not possible to decrease the error by going further inside the constraint ellipsoid. This implies that LSQI and Least Squares Quadratic Equality (LSQE) are identical if there exists no solution such that LS already satisfies the constraint, i.e., already lies inside the boundary. Following the discussion after (8.3.13), the energy of the t_1 solution almost certainly exceeds that from the t_0 solution, so the LS solution lies outside the constraint boundary.

Thus, the best-fit solution for the physical displacements and velocities is the solution with the maximum energy permitted by the constraint. This raises the question of whether an inequality is really the most appropriate constraint – we would arrive at the same answer using an equality constraint, i.e., (8.3.18) could be written

$$\min \left\| \boldsymbol{\Psi}_1 \dot{\boldsymbol{\eta}}_1 - \dot{\mathbf{x}}_0 \right\|_2 \text{ subject to } \left\| \dot{\boldsymbol{\eta}}_1 \right\|_2 = \sqrt{2T_0} \tag{8.3.19}$$

The practical answer to this question is that a set of well-tested computational tools using the inequality is available for this problem, and while some computational inefficiency may result, this is not a serious issue since much of the computation can be done offline. A conceptual illustration of the method is shown in Fig. 8.6 where the hypothetical average kinetic energy $\bar{V}(t)$ is shown with respect to time. While the LS solution will provide an essentially identical match to the required displacements, it does so at the expense of an arbitrarily large jump in the potential energy. The LSQI solution does so while maintaining energy continuity, with a small error in the physical approximation. With a sufficiently large number of modes, this small error can be made arbitrarily small.

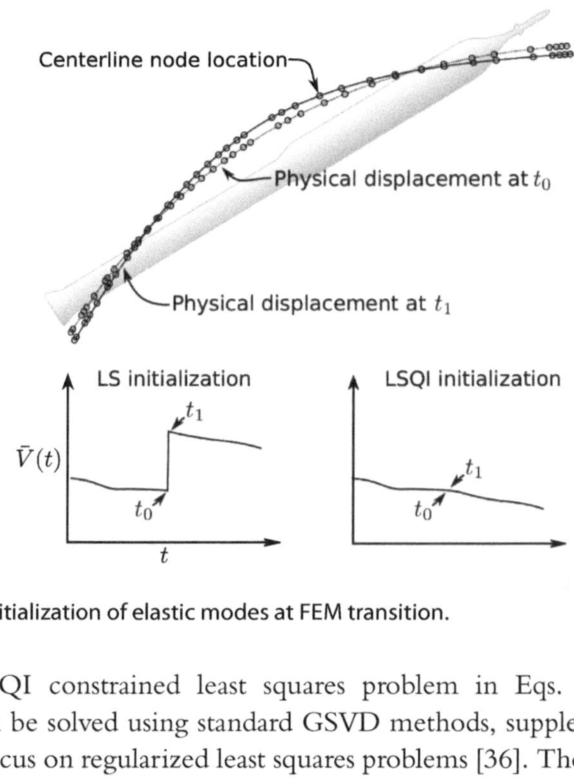

Figure 8.6 Initialization of elastic modes at FEM transition.

The LSQI constrained least squares problem in Eqs. (8.3.17) and (8.3.18) can be solved using standard GSVD methods, supplemented with tools that focus on regularized least squares problems [36]. The numerically intensive GSVD computations can be performed using the modal data before the simulation is run; the result is *not dependent on the current states.* Therefore, the initialization technique is well-suited to simulation implementation since the GSVD matrices associated with each model can be stored in a lookup table.

Direct shape matching (DSM)

LSQI is based on the assumption that energy continuity is the correct objective. In fact, when truncated models are used, and there is a large steady state strain, this may not be the case. Under thrusting conditions, such a steady state strain may result if the selected modes have a large axial component, or there is a center of mass offset. In such situations, methods expecting constant energy will not produce smooth transitions. This is easy to see if one imagines that the only excitation is a static load, but there is still a transition in the model used to describe the structure. Even when the load on the structure remains fixed, the energy computed from (8.3.8) will jump during transitions. This jump is supposed to happen. If this jump is not accommodated, a "twang" results in the response, although at a smaller amplitude than without LSQI. Jurenko et al. [37] separate the response into a steady-state portion that matches the applied forces at the time of transition and a homogeneous portion that results in a good match to the desired physical shape. They call this technique LSQI-DSM. The velocities are still obtained using LSQI.

A modification of LSQI initialization is necessary for discontinuous changes in the vehicle mass or configuration, such as a panel jettison or staging event. This requires knowledge of the fraction of kinetic and potential energy that is retained after the transition so that the remaining structure to be propagated forward is properly initialized. For a large change, such as staging, the alternative is to simply set all the generalized displacements and velocities $\eta, \dot{\eta}$ to zero. However, doing so ignores the release of strain energy that can occur when major load-bearing vehicle elements, such as solid rocket boosters, are suddenly removed from the vehicle load path. Such transients may affect control system and navigation system response during staging events. The conservatism of this approach is application-specific and should be evaluated on a case-by-case basis.

Conclusions to Section 8.3

The concept of interpolating between FEMs has been described in Section 8.2. The LSQI initialization technique described in the present section obviates any need for interpolation. The idea is that small mismatches between the rocket mass at any given time and the mass in the FEM can be tolerated as long as there is a smooth and physically realistic transition from one FEM to the next. This is especially true if the time interval covered by each FEM is short, which becomes all the more feasible since the manual or automated mode shape correlation described in Section 8.2 can be eliminated, thus decreasing the effort required to incorporate a larger number of FEMs.

Some judgment may be required to determine if LSQI is actually required. For small or relatively simple rockets, resetting the states to zero or using a simple pseudoinverse may be adequate if the discontinuity at transitions is not too disruptive.

CHAPTER 9

Stability and control

In this chapter we present an introduction to the design of flight control systems for launch vehicles, with a focus on those techniques that are required for very large boosters. While the present treatment is abbreviated, it provides a basis for the modeling and simulation engineer to become familiar with algorithm principles typically delivered from a GN&C software group for analysis. Since the rocket modeling, simulation, and control design problems are tightly coupled in practice, it also provides a foundation from which a novice control engineer can understand how the models presented in the previous chapters are to be used. Similarly, the control system design concepts can be used by a modeling and simulation analyst to build a representation of a launch vehicle flight control system within a simulation environment, avoiding the need to implement the full suite of GN&C software in order to simply have the simulation produce useful output.

Compared with smaller aerospace vehicles, large rockets require a more integrated approach to control design. Linear frequency-domain, e.g., "classical," methods are standard because they allow fairly arbitrary representations of the vehicle dynamics with fewer theoretical limitations. Restricting the dynamic analysis to accommodate more sophisticated control design techniques risks the later realization that such restrictions would have to be lifted and would invalidate the control design. As a result, classical techniques are the established and accepted practice in the launch vehicle design industry.

Proportional-derivative control, elementary at its core, is the basis of practically all launch vehicle control systems. The task of the inner-loop flight control algorithm, classically referred to as the "autopilot", is to stabilize the angular dynamics of the rocket, point the undeformed centerline according to a commanded profile, and either ignore or suppress the flexible-body dynamics.

The problem of controlling an ascending launch vehicle is dominated by parameter uncertainty and structural flexibility. During the design process, the booster structural mass is systematically reduced and optimized in order to achieve performance goals (e.g., maximize payload to orbit). Necessarily, the compliance of the structure is increased to essentially the limit of the structural loads. In addition, many orbital space boosters are more than 90%

Dynamics and Simulation of Flexible Rockets
https://doi.org/10.1016/B978-0-12-819994-7.00014-5

liquid by mass; the dynamics of sloshing propellant, discussed in Chapters 2 and 3, must be directly considered in the stability and control design. Finally, the use of high-power thrust vector control introduces complicated control challenges such as engine servoelastic interactions (Chapter 6) that may also couple with the control system.

9.1 Problem formulation

The fundamental task of a launch vehicle autopilot is control of the average thrust vector, via pointing of the body attitude, along a prescribed direction in inertial space. Generally, this direction is provided by a guidance algorithm, which may either be a closed-loop steering algorithm or an open-loop attitude profile. In both cases, the desired attitude command of the vehicle can be expressed as a time-varying attitude quaternion $\mathbf{q}_c(t)$ or Euler angles $\Theta(t)$, possibly with an associated attitude rate command $\boldsymbol{\omega}_c(t)$. These attitude and rate commands are issued to the autopilot by some guidance function, and compared with sensed outputs in order to generate an error signal. These error signals are then used to generate torque or acceleration commands, which are distributed to actuators to effect the proper orientation of the vehicle while maintaining the required performance and suppressing undesirable dynamics such as bending and slosh.

It is helpful to the control engineer to know the nature of the commands that he or she might expect from the guidance algorithm. Ascent guidance is a complex topic that is founded in the theory of optimal control. Usually the designer is interested in satisfying a two-point boundary value problem for the three degrees of freedom of the vehicle moving from the launch site into orbit. This 3-DoF problem uses an inertial coordinate frame. Using the nomenclature of Chapters 2 and 5, the simplified equations of motion are given by

$$m\dot{\mathbf{v}}_I = \mathbf{f}_{\text{grav}} + \mathbf{f}_{\text{aero}} + F(t)\mathbf{u}(t) \tag{9.1.1}$$

$$\dot{\mathbf{r}}_I = \mathbf{v}_I \tag{9.1.2}$$

$$\dot{m} = -\frac{F(t)}{g_0 I_{sp}}. \tag{9.1.3}$$

In this problem, the vehicle begins at time $t = t_0$ at the launch site, and ends at some terminal state $\mathbf{r}(t_f)$, $\mathbf{v}(t_f)$ that satisfies the mission objectives. The optimal solution to this problem maximizes the terminal mass $m(t_f)$. The control variables are the thrust direction in inertial space, $\mathbf{u}$, and if throttling is possible, the thrust F. Such an optimization problem is complicated by aerodynamics, accommodation of multiple stages, constraints on aerodynamic and aerothermal loads, and so on. A rich literature of semi-analytical techniques has been developed to solve the general problem of trajectory optimization, both on the ground before flight and in the onboard software. The solution of this general problem is not treated in this book.

Moments on the vehicle can be provided by a variety of actuation systems, including moveable control surfaces, reaction control thrusters, and thrust vector control (TVC). Control surfaces are generally reserved for highly maneuverable tactical missiles, and reaction controls are seldom used within the atmosphere except for roll stabilization. The design of these systems will not be treated extensively herein but the required concepts follow readily from those presented for the most common form of launch vehicle control, a thrust vector control system.

A secondary task of many launch vehicle autopilots is the reduction of bending loads or reduction of lateral path deviations. Designing a control system that rigidly follows a prescribed guidance attitude profile while in the atmosphere may increase structural loading in the presence of uncertain winds. As such, many launch vehicle autopilots include a load relief function, which uses sensed or estimated aerodynamic incidence angles to allow the vehicle to turn into or drift with the wind, depending on the relative weighting of control objectives. Particularly for systems using thrust vector control, a large fraction of the induced bending load may come from the thrust vectoring moments. Autopilot designers must carefully consider limits on control authority and system bandwidth to avoid exceeding maneuvering loads.

Finally, the goal of pointing the thrust vector in a prescribed direction can be realized more effectively if the system is able to compensate for deviations from assumed conditions. Examples include center of mass uncertainty, engine-to-engine mounting misalignments, and aerodynamic disturbance torques. Classical integral control is one form of a disturbance compensation algorithm.

An example of a general guidance, navigation, and control (GN&C) architecture for a large liquid-propelled launch vehicle, the Saturn V, is shown in Fig. 9.1. This architecture consisted of the ST-124M-3 inertial measurement unit (IMU), a separate flight control rate gyro assembly (RGA), the Launch Vehicle Data Adapter (LVDA), the Launch Vehicle Digital Computer (LVDC), and the Flight Control Computer (FCC). During first and second-stage flight, moments on the vehicle were provided by eight TVC actuators that moved the four outboard engines in pitch and yaw. The motions of the vehicle were sensed by a combination of the body-mounted rate gyro and the IMU. The IMU also provided estimates of attitude, position, velocity, and time to an ascent guidance function, which was executed in the LVDC. The ascent guidance function generated the attitude command profile; this profile was used to generate *attitude errors*, expressed as a set of small angles ϕ_e, θ_e, ψ_e for processing by the autopilot.

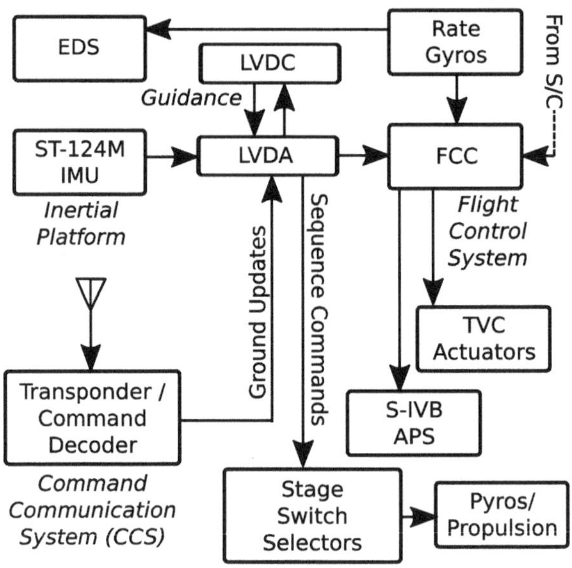

Figure 9.1 Saturn IB/V Guidance, Navigation, and Control Architecture (*adapted from NASA MSFC IV-4-401-1*).

The gravity turn

In practice, the atmospheric phase of the ascent guidance program is almost always a gravity turn. Frequently, the 3-DoF equations of motion (9.1.1) through (9.1.3) are inserted into a numerical optimization scheme, and the trajectory designer finds that, given constraints on loads, the resultant optimal trajectory is very nearly one of zero angle of attack. Consequently, the optimal trajectory through the atmosphere has little to gain from closed-loop guidance, and the gravity turn ascent from liftoff to approximately the end of the first stage burn can be implemented in an open-loop guidance program called a χ *(chi) table*. The χ table is an explicit attitude of the

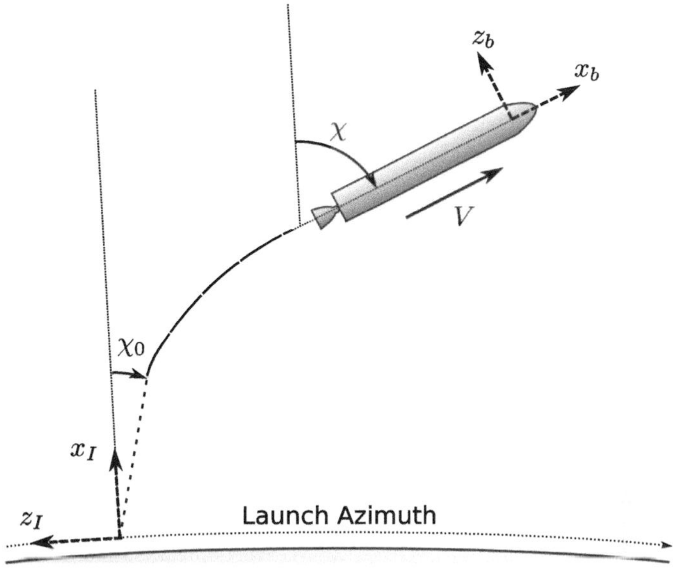

Figure 9.2 Gravity turn geometry in the pitch plane.

vehicle defined as a function of a monotonically increasing independent variable, such as time, altitude, or velocity. The angle χ is usually referenced to the launch site local vertical at the time of liftoff, as shown in Fig. 9.2. The symbol χ thus represents the assigned pitch angle relative to whatever launch site reference direction is used to define $\chi = 0$. It should be noted that the direction defined by χ, as shown in Fig. 9.2, can serve as the x axis of the trajectory reference frame defined in Chapter 7. Thus ϕ_y is the pitch perturbation from this reference.

A simple illustration of a gravity turn is a vehicle in a circular orbit, with the x axis continually aligned with the velocity vector. In this motion, gravity imposes a continuous lateral acceleration, but the body frame rotates in such a way as to keep the y and z components of velocity at zero. This is a characteristic of gravity turns in general – lateral acceleration but no body axis lateral velocity. Thus

$$\mathbf{v} = \begin{bmatrix} V & 0 & 0 \end{bmatrix}^T \tag{9.1.4}$$

The objective during the gravity turn phase of ascent is to steer the rocket to maintain this condition. Because of the aforementioned characteristic, it becomes convenient to solve the gravity turn equations using the body frame, in contrast to the inertial frame used by the 3-DoF equations of motion defined above. To the extent that (9.1.4) is achieved, and in the absence of wind, the only aerodynamic force is drag. Assuming that the only other forces acting on the rocket are thrust and gravity, the inertial acceleration expressed in the body frame is

$$\mathbf{a}_b = \mathbf{C}_{bI}\mathbf{g}_I + \begin{bmatrix} \bar{g} & 0 & 0 \end{bmatrix}^T \tag{9.1.5}$$

where $\mathbf{g}_I$ is the gravity vector in the inertial frame. Recall from Chapter 2 that the acceleration in the body frame is given by $\mathbf{a}_b = \dot{\mathbf{v}} + \boldsymbol{\omega}^{\times}\mathbf{v}$, thus

$$\dot{\mathbf{v}} + \boldsymbol{\omega}^{\times}\mathbf{v} = \mathbf{C}_{bI}\mathbf{g}_I + \begin{bmatrix} \bar{g} & 0 & 0 \end{bmatrix}^T \tag{9.1.6}$$

It should be noted that all of the quantities in this equation are functions of time. Assuming that the launch site frame has the local vertical aligned to the body x axis at time of launch, as in Fig. 9.2, we have

$$\mathbf{g}_I = \begin{bmatrix} -g & 0 & 0 \end{bmatrix}^T \tag{9.1.7}$$

One obvious solution to Eq. (9.1.6) occurs when $\boldsymbol{\omega}(t) = \mathbf{0}$, so that $\mathbf{C}_{bI}$ is constant and equal to the identity matrix. Thus $\mathbf{C}_{bI}\mathbf{g}_I = \begin{bmatrix} -g & 0 & 0 \end{bmatrix}^T$, and $\dot{\mathbf{v}} = \begin{bmatrix} (\bar{g} - g) & 0 & 0 \end{bmatrix}^T$. This is the case if the rocket ascends straight up. A more useful case is a turning trajectory which allows the rocket to gain a downrange velocity necessary for achieving orbit.

Assume that the inertial frame is defined such that the $-z$ axis is aligned to the launch azimuth (Fig. 9.2). Under the foregoing assumptions, and neglecting the rotation of the earth, this launch site frame is approximately an inertial frame. Since the z_I axis is antiparallel to the launch azimuth, the vehicle needs to turn only about the pitch axis.[1] Thus

$$\boldsymbol{\omega} = \begin{bmatrix} 0 & \dot{\chi} & 0 \end{bmatrix}^T \tag{9.1.8}$$

and

$$\mathbf{C}_{bI} = \begin{bmatrix} \cos\chi & 0 & -\sin\chi \\ 0 & 1 & 0 \\ \sin\chi & 0 & \cos\chi \end{bmatrix} \tag{9.1.9}$$

[1] An initial rolling maneuver called the *roll program* is performed to align the body z axis to the launch azimuth before beginning the pitching maneuver.

Inserting (9.1.4), (9.1.7), (9.1.8), and (9.1.9) into (9.1.6) yields one equation from the x component and another from the z component

$$\dot{V} = \bar{g} - g\cos\chi \tag{9.1.10}$$

$$\dot{\chi} = \frac{g\sin\chi}{V} \tag{9.1.11}$$

These are simultaneously integrated numerically. Depending on the initial velocity V_0 and the initial *kick angle* χ_0, a solution for $\chi(t)$ will be generated. In order to calculate $\bar{g}(t)$, it is necessary to compute the altitude (in order to get D), and also the mass m. These are obtained using (9.1.2) and (9.1.3). The solution must be iterated to find the appropriate V_0, χ_0 that satisfy constraints. There may also be iterations with the design of the rocket to adjust the propellant loading or structural mass.

The reader will note that the z_b axis points up in Fig. 9.2, as opposed to pointing down in the figures of the preceding chapters. The convention of having the z axis point down arises from aircraft dynamic analysis. With this convention, the y axis points to the right. Using the right hand rule to define rotations, this results in a positive pitch angle being nose-up. This desirable feature is the reason why virtually all airplane dynamic analyses have the z axis pointing down.

For a crewed rocket, the problem has additional considerations. With the astronauts in a seated position, the relative orientations of their heads and feet help define what is meant by up or down in the body frame. The astronauts have a desire to be able to see the earth's horizon during the ascent to orbit so as to have a visual reference of the vehicle's attitude. Many crewed launch vehicles, therefore, fly "heads down." The body frame is attached to the rocket in the same way as it is for an airplane, with the z axis pointing in the down direction defined by the astronauts' seats. The rocket is therefore flying "upside down", so the z axis points upward as shown in Fig. 9.2.

Consider the example of a gravity turn shown in Fig. 9.3. These data are representative of a single-stage suborbital rocket having a liftoff thrust-to-weight ratio of about 1.4 and a peak acceleration of about 4 g at burnout. Allowing for a brief period of initial vertical rise and azimuth alignment, the initial pitch angle χ_0 is 8 degrees, yielding a burnout χ of about 45 degrees. The angular rates along the trajectory are quite benign, averaging less than 1 degree per second. This type of profile is typical of an ascending booster.

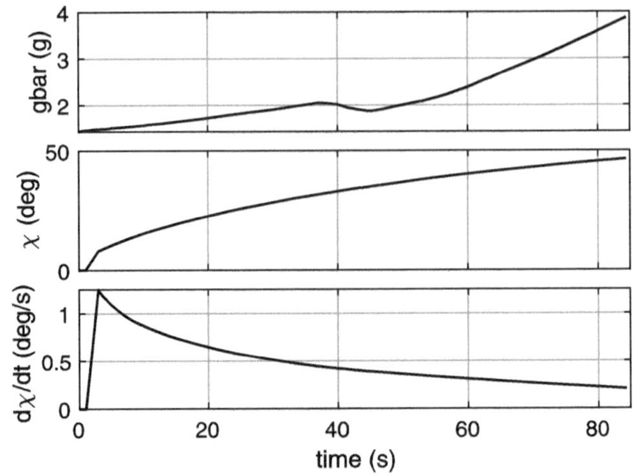

Figure 9.3 Gravity turn for a single-stage rocket.

9.2 Design methods

The approach to the launch vehicle stability analysis and control design problem is tightly integrated with the modeling and simulation of the system, since it is difficult to incrementally test a flight control system on a rocket. The control designer should first choose a modeling framework of adequate fidelity. For initial studies, it is sufficient to model the dynamics of the vehicle as a rigid body in a single plane. Subsequent studies can include pitch/yaw motion, fuel slosh, and flexibility effects using the more detailed linear model presented in Chapter 7. When the stability characteristics of the linear model are shown to be adequate along the entire expected trajectory, the rocket flight is simulated using nonlinear equations of motion, and the performance of the control system is evaluated.

The accuracy and fidelity limitations of linear models must be accounted for in the design process by allowing for model and parameter uncertainty. For linear, classical feedback control, this takes the form of *stability margins*. As will be discussed later in this chapter, stability margins both relate to physical system uncertainties (such as thrust variations) and model uncertainties (such as unmodeled dynamics). In addition to uncertainties in the model fidelities, there are certain parameters that cannot be known practically or accurately prior to flight, for example, temperature at launch or winds aloft. Analysis methods are used to show that the design is satisfactory for both the design reference parameters, as well as some expected range of uncertain parameters, plus margin. The common method is Monte Carlo analysis, discussed briefly in Chapter 10.

It is convenient to assume the vehicle control problem is single-input, single-output (SISO), having the description

$$\dot{\mathbf{x}} = \mathbf{A}\mathbf{x} + \mathbf{b}u \tag{9.2.1}$$

$$y = \mathbf{c}^T\mathbf{x} + du \tag{9.2.2}$$

where $\mathbf{x}$ is the system state vector, y is the sensor output, and u is the control input. This is a useful assumption for many aerospace systems where the controlled degrees of freedom are approximately decoupled, and is a special case of the linear equations presented in Chapter 7.

9.2.1 Classical control

Classical control techniques are applied extensively in industry for rocket design problems. These methods' ability to easily handle general linear systems, especially those that are inherently unstable and with flexible dynamics, turns out to be a major advantage in practical application. The fundamentals of classical feedback control theory are given only a cursory treatment herein. Two comprehensive references on the subject are found in the textbooks by D'Azzo and Houpis [38] as well as Hsu and Meyer [39].

Classical control techniques originated during a time when numerical analysis capabilities were very limited. Prior to the advent of digital computers, determining the eigenvalues - and therefore the stability properties - of a system greater than about order six was an extremely tedious exercise. However, determining a system's frequency response is often more straightforward, which encouraged the development of a rigorous and practical frequency-domain theory. So-called *modern* control emerged in the early 1960s, a time when computers were becoming more available to the control engineer. This development of control theory uses state variable and variational methods to develop optimal control laws, state observers, and so on. These procedures are very powerful when combined with classical, frequency-domain techniques.

Laplace transforms and transfer functions

This section will be most valuable for the engineer with some prior exposure to Laplace transforms, and can serve as a refresher on the fundamentals.

The single-sided Laplace transform

$$F(s) = \mathcal{L}\{f(t)\} = \int_0^\infty e^{-st} f(t)dt \tag{9.2.3}$$

is used to map time-domain signal functions into the complex domain, so that differential equations can be expressed as polynomial expressions of $s = \sigma + j\omega$. The Laplace transform is a convenient operator that simplifies many common operations on constant-coefficient linear differential equations. Since $\mathcal{L}$ is a linear operator, for two arbitrary constants a and b,

$$af(t) + bg(t) \rightarrow aF(s) + bG(s). \tag{9.2.4}$$

Most importantly, the Laplace transform allows the mapping of time derivatives of a function $f(t)$ as follows;

$$\dot{f}(t) \rightarrow sF(s) - f(0), \ddot{f}(t) \rightarrow s^2F(s) - sf(0) - f'(0) \tag{9.2.5}$$

and so on. For many problems in flight mechanics, the derivatives are of order two or less.

Using these basic properties, any linear, constant coefficient system of ordinary differential equations can be transformed into a transfer function or a transfer function matrix. In terms of its output y and its input u, an nth order SISO system can be described by

$$a_k y^{(k)} + \cdots + a_1\dot{y} + a_0 y = b_j u^{(j)} + \cdots + b_1\dot{u} + b_0 u. \tag{9.2.6}$$

Using the Laplace transform property given in Eq. (9.2.5) and assuming $y(0) = \dot{y}(0) = \cdots = 0$,

$$a_k s^k Y(s) + \cdots + a_1 sY(s) + a_0 Y(s) = b_j s^j U(s) + \cdots + b_1 sU(s) + b_0 U(s). \tag{9.2.7}$$

Solving for the transfer function $G(s) = Y(s)/U(s)$, it follows that

$$G(s) = \frac{b_j s^j + \cdots + b_1 s + b_0}{a_k s^k + \cdots + a_1 s + a_0}. \tag{9.2.8}$$

Such a transfer function is called *proper* if $k > j$ and *strictly proper* if $k = j$. Transfer functions are usually normalized so that $a_k = 1$. Note that systems having $j > k$ are non-causal systems and are physically unrealizable.

For linear, constant-coefficient systems of ordinary differential equations, the roots of the characteristic polynomial equation

$$a_k s^k + a_{k-1} s^{k-1} \cdots + a_2 s^2 + a_1 s + a_0 = D(s) = 0 \tag{9.2.9}$$

completely determine the stability properties of the system. Here, D stands for "denominator." These roots or *poles* are equivalent to the eigenvalues of the matrix $\mathbf{A}$ in Eq. (9.2.1) and will be denoted by p_i. Once the poles have been found, the denominator can be factored as

$$D(s) = (s - p_1)(s - p_2) \cdots (s - p_k) \tag{9.2.10}$$

Conversely, the roots z_i of the numerator polynomial equation

$$b_j s^j + b_{j-1} s^{j-1} \cdots + b_2 s^2 + b_1 s + b_0 = N(s) = 0 \tag{9.2.11}$$

are called the *zeros* of the system and do not affect the stability properties, unless the system is placed in a feedback configuration.

Many practical control design problems can be reduced to an assembly of first and second order systems. For example, a first-order system might describe the sensing of velocity or attitude rate with respect to a force or torque input. A second-order system can be used to represent the response of position with respect to force, or elastic displacement with respect to generalized force.

One important second-order system that has appeared in the preceding chapters is the second-order low-pass system

$$G(s) = \frac{k\omega_0^2}{s^2 + 2\zeta\omega_0 s + \omega_0^2} \tag{9.2.12}$$

which occurs as a model of physical systems having a single degree of freedom, a restoring force, and a damping force proportional to a velocity. The parameter ω_0 is the called the *natural frequency* and the parameter ζ is called the damping ratio. For $|\zeta| < 1$, the denominator has complex roots and cannot be decomposed into the product of two real factors. Its roots represent the important case of an underdamped time response for $0 < \zeta < 1$ and an unstable response for $\zeta < 0$.

Second-order (quadratic) systems with $\frac{\sqrt{2}}{2} \leq \zeta < 1$ have desirable properties in both the time and frequency domain, and therefore can be used as model systems for control design. As a model system, a designer develops

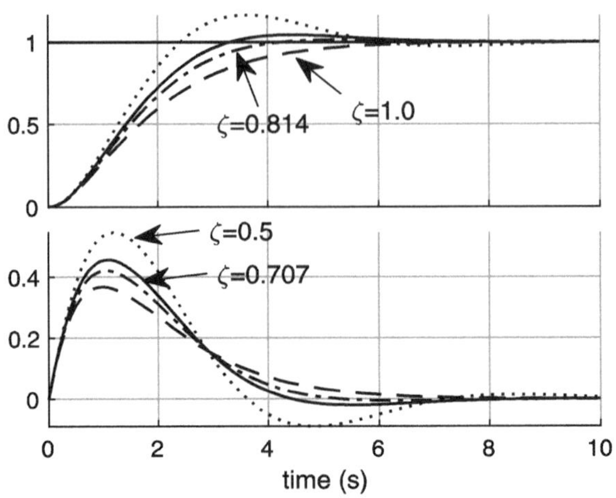

Figure 9.4 Time response of quadratic factor (top: unit input, bottom: unit initial velocity).

a feedback control law such that the closed-loop system approximates the behavior of a simpler, second-order system with a desired natural frequency and damping ratio.

The time responses of a second-order system $\omega_0^2\left(s^2+2\zeta\omega_0 s+\omega_0^2\right)^{-1}$ with $\omega_0=1$ and several values of ζ are shown in Fig. 9.4. The upper portion of the figure depicts the response of the system to a unit input or initial displacement, such as an initial attitude error. The lower portion shows the response to an initial velocity, such as an initial angular rate.

For a second-order system of this form, a value of $\zeta=0.753$ corresponds to the time response which minimizes the integral of the product of time and absolute error (ITAE)[2] in response to a unit command or initial condition [41], and a value of $\zeta=0.814$ minimizes the related cost function

$$J=\int_0^\infty t^a\,|e|\,dt \tag{9.2.13}$$

for $a=2$. As a increases, the response becomes more "deadbeat," which means that overshoot is minimized.

[2] The reader is cautioned that the ITAE system coefficients published in many controls textbooks are reproduced from a 1953 technical report [40] and contain small errors; see [41].

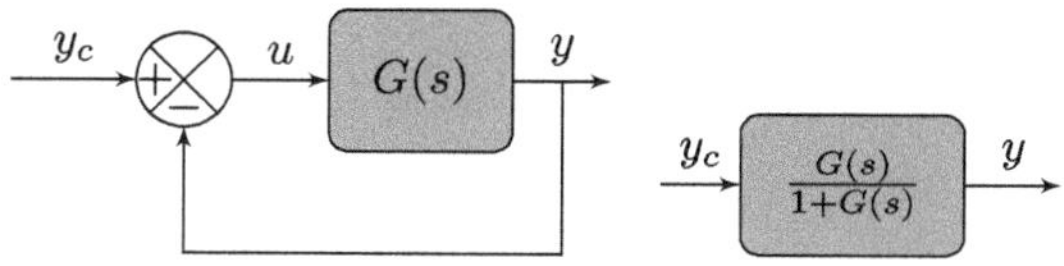

Figure 9.5 Equivalent unity feedback configurations.

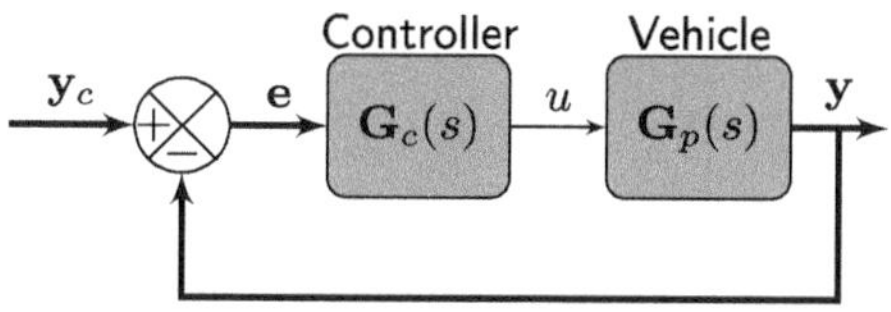

Figure 9.6 Compensator structure for classical control problems.

Feedback configurations

If a transfer function $G(s)$ is placed into a unity feedback configuration (Fig. 9.5), the closed-loop transfer function $G_{CL}(s)$ is given by

$$G_{CL}(s) = \frac{G(s)}{1 + G(s)}. \tag{9.2.14}$$

The open-loop transfer function is therefore $G(s)$. The denominator of the closed-loop transfer function, $B(s) = 1 + G(s)$, is called the *return difference function* and is important in stability and performance. If $G_{CL}(s)$ is written in terms of the polynomials $N(s)$ and $D(s)$ of $G(s)$, it follows that

$$G_{CL}(s) = \frac{N(s)}{D(s) + N(s)}. \tag{9.2.15}$$

It is important to realize that while the zeros of $G(s)$ have no influence on the open-loop stability of $G(s)$, they contribute directly to the poles of the closed-loop transfer function $G_{CL}(s)$.

The examples in this chapter can usually be explained using the *forward loop* or *compensator* configuration for feedback control (Fig. 9.6). In this configuration, vehicle dynamics and control law form a chain that is closed using unity negative feedback. For analysis, it is desired to compute a suitable open-loop transfer function that is SISO.

The controller and vehicle blocks in Fig. 9.6 are depicted using bold symbols and interconnections, indicating that these elements usually have multiple inputs and outputs. The sensed output $\mathbf{y}$ is subtracted from the commanded output $\mathbf{y}_c$ to produce an error signal $\mathbf{e}$, which is the input to a controller or compensator. The output of the controller (in this case, a scalar, u) is the input to the vehicle dynamics, sometimes called the *plant*.

A transfer function matrix $\mathbf{G}(s)$ is simply a mapping of a set of inputs u_i to a set of outputs y_j using a matrix containing scalar transfer functions. For example, for a two-input, one-output system, the transfer function matrix $\mathbf{G}(s)$ can be used to write

$$y = \begin{bmatrix} G_1(s) & G_2(s) \end{bmatrix} \begin{bmatrix} u_1 \\ u_2 \end{bmatrix} \tag{9.2.16}$$

which means that $y = G_1(s)u_1 + G_2(s)u_2$. If $G_1(s)$ and $G_2(s)$ are transfer functions representing the same linear system of ordinary differential equations, G_1 and G_2 may have a common denominator polynomial but different numerator polynomials. In this case, Eq. (9.2.16) can be written as

$$y = \frac{\begin{bmatrix} N_1(s) & N_2(s) \end{bmatrix}}{D(s)} \begin{bmatrix} u_1 \\ u_2 \end{bmatrix}. \tag{9.2.17}$$

Transfer function matrices play only a cursory role in framing the problem. Consider for example a rocket with an input equal to the commanded gimbal angle β_c and outputs from perfect sensors consisting of the measured trajectory-relative angle and body rate, ϕ_y and $\dot{\phi}_y$, respectively. The vehicle can be represented by one-input, two-output transfer function matrix $\mathbf{G}_p(s)$ that takes as its input the gimbal command β_c^-. Likewise, the controller $\mathbf{G}_c(s)$ takes as its inputs the values ϕ_y and $\dot{\phi}_y$, producing a scalar control command $u = \beta_c^+$ which is compatible with the input of $\mathbf{G}_p$. For analysis, there exists only one signal path in the system that is a scalar. If the commanded values are assumed to be zero, Fig. 9.6 can be redrawn as in Fig. 9.7. Note that due to unity negative feedback,

$$\beta_c^- = -\beta_c^+. \tag{9.2.18}$$

Here, the open-loop transfer function $G(s) = \mathbf{G}_c\mathbf{G}_p$ is SISO, while the complexity of the sensor signal paths $\mathbf{y}$ is absorbed into the system. This feedback loop broken at the gimbal output-to-input path is called the β *break* and is standard in rocket autopilot analysis.

The reader will note that $\mathbf{y}_c$ has been omitted from Fig. 9.7. The structure of Fig. 9.7 with $\mathbf{y}_c = \mathbf{0}$ is referred to in the controls literature as the *regulator* problem, as opposed to the *servo* problem. The regulator problem

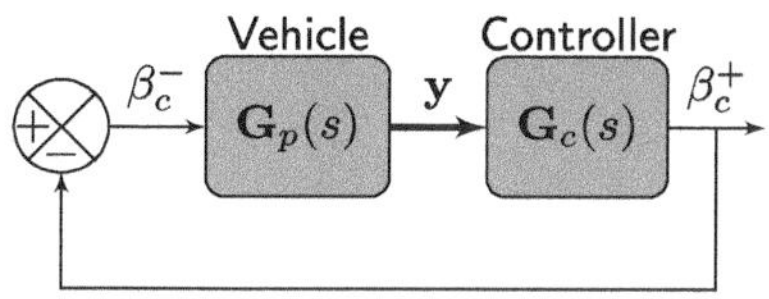

Figure 9.7 Compensator structure for a simple rocket problem.

is concerned with controlling the states of the system so that they return to equilibrium when perturbed, whereas the servo problem attempts to control the states of the system to follow a prescribed trajectory. Since the linear rocket model uses ϕ_Y to represent a perturbation from the gravity turn trajectory, ϕ_Y is, in fact, already the attitude error, and the control design can be cast as a regulator problem without introducing further complications. Because of this structure, traditional servocontrol metrics such as step response are not particularly useful. A step attitude command is equivalent to commanding the vehicle to fly at a nonzero angle of attack with respect to the nominal gravity turn trajectory, which is seldom done in practice. A more useful metric for time-domain transient response are responses to initial conditions, such as in the lower portion of Fig. 9.4.

9.2.2 Plant dynamic equations

The subset of the linear equations that can be used for design, derived in Chapter 7, will be restated here, assuming for simplicity that the rocket center of mass and origin of the coordinate system are coincident, there is no motion in the yaw plane, and a prescribed motion model of the engine is used. The rotation equation (7.2.7) is

$$
\begin{aligned}
I_{YY}\dot{\omega}_Y - \sum_j m_{sj} X_{sj} \ddot{z}_{sj} = &-\sum_j m_{sj} \bar{g}\, z_{sj} + C_{N\alpha}\bar{q} S_{ref} X_{cp}\left(\phi_Y + \frac{\dot{z}_I}{V}\right) \\
&+ X_G\left(F_R \beta_{EY} + F\sum \sigma_{Y\beta\, i}\eta_i\right) + F\sum \psi_{z\beta i}\eta_i \\
&+ \left(X_G S_E - I_{EG}\right)\ddot{\beta}_{EY} - X_{cp}\frac{C_{N\alpha}\bar{q}S_{ref}}{V} V_{wz} \qquad (9.2.19)
\end{aligned}
$$

and the translation equation (7.2.8) is given by

$$
\begin{aligned}
m_T \ddot{z}_I + \sum m_{sj}\ddot{z}_{sj} = &-m_T \bar{g}\phi_Y - C_{N\alpha}\bar{q}S_{ref}\left(\phi_Y + \frac{\dot{z}_I}{V}\right) \\
&- F\sum \sigma_{Y\beta\, i}\eta_i - F_R \beta_{EY} + S_{Ex}\ddot{\beta}_{EY} + \frac{C_{N\alpha}\bar{q}S_{ref}}{V} V_{wz} \qquad (9.2.20)
\end{aligned}
$$

The slosh dynamics (9.2.21) are given by

$$\ddot{z}_{sj} - X_{sj}\ddot{\phi}_y + \ddot{z}_I + \sum_i \psi_{zji}\ddot{\eta}_i = -\Omega_{sj}^2 z_{sj} - 2\zeta_{sj}\Omega_{sj}\dot{z}_{sj} - \bar{g}\phi_y \tag{9.2.21}$$

and the bending dynamics (7.2.10) are

$$\begin{aligned}\ddot{\eta}_i + \sum_j m_{sj}\psi_{zji}\ddot{z}_{sj} = {} & -\bar{q}S_{ref}C_{N\eta i}\left(\phi_y + \frac{\dot{z}_I}{V}\right) \\ & - \Omega_{Bi}^2\eta_i - 2\zeta_{Bi}\Omega_{Bi}\dot{\eta}_i - F_R\psi_{z\beta i}\beta_{Ey} \\ & + \left(\psi_{z\beta i}S_{Ex} - \sigma_{y\beta i}I_{EG}\right)\ddot{\beta}_{Ey} + \frac{\bar{q}S_{ref}C_{N\eta i}}{V}V_{wz}\end{aligned} \tag{9.2.22}$$

In developing the design techniques, these equations will first be simplified to represent only the pitching of the rigid body with aerodynamics. Winds and slosh will be added back to demonstrate their effects on the control system. Since the dynamics with only the rigid body and slosh are relatively simple, they can be analyzed using transfer functions.

In Section 9.3, the dynamics of a TVC actuator are introduced into the rocket dynamic equations. At this juncture, the equations become too complex to handle easily with transfer functions, so the analyses introduced in Section 9.4, which reintroduce the flexible body effects, will use state variable and frequency-domain techniques.

9.2.3 Proportional-derivative control

The stabilization and control of the bare airframe with aerodynamics is the first step to the design of the flight control system. Some initial simplifications are necessary to assess the design problem, and these effects will subsequently be added back. If the wind, flexible dynamics, and slosh are eliminated and the TWD effects are ignored, it follows that

$$I_{yy}\dot{\omega}_y = C_{N\alpha}\bar{q}S_{ref}X_{cp}\left(\phi_y + \frac{\dot{z}_I}{V}\right) + X_G F_R \beta_c^- \tag{9.2.23}$$

where it has been assumed that the actuator is perfect; that is, β_{Ey} is always equal to its commanded value, β_c^-. Finally, let the plunging velocity $\dot{z}_I$ be zero. This is equivalent to stating that the lateral or plunging motion's effect on the rocket's angle of attack is slow compared with the change in the freestream velocity direction with respect to the body. If this the case, the translation equation can be eliminated entirely, and the remaining pitch dynamics equation is

$$I_{yy}\ddot{\phi}_Y = C_{N\alpha}\bar{q}S_{ref}X_{cp}\phi_Y + X_G F_R \beta_c^- . \tag{9.2.24}$$

The substitution $\ddot{\phi}_Y = \dot{\omega}_Y$ has been used. This fundamental form of the pitching dynamics with aerodynamic moments is sufficient for preliminary design. Taking the Laplace transform of (9.2.24), the trajectory-relative attitude error transfer function is

$$G_p(s) = \frac{\phi_Y(s)}{\beta_c^-(s)} = \frac{X_G F_R}{I_{yy}\left(s^2 - C_{N\alpha}\bar{q}S_{ref}X_{cp}\right)} \tag{9.2.25}$$

This represents the uncontrolled airframe, and for $X_{cp} > 0$, the rocket is statically unstable. In this case, the roots of the characteristic polynomial in (9.2.25) are real and are located at

$$\lambda_i = \pm\sqrt{\frac{-C_{Z\alpha}\bar{q}S_{ref}X_{cp}}{I_{YY}}}. \tag{9.2.26}$$

The positive real root of the uncontrolled vehicle represents the divergent aerodynamic instability resulting from a center of pressure forward of the center of mass.

The simplest control law required to stabilize this system, assuming the rocket has no pitch damping, is called *proportional-derivative* or PD control. A PD control law has the form

$$\beta_c^+ = -k_p\phi_Y - k_D\dot{\phi}_Y \tag{9.2.27}$$

where the proportional and derivative gains $k_P, k_D > 0$ are constants.

Note that Eq. (9.2.25) has as its output the attitude error ϕ_Y. While in general we might use a matrix transfer function (or state-space model) to represent the differences in the outputs ϕ_Y and $\dot{\phi}_Y$, in this simplified case they are related by a time derivative and therefore we can just write the control law as

$$G_c(s) = \frac{\beta_c^+(s)}{\phi_Y(s)} = -k_P - k_D s. \tag{9.2.28}$$

The open-loop transfer function $H(s) = G_c(s)G_p(s)$ for the rocket with a PD control law is

$$\frac{\beta_c^+(s)}{\beta_c^-(s)} = G_c(s)G_p(s) = \frac{-X_G F_R\left(k_P + k_D s\right)}{I_{yy}\left(s^2 - C_{N\alpha}\bar{q}S_{ref}X_{cp}\right)}. \tag{9.2.29}$$

Using Eq. (9.2.14), the closed-loop transfer function is

$$G_{CL}^{\beta}(s) = \frac{\beta_c^+(s)}{\beta_c(s)} = \frac{-X_G F_R\left(k_P + k_D s\right)}{I_{YY}\left(s^2 - C_{N\alpha}\bar{q}S_{ref}X_{cp}\right) - X_G F_R\left(k_P + k_D s\right)} \tag{9.2.30}$$

where the superscript β has been used to emphasize that this transfer function has its loop closure at the β signal path, as in Fig. 9.7.

The transfer function for $G_{CL}^{\beta}(s)$ can be converted into the transfer function $G_{CL}^{\phi}(s)$ by noting that

$$G_{CL}^{\phi}(s) = \frac{\phi_Y(s)}{\beta_c^+(s)} = G_{CL}^{\beta}(s)G_p^{-1}(s) = -\frac{\beta_c^+(s)}{\beta_c(s)}\frac{\phi_Y(s)}{\beta_c^-(s)}. \tag{9.2.31}$$

Grouping terms and simplifying, the closed-loop transfer function in terms attitude error is given by

$$G_{CL}^{\phi}(s) = \frac{\frac{X_G F_R}{I_{YY}}}{s^2 - \frac{X_G F_R}{I_{YY}}k_D s - \frac{1}{I_{YY}}\left(X_G F_R k_P + C_{N\alpha}\bar{q}S_{ref}X_{cp}\right)}. \tag{9.2.32}$$

The real parts of the roots of the characteristic polynomial equation

$$s^2 + a_1 s + a_0 = D(s) = 0 \tag{9.2.33}$$

are negative if and only if a_1, a_0 are positive, which is required for stability of the closed-loop transfer function (9.2.32). Since X_G is conventionally negative (gimbal aft of CG) and $k_D > 0$, a_1 is positive. The proportional gain k_P must satisfy

$$k_P \geq \frac{C_{N\alpha}\bar{q}S_{ref}X_{cp}}{-X_G F_R}. \tag{9.2.34}$$

This inequality is fundamental for statically unstable rockets; the critical control gain for stability must equate the unit aerodynamic moment due to pitching ($C_{N\alpha}\bar{q}S_{ref}X_{cp}$) with the unit thrust vectoring moment due to pitching ($-X_G F_R k_P$). Additional insight can be gained by considering the characteristic polynomial in (9.2.32) as a second-order factor. From this, we see that the closed-loop frequency and damping ratio of the pitch dynamics are related by

$$\omega_c^2 = -\frac{1}{I_{YY}}\left(X_G F_R k_P + C_{N\alpha}\bar{q}S_{ref}X_{cp}\right) \tag{9.2.35}$$

$$2\zeta_c\omega_c = -\frac{X_G F_R}{I_{YY}}k_D. \tag{9.2.36}$$

These yield the design equations

$$k_P = \frac{I_{yy}\omega_c^2 + C_{N\alpha}\bar{q}S_{ref}X_{cp}}{-X_G F_R} \tag{9.2.37}$$

and

$$k_D = \frac{I_{yy}2\zeta_c\omega_c}{-X_G F_R} \tag{9.2.38}$$

where the designer chooses a desired ω_c, ζ_c that meets the design criteria. The frequency ω_c is called the *control frequency*, and the control damping ratio ζ_c can be selected based on optimizing a performance index such as (9.2.13). If the time profiles of the quantities in Eqs. (9.2.37) and (9.2.38) are known before flight with reasonable certainty, these provide a useful basis for an initial gain schedule for many large rockets.

The ratio of the quantities in parentheses on the RHS of (9.2.35) is important for handling uncertainty. Since this quantity must be positive for stability, the designer must ensure that $\omega_c > 0$ for all variations of the uncertain parameters, such as the gimbal moment arm, thrust, and dynamic pressure. The *aerodynamic gain margin*, abbreviated GM_a, is approximately

$$\mathrm{GM}_a \approx 20\log_{10}\frac{X_G F_R k_P}{C_{N\alpha}\bar{q}S_{ref}X_{cp}}. \tag{9.2.39}$$

Note that this is a negative quantity, and corresponds to the amount of gain reduction (in dB) that can be tolerated without a loss of stability. This is, however, only a necessary condition, and other factors may make the control design unsatisfactory even if the aerodynamic gain margin is less than zero. Standard practice is to provide for a minimum of 6 dB or a factor of 2 in selecting k_P to ensure robustness to uncertainty.

9.2.4 Integral control

The shortcoming of the proportional-derivative controller is its inability to compensate for external disturbances. Consider again the simplified pitch dynamics, but suppose there is an additional thrust vector misalignment $\Delta\beta_{Ey}$. With the control law (9.2.27) included, the result is

$$I_{yy}\ddot{\phi}_y = C_{N\alpha}\bar{q}S_{ref}X_{cp}\phi_y + X_G F_R\left(k_P\phi_y + k_D\dot{\phi}_y\right) + X_G F_R\Delta\beta_{Ey}. \tag{9.2.40}$$

Taking the Laplace transform of Eq. (9.2.40) and solving for the transfer function of the attitude error with respect to the external disturbance, it follows that

$$\frac{\phi_y(s)}{\Delta\beta_{Ey}(s)} = S_\beta(s) = \frac{X_G F_R}{I_{yy}s^2 - X_G F_R k_D s - \left(X_G F_R k_P + C_{N\alpha}\bar{q}S_{ref}X_{cp}\right)}. \quad (9.2.41)$$

This is known as the β sensitivity function, and relates the frequency response of the attitude error to an external moment caused by a misaligned gimbal $\Delta\beta_{Ey}$. The importance of the sensitivity function can be shown using the *final value theorem* (FVT). The final value theorem determines the steady-state value of the time response of a transfer function model without requiring computation of the inverse transform. The final value theorem states that

$$\lim_{t\to\infty} f(t) = \lim_{s\to 0} sF(s) \quad (9.2.42)$$

where $F(s)$ is the product of the closed-loop transfer function and the Laplace transform of the input, $W(s)$. Letting $w(t)$ be the unit step function

$$w(t) = \mathcal{U}(t) = \begin{cases} 0 & t < 0 \\ 1 & 0 \le t \end{cases} \quad (9.2.43)$$

with the Laplace transform W

$$W(s) = \mathcal{L}\{\mathcal{U}(t)\} = \frac{1}{s} \quad (9.2.44)$$

the steady state value of the sensitivity function (9.2.41) to a unit thrust vector misalignment is

$$\lim_{s\to 0} sW(s)S_\beta(s) = \lim_{s\to 0} S_\beta(s) = \frac{-X_G F_R}{\left(X_G F_R k_P + C_{N\alpha}\bar{q}S_{ref}X_{cp}\right)}. \quad (9.2.45)$$

This steady state error may not be acceptable from a design standpoint. In the limiting case of vacuum flight where $\bar{q} \to 0$, the steady-state error is

$$\lim_{\substack{s\to 0\\ \bar{q}\to 0}} S_\beta(s) = -\frac{1}{k_P}. \quad (9.2.46)$$

This situation can be remedied by the introduction of integral control, which puts an additional pole into the open-loop transfer function

$G_c(s)G_p(s)$. Let the control law be given by[3]

$$\beta_c^+ = -k_P\phi_y - k_D\dot{\phi}_y - \int_0^t k_I\phi_y d\tau \tag{9.2.47}$$

with the pitch dynamics

$$I_{yy}\ddot{\phi}_y = C_{N\alpha}\bar{q}S_{ref}X_{cp}\phi_y + X_G F_R\left(k_P\phi_y + k_D\dot{\phi}_y + \int_0^t k_I\phi_y d\tau\right) + X_G F_R \Delta\beta_{Ey}. \tag{9.2.48}$$

The Laplace transform of Eq. (9.2.48) is

$$S_\beta(s) = \frac{sX_G F_R}{I_{yy}s^3 - X_G F_R k_D s^2 - \left(X_G F_R k_P + C_{N\alpha}\bar{q}S_{ref}X_{cp}\right)s - X_G F_R k_I}. \tag{9.2.49}$$

Note the factor of s now appearing in the numerator. The steady-state value of the attitude error with respect to a constant thrust vector misalignment is

$$\lim_{s\to 0} sW(s)S_\beta(s) = 0. \tag{9.2.50}$$

The key feature of integral control is that the steady state error with respect to a constant disturbance will converge to zero asymptotically. However, the inclusion of integral control can degrade transient performance, since the denominator polynomial is now third order. Assuming that Eq. (9.2.37) is still to be used to set the control frequency, let the transfer function in Eq. (9.2.49) be written in the normalized form

$$S_\beta(s) = \frac{k_0\omega_1^3 s}{s^3 + \gamma_2\omega_1 s^2 + \gamma_1\omega_1^2 s + \omega_1^3} \tag{9.2.51}$$

where

$$\gamma_1\omega_1^2 = -\frac{1}{I_{yy}}\left(X_G F_R k_P + C_{N\alpha}\bar{q}S_{ref}X_{cp}\right) = \omega_c^2 \tag{9.2.52}$$

$$\gamma_2\omega_1 = -\frac{1}{I_{yy}}X_G F_R k_D \tag{9.2.53}$$

[3] In practice, it is important that the integral gain k_I be placed inside the integral. If the gain is varying in time, the rate of integration is modulated and not the total value of the integral. If the gain is constant there is no distinction.

$$\omega_1^3 = -\frac{1}{I_{yy}} X_G F_R k_I. \tag{9.2.54}$$

The values γ_i are model coefficients, and k_0 is an arbitrary constant. Eq. (9.2.13) introduced the generalized integral performance index J, which was used to find that for $a = 2$ and a second-order system, $\zeta_c = 0.814$ for a near-deadbeat response. Using the same approach (minimize J using a numerical search) on the third-order system in Eq. (9.2.51), the optimal coefficients can be found to be

$$\gamma_1 = 2.369 \tag{9.2.55}$$

$$\gamma_2 = 2.205. \tag{9.2.56}$$

It should be noted that the optimal parameters γ_i depend on the numerator polynomial. With γ_i given and the proportional gain already determined using (9.2.37), the integral and derivative gains can be determined explicitly. The third-order natural frequency is

$$\omega_1^2 = \frac{\omega_c^2}{\gamma_1} \tag{9.2.57}$$

giving

$$k_D = -\frac{\gamma_2 \omega_c I_{yy}}{\sqrt{\gamma_1} X_G F_R} \tag{9.2.58}$$

and

$$k_I = -I_{yy} (X_G F_R)^{-1} \left(\frac{\omega_c^2}{\gamma_1} \right)^{\frac{3}{2}}. \tag{9.2.59}$$

Control laws with proportional, integral, and derivative components are ubiquitous. This is commonly called PID control.

Design example

A design example can be used to illustrate the effects of PD and PID control on a typical launch vehicle configuration. The vehicle parameters are representative of a large rocket in the atmospheric boost phase near maximum dynamic pressure. The parameters at this flight condition are given in Table 9.1.

The vehicle dynamics are simulated using both the simplified transfer functions in Eqs. (9.2.32), (9.2.41), and (9.2.49) along with the detailed linear model presented in Chapter 7. The detailed model uses an idealized

Table 9.1 Example vehicle parameters.

Parameter	Name	Value
m_T	Mass	40000 slug
I_{yy}	Pitch moment of inertia	2×10^8 slug · ft^2
X_G	Gimbal moment arm	−100 ft
X_{cp}	Center of pressure	90 ft
S_{ref}	Aerodynamic reference area	100 ft^2
V	Relative velocity	2000 ft/s
$\bar{q}$	Dynamic pressure	800 lbf/ft^2
$C_{N\alpha}$	Normal force coefficient	9.0/rad
C_A	Drag coefficient	2.0
$F_R = F_0$	Gimbaled thrust	3×10^6 lbf
S_E	Gimbal first moment	500 slug · ft
I_E	Gimbal second moment	10000 slug · ft

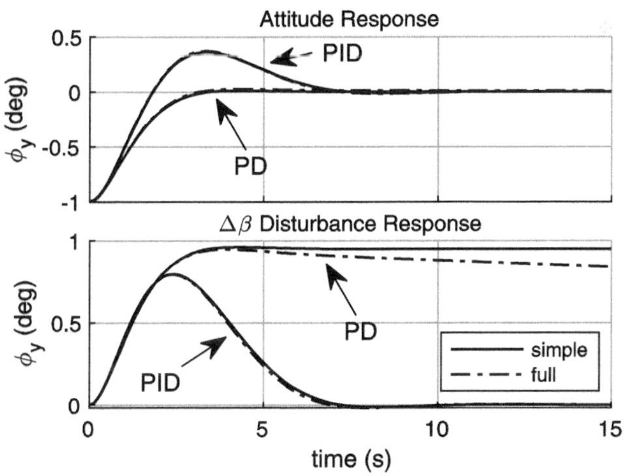

Figure 9.8 Transient vehicle response with PD and PID control (top: unit initial attitude error; bottom: unit thrust vector misalignment).

actuator. The actuator dynamics will be discussed in Section 9.3. The proportional, derivative, and integral gains are designed using $\omega_c = 1.57$ rad/s and either (9.2.37) and (9.2.38), or (9.2.38), (9.2.58) and (9.2.59).

A comparison of the time response of the models is shown in Fig. 9.8. The simplified transfer functions provide a very good approximation for the transient response. The tradeoff between improved transient performance and steady-state error rejection is clear in comparing the vehicle responses with and without integral control. While the PID control law increases overshoot, the ability to converge to zero error in the case of a thrust vector misalignment disturbance is markedly improved.

Integral control is the simplest form of steady-state disturbance rejection. Other approaches, such as simple linear state observers [42], can provide similar disturbance attenuation without significant transient performance penalties.

9.2.5 Load relief

During flight through the atmosphere, the most stressing environment for the vehicle structure is often the region of maximum dynamic pressure. The detailed vehicle bending loads can be derived using a variety of approaches, and complex static and dynamic load indicators have been used to help with trajectory shaping and launch availability decisions [43]. A simpler metric used for control design and performance analysis is the product of dynamic pressure and total angle of attack, $\bar{q}\alpha_T$. Since for small perturbations $\bar{q}$ is not affected by maneuvering, the goal of any load relief algorithm is short-term reduction of the total angle of attack.

Load relief has been explored as an option on many large launch vehicles, including the Saturn I and Saturn V. The formulations presented by Hoelker [44] and Garner [3] which rely on pure normal acceleration feedback are difficult to stabilize in practice and are based on quasi-steady state conditions, derived via the final value theorem. For practical values of the accelerometer feedback gain, the steady state conditions cannot be reached in a time that is relevant relative to the vehicle's trajectory. Since online algorithms or air data systems can either estimate or measure the angle of attack, respectively, the use of α feedback can be considered in the development of a preliminary load relief design.

By definition, following the gravity turn trajectory with $\phi_y = 0$ also has $\alpha_y = 0$ for zero wind, i.e., $V_{wz} = 0$. In the presence of wind, $\phi_y = 0$ does not ensure that $\alpha_y = 0$, and there will be a transient or steady-state load. Immediate reduction of the angle of attack can be of great benefit, but this must also be balanced with the induced loads due to gimbaling the engine. Only the former is considered in the present discussion.

A significant issue that arises is that the trajectory relative velocity state $\dot{z}_I$ is unstable for some level of α gain. This is because in the presence of a wind, the vehicle will turn into the wind to reduce α, but may diverge from the nominal path as a result. For the linearized model, this is represented by the third-order closed-loop transfer function having two damped complex poles and one real pole with a small positive value. This real pole is known as the *drift root*.

It is permissible to allow the α feedback gain to exceed the critical value temporarily, allowing path deviations near maximum dynamic pressure in order to reduce loads. For simplicity, the approach we will use here is to set the α gain at the critical level, which will reduce the loads considerably but not affect the fundamental gain design assumptions used in Section 9.2.3.

In order to introduce α feedback, we will let the control law take the form

$$\beta_c^+ = -(1-\kappa)\,k_P\phi_y - k_D\dot{\phi}_y - \kappa k_P\alpha_y \tag{9.2.60}$$

where α_y is the sensed angle of attack and $0 \le \kappa \le 1$.

The α-feedback control law (9.2.60) amounts to dividing the proportional gain into an attitude error component and an α component. Note that if we assume the plunging velocity is zero, the control equation becomes equivalent to (9.2.27). Thus, the same equations ((9.2.37) and (9.2.38)) can be used to determine the gains k_P and k_D. The short-term attitude response (and aerodynamic gain margin) will be largely insensitive to κ.

The plunging velocity $\dot{z}_I$ must be included in order to determine the effect of α feedback on the drift root and on the angle of attack. In order to simplify the analysis, the rigid-body perturbation equations will first be expressed in a state-space format consistent with Chapter 7, and the inertial coupling terms (coefficients of $\ddot{\beta}_{Ey}$) will be ignored. We will also use the term $N_\alpha = \bar{q}S_{ref}C_{N\alpha}$ to simplify the equations.

The state matrices associated with the rigid pitching and lateral motion, using Eqs. (7.2.14), (7.2.15) and (7.2.24) are

$$\tilde{\mathbf{A}}_r = \begin{bmatrix} 0 & 1 & 0 \\ \frac{N_\alpha X_{cp}}{I_{yy}} & 0 & \frac{N_\alpha X_{cp}}{I_{yy}V} \\ -\frac{\bar{g}m_T + N_\alpha}{m_T} & 0 & -\frac{N_\alpha}{m_T V} \end{bmatrix} \tag{9.2.61}$$

$$\tilde{\mathbf{B}}_r = \begin{bmatrix} 0 & 0 \\ \frac{F_R X_G}{I_{yy}} & -\frac{N_\alpha X_{cp}}{I_{yy}V} \\ -\frac{F_R}{m_T} & \frac{N_\alpha}{m_T V} \end{bmatrix} \tag{9.2.62}$$

for the system described by

$$\dot{\mathbf{x}} = \tilde{\mathbf{A}}_r\mathbf{x} + \tilde{\mathbf{B}}_r\mathbf{u} \tag{9.2.63}$$

with the state vector $\mathbf{x} = \begin{bmatrix} \phi_Y & \dot{\phi}_Y & \dot{z}_I \end{bmatrix}^T$. The inputs are the feedback gimbal command β_c^- and the wind velocity V_w,

$$\mathbf{u} = \begin{bmatrix} \beta_c^- & V_w \end{bmatrix}^T. \tag{9.2.64}$$

The output equation is

$$\mathbf{y} = \begin{bmatrix} \phi_Y & \dot{\phi}_Y & \alpha_Y \end{bmatrix}^T = \mathbf{C}\mathbf{x} + \mathbf{D}\mathbf{u} \tag{9.2.65}$$

where

$$\mathbf{C} = \begin{bmatrix} 1 & 0 & 0 \\ 0 & 1 & 0 \\ 1 & 0 & \frac{1}{V} \end{bmatrix} \quad \mathbf{D} = \begin{bmatrix} 0 & 0 \\ 0 & 0 \\ 1 & \frac{1}{V} \end{bmatrix} \tag{9.2.66}$$

In order to make the control feedback equation compatible with Eq. (9.2.63), let the control $\mathbf{u}_0 = \mathbf{K}\mathbf{y}$ where

$$\mathbf{K} = \begin{bmatrix} (1-\kappa)\,k_P & k_D & \kappa k_P \\ 0 & 0 & 0 \end{bmatrix}. \tag{9.2.67}$$

The closed-loop transfer function matrix can be found by substituting the expression

$$\mathbf{u} = \mathbf{u}_0 + \Delta\mathbf{u} \tag{9.2.68}$$

into Eqs. (9.2.63), taking the Laplace transform, and solving for $\mathbf{y}(s)$ as a function of $\Delta\mathbf{u}(s)$. The details are rather arduous and are omitted. The resulting transfer function matrix contains the following entries:

$$\mathbf{G}_{CL}(s) = \begin{bmatrix} \frac{\phi_Y(s)}{\beta_c^-(s)} & \frac{\phi_Y(s)}{V_{wz}(s)} \\ \frac{\dot{\phi}_Y(s)}{\beta_c^-(s)} & \frac{\dot{\phi}_Y(s)}{V_{wz}(s)} \\ \frac{\alpha_Y(s)}{\beta_c^-(s)} & \frac{\alpha_Y(s)}{V_{wz}(s)} \end{bmatrix} \tag{9.2.69}$$

The primary transfer function of interest is the closed-loop response of angle of attack with respect to wind. This is given by

$$S_w(s) = \frac{\alpha_Y(s)}{V_{wz}(s)} \tag{9.2.70}$$

and has the form

$$S_w(s) = \frac{s\left(b_3 s^2 + b_2 s + b_1\right)}{s^3 + a_2 s^2 + a_1 s + a_0}. \tag{9.2.71}$$

If the solutions for k_p and k_D from Eqs. (9.2.37) and (9.2.38) are inserted into Eq. (9.2.70), the numerator coefficients are given by

$$b_3 = -\frac{1}{V} \tag{9.2.72}$$

$$b_2 = -\frac{2\omega_c\zeta_c}{V} \tag{9.2.73}$$

$$b_1 = (\kappa - 1)\frac{\left(I_{yy}\omega_c^2 + N_\alpha X_{cp}\right)}{I_{yy}V}. \tag{9.2.74}$$

The denominator coefficients are

$$a_2 = 2\omega_c\zeta_c - \frac{I_{yy}\kappa\omega_c^2 - N_\alpha X_G + N_\alpha X_{cp}\kappa}{VX_G m_T} \tag{9.2.75}$$

$$a_1 = \omega_c^2 + \frac{2N_\alpha\omega_c\zeta_c\left(X_G - X_{cp}\right)}{VX_G m_T} \tag{9.2.76}$$

and

$$a_0 = -\frac{\bar{g}\kappa\omega_c^2}{V} + (1-\kappa)\frac{N_\alpha\omega_c^2\left(X_G - X_{cp}\right)}{VX_G m_T}$$
$$(1-\kappa)\frac{N_\alpha X_{cp}\left(N_\alpha X_G - N_\alpha X_{cp} + X_G\bar{g}m_T\right)}{I_{yy}VX_G m_T}. \tag{9.2.77}$$

Eq. (9.2.77) is also a necessary condition for stability of the drift root. That is, for the third order polynomial in the denominator of (9.2.71), $a_0 \geq 0$ is a necessary condition for its roots to have zero or negative real parts. The maximum α feedback gain occurs for the κ where $a_0 = 0$ and the drift root is identically zero, which is given by

$$\kappa_c = \frac{N_\alpha}{I_{yy}\omega_c^2 + N_\alpha X_{cp}}\left\{X_{cp} + \frac{I_{yy}\omega_c^2\left(X_G - X_{cp}\right)}{N_\alpha\left(X_G - X_{cp}\right) + X_G\bar{g}m_T}\right\}. \tag{9.2.78}$$

If $\kappa = \kappa_c$, $a_0 = 0$ and the response to wind becomes finite since Eq. (9.2.70) now has the form

$$S_w(s) = \frac{b_3 s^2 + b_2 s + b_1}{a_3 s^2 + a_2 s + a_1} \tag{9.2.79}$$

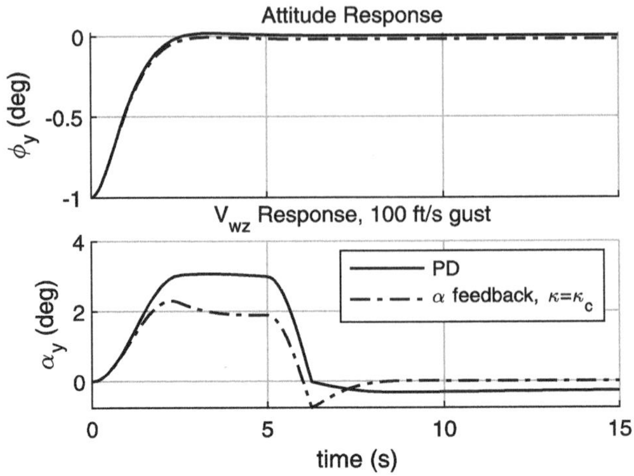

Figure 9.9 Vehicle α response with PD and optimal α feedback control (top: unit initial attitude error; bottom: 100 m/s gust).

and can be evaluated using the FVT. The steady-state response to a wind is equal to b_1/a_1.

For implementation, the reader is cautioned that estimates or measurements of angle of attack are inherently prone to errors, and the parameters appearing in Eq. (9.2.78) cannot be known exactly prior to flight. Therefore, it can be expected that the limiting α gain can be realized only in the ideal case. Whether a slightly unstable drift root is permissible should be thoroughly evaluated in nonlinear time-domain simulation. Alternatively, a positive value of $0 < \kappa < \kappa_c$ can be selected that has the desired effect of reducing transient loads but provides sufficient margin for parameter dispersions. The use of a load relief scheme can be scheduled to correspond with the region of maximum dynamic pressure.

An example of the effect of α feedback load relief is shown in Fig. 9.9, using the vehicle parameters given in Table 9.1. The proportional and derivative gains have the same values as in the example in Fig. 9.8, and using Eq. (9.2.78), the α gain $\kappa = \kappa_c \approx 0.4$. That is, about 60% of the proportional feedback is attitude error, and the drift root is at the stability limit.

As expected, the response to an attitude error or command is essentially unchanged by α feedback. The vehicle response to a gust input having a period of about 5 seconds and a magnitude of $V_{wz} = -100$ m/s is greatly improved. This would provide a significant reduction in loads, although the vehicle will deviate from the nominal trajectory in the presence of a steady-state wind. It is common to balance the effects of load relief with a biasing of the attitude commands based on predicted winds so that the commanded trajectory is close to zero angle of attack.

9.2.6 Slosh stability

The two most important issues affecting a feedback control system for a large rocket are the effects of bending and sloshing propellant. While the former can usually be addressed via filtering and frequency-domain techniques, sloshing propellant can create a major design hurdle. The relationships of acceleration, vehicle diameter, propellant density, and control performance requirements often line up in such a way that the sloshing natural frequencies are close to the control frequency. If this occurs, there is a possibility that the liquid response will phase unfavorably with the control system, leading to instability.

Bauer [7] derived a simple criterion that is useful for vehicles with a single dominant sloshing mode, or multiple sloshing modes that are separated in frequency. The criterion determines the conditions under which sloshing dynamics are unfavorable with respect to the control system. Although the effects of slosh can be predicted more accurately using frequency-domain methods and time-domain simulation, the Bauer criterion is still very useful for preliminary analyses and checks on the feasibility of a design. If the criterion predicts that a sloshing mode will be unstable, the propellant tank in question is likely to require ring baffles to add sufficient damping for stability.

Again using the approach developed in Chapter 7, consider a rocket *in vacuo* with a single sloshing propellant mode having zero damping. If the aerodynamics, bending, and engine inertial coupling effects are neglected, Eqs. (9.2.19), (9.2.20), and (9.2.21) can be simplified as

$$I_{yy}\dot{\omega}_Y - m_s X_s \ddot{z}_s = -m_s \bar{g} z_s + X_G F_R \beta_{EY} \tag{9.2.80}$$

$$m_T \ddot{z}_I + m_s \ddot{z}_s = -m_T \bar{g} \phi_Y - F_R \beta_{EY} \tag{9.2.81}$$

$$\ddot{z}_s - X_s \ddot{\phi}_Y + \ddot{z}_I = -\Omega_s^2 z_s - \bar{g}\phi_Y. \tag{9.2.82}$$

Using Eq. (7.2.6) for the system described by

$$\dot{\mathbf{x}} = \tilde{\mathbf{A}}\mathbf{x} + \tilde{\mathbf{b}}u \tag{9.2.83}$$

with the state vector $\mathbf{x} = \begin{bmatrix} \phi_y & \dot{\phi}_y & \dot{z}_I & z_s & \dot{z}_s \end{bmatrix}^T$, the system matrix is

$$\tilde{\mathbf{A}} = \begin{bmatrix} 0 & 1 & 0 & 0 & 0 \\ 0 & 0 & 0 & -\frac{m_s\left(X_s m_T \omega_s^2 + \bar{g}(m_T - m_s)\right)}{I_{yy}(m_T - m_s) - X_s^2 m_T m_s} & 0 \\ -\bar{g} & 0 & 0 & -\frac{m_s\left(I_{yy}\omega_s^2 + \bar{g} X_s m_s\right)}{I_{yy} m_s - m_T\left(I_{yy} - X_s^2 m_s\right)} & 0 \\ 0 & 0 & 0 & 0 & 1 \\ 0 & 0 & 0 & \frac{m_T\left(I_{yy}\omega_s^2 + \bar{g} X_s m_s\right)}{m_T m_s X_s^2 - I_{yy} m_T + I_{yy} m_s} & 0 \end{bmatrix} \tag{9.2.84}$$

with the input matrix

$$\tilde{\mathbf{b}} = \begin{bmatrix} 0 \\ \frac{F_R\left(X_G(m_T - m_s) + X_s m_s\right)}{I_{yy}(m_T - m_s) - X_s^2 m_T m_s} \\ \frac{F_R\left(-m_s X_s^2 + X_G m_s X_s + I_{yy}\right)}{I_{yy} m_s - m_T\left(I_{yy} - X_s^2 m_s\right)} \\ 0 \\ \frac{F_R\left(I_{yy} + X_G X_s m_T\right)}{I_{yy} m_T - m_s\left(m_T X_s^2 + I_{yy}\right)} \end{bmatrix}. \tag{9.2.85}$$

It is of course possible to include a nonzero slosh damping ζ_s, but this complicates the expressions substantially. Some basic and very useful results can be derived for the system without damping. Using the same approach as in the discussion on load relief, we assume proportional-derivative feedback control and let the control $u_0 = \mathbf{Ky}$ where

$$\mathbf{K} = \begin{bmatrix} k_P & k_D \end{bmatrix}. \tag{9.2.86}$$

The output equation is

$$\mathbf{y} = \begin{bmatrix} \phi_y & \dot{\phi}_y \end{bmatrix}^T = \mathbf{Cx} \tag{9.2.87}$$

where

$$\mathbf{C} = \begin{bmatrix} \mathbf{1}_{2\times 2} & \mathbf{0}_{2\times 3} \end{bmatrix}. \tag{9.2.88}$$

It is helpful to choose values for k_P and k_D according to Eqs. (9.2.37) and (9.2.38), which in vacuum flight give

$$k_P = \frac{I_{yy}\omega_c^2}{-X_G F_R} \tag{9.2.89}$$

$$k_D = \frac{I_{yy} 2\zeta_c \omega_c}{-X_G F_R} \tag{9.2.90}$$

where ω_c is the control frequency, and ζ is the desired closed-loop damping ratio. For the analysis of stability, it is only necessary to compute the characteristic polynomial equation

$$D(s) = \det(s\mathbf{1} - \mathbf{A}_{CL})^{-1} = 0 \tag{9.2.91}$$

where $\mathbf{A}_{CL}$ is the closed-loop system matrix. This matrix can be obtained by substituting the output and control relationships into Eq. (9.2.83), giving

$$\mathbf{A}_{CL} = \tilde{\mathbf{A}} + \tilde{\mathbf{b}}\mathbf{KC} \tag{9.2.92}$$

which then can be used in Eq. (9.2.91) to yield the closed-loop characteristic polynomial. For this system, the polynomial is 5th order and has the coefficients

$$a_4 = 1 \tag{9.2.93}$$

$$a_3 = \frac{2\zeta_c}{\omega_c} \tag{9.2.94}$$

$$a_2 = -\frac{X_G\left(m_T\omega_s^2 + \omega_c^2(m_T - m_s)\right) + X_s m_s \omega_c^2 + I_{yy}^{-1} X_G X_s \bar{g} m_T m_s}{\omega_c^2\left(\bar{g}m_s - X_G m_T \omega_s^2\right)} \tag{9.2.95}$$

$$a_1 = -\frac{2\zeta\left(X_G m_T + m_s(X_s - X_G)\right)}{\omega_c\left(\bar{g}m_s - X_G m_T \omega_s^2\right)} \tag{9.2.96}$$

$$a_0 = \frac{X_G\left(m_T m_s X_s^2 - I_{yy}(m_T - m_s)\right)}{I_{yy}\omega_c^2\left(\bar{g}m_s - X_G m_T \omega_s^2\right)} \tag{9.2.97}$$

For a characteristic polynomial of degree k, a necessary and sufficient condition for stability (i.e., all poles of $D(s)$ have negative real parts) is that the well-known Routh stability criterion be satisfied. A general statement of the Routh criterion is not given here; the reader is referred to Reference [39]. The Routh criterion first requires that all coefficients a_i be positive. For a system of order $k = 5$ with $a_4 = 1$, it is also necessary to satisfy the inequalities

$$a_3 > 0 \tag{9.2.98}$$

$$\frac{a_2 a_3 - a_1}{a_3} > 0 \tag{9.2.99}$$

$$\frac{a_1^2 - a_1 a_2 a_3 + a_0 a_3^2}{a_1 - a_2 a_3} > 0. \tag{9.2.100}$$

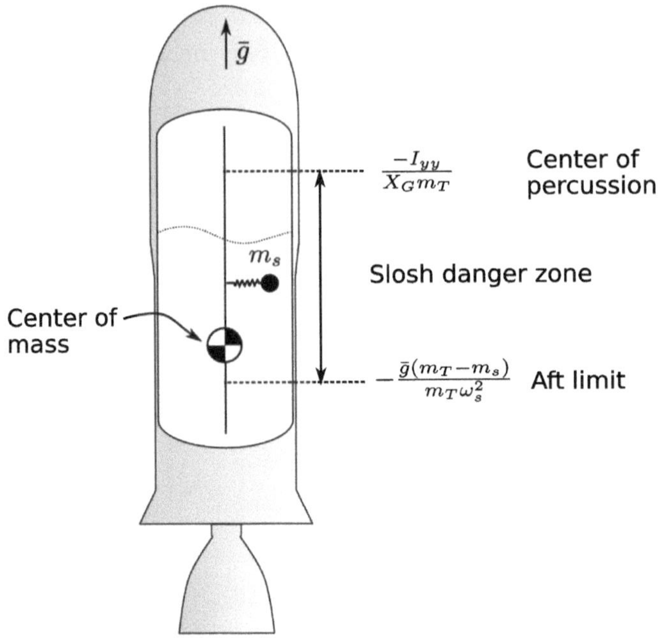

Figure 9.10 Slosh danger zone.

After substitution of the coefficients a_i into the Routh equations, it can be shown that if the slosh mass X_s is located in the region

$$-\frac{\bar{g}\,(m_T - m_s)}{m_T \omega_s^2} \le X_s \le \frac{-I_{\gamma\gamma}}{X_G m_T}, \tag{9.2.101}$$

the closed-loop system is *unstable*. This result was first derived by Bauer [7] and is called the *danger zone*. The danger zone is depicted in Fig. 9.10. The quantity

$$X_P = \frac{-I_{\gamma\gamma}}{X_G m_T} \tag{9.2.102}$$

is called the *center of percussion* and is the point on the vehicle where the instantaneous translation and rotation motions in response to a gimbal motion are equal and opposite. The danger zone lies approximately between the center of mass and the center of percussion. The aft limit given in Eq. (9.2.101) is nearly zero for some vehicle configurations, corresponding with the location of the CG.

The danger zone is important for vehicle sizing and design. A sloshing propellant mode whose slosh mass lies within the danger zone will be inherently unstable under feedback control, and will require some level of additional damping ζ_s in order to ensure stability. Since the vehicle mass properties and liquid levels vary as propellant is burned, it is important that the designer evaluate the characteristics of the sloshing modes for those times of flight where the slosh mass is located in this region. In many cases, baffles are required. The required damping depends on the system parameters, and can be determined analytically using the same method as used to determine the limits on X_s. Bauer presented curves of ζ_s with respect to X_s in Reference [7].

The present analysis is limited to the case of a single tank. If the sloshing mode frequencies are well separated, then the analysis is generally valid when considering a single tank at a time. However, two or more sloshing modes having similar frequencies can lead to incorrect conclusions if treated individually. The best approach in this case is to investigate the location of the eigenvalues numerically using the linear model of Chapter 7.

9.3 Actuation systems

As discussed in the preceding chapters, the flight control actuation systems of large space boosters employ almost exclusively a thrust vector control (TVC) scheme, where, for liquid engines, the entire engine and thrust chamber are rotated in two degrees of freedom (pitch and yaw) about a gimbal so as to point the thrust vector in the required direction. While the actuator systems are somewhat heavy and complex, the performance benefits have made TVC a standard actuation scheme for large rockets.

Traditional methods for modeling TVC dynamics have relied upon simplified models anchored to test data obtained in a laboratory. An additional detailed, or "complex" model, is used for subsystem-level requirements verification. The simplified model is sometimes called a "simplex" model, and for hydraulic actuators, can be reduced to a fourth-order linear model per engine degree of freedom. Two of the states are used to describe the engine angle and rate, and two states of model are used to model the actuator feedback dynamics. One of the most significant considerations in modeling the actuator-engine system is to ensure that the inertial effects discussed in Chapter 6 are correctly modeled.

In the terminology of actuation systems, the moveable mass of the engine is called the *load*. For large rockets the engine and its attachments to

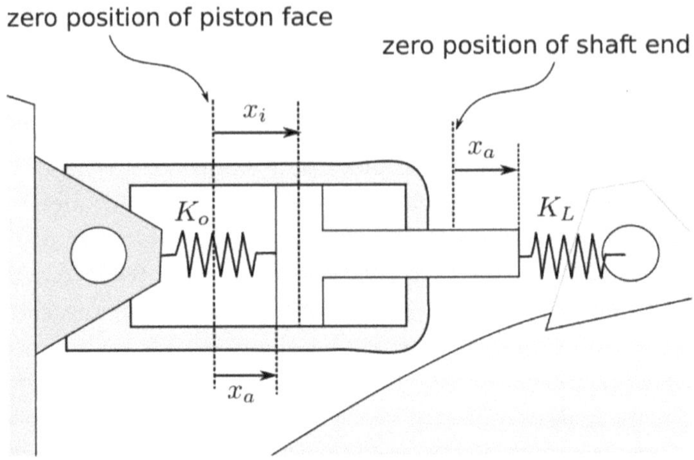

Figure 9.11 TVC actuator with load compliance.

the vehicle cannot be considered rigid. The natural frequency of the engine on its compliant mounting is called the *load resonance*. One of the goals of the TVC actuator is to stabilize this motion. For this reason, almost all TVC actuators use some type of load feedback. This method senses the dynamics of the load and provides a feedback signal that is proportional to the load force. The details of this type of load control system are detailed in the work by Thompson and Hung [45,46].

Consider the notional diagram of a hydraulic actuator given in Fig. 9.11. In this simplified model, the actuator components are massless and the only compliances are in the load (K_L) and the hydraulic oil (K_o). The load compliance is a single spring that represents all of the compliance in the system, including the actuator case, engine structure, and the backup structure. As a simplification, the diagram only shows the bare minimum necessary to represent motion in one direction. A more realistic diagram would be more symmetrical, with the zero position of the piston face close to the center of the cylinder.

The total compliance from the rocket stage to the engine interface is

$$K_T = \left(\frac{1}{K_L} + \frac{1}{K_o}\right)^{-1} \tag{9.3.1}$$

and the displacement of the engine is given by

$$x_E = d_{arm}\beta_E \tag{9.3.2}$$

where d_{arm} is the moment arm from the actuator attach point to the engine pivot point (see Fig. 6.3). It is helpful to introduce the ideal piston position x_i which can be thought of as the displacement of the piston if the oil were not compressible. The ideal piston position is related to the flow by

$$\dot{x}_i = \frac{1}{A_p} Q(t) \tag{9.3.3}$$

where A_p is the piston area and $Q(t)$ is the input flow. The actual displacement of the piston x_a is less than x_i due to the compliance of the oil. The displacement of the engine, x_E, is less than x_a due to the fact that K_L is not infinite. Thus we can write

$$x_E = x_a - \frac{f_a}{K_L} \tag{9.3.4}$$

where

$$f_a = K_T (x_i - x_E) \tag{9.3.5}$$

is the force developed in the actuator. Combining these expressions,

$$x_a = \left(1 - \frac{K_T}{K_L}\right) d_{arm}\beta_E + \frac{K_T}{K_L} x_i. \tag{9.3.6}$$

Now imagine that there is a torsional spring K_E and damper C_E at the engine gimbal. The former is a viscous damping approximation to the engine gimbal running friction, and the latter is the sum of the gravitational stiffness and any loads about the engine gimbal due to flexible bearings or pressurized propellant ducts. The engine rotation is accelerated by the actuator torque and decelerated by the torsional spring and damper:

$$I_E \ddot{\beta}_E = f_a d_{arm} - C_E \dot{\beta}_E - K_E \beta_E$$

Substituting from (9.3.5) and then from (9.3.2) we obtain

$$I_E \ddot{\beta}_E = K_T d_{arm} x_i - C_E \dot{\beta}_E - \left(K_E + K_T d_{arm}^2\right) \beta_E \tag{9.3.7}$$

The engine natural frequency for a stationary actuator ($x_i = 0$) is

$$\omega_p^2 = \frac{\left(K_E + K_T d_{arm}^2\right)}{I_E} \tag{9.3.8}$$

This is called the *pendulum mode* frequency, which was discussed in Chapter 6. It represents the natural frequency of the engine-actuator system when only the compliance of the structure and oil are included and the actuator is inoperative.

In order to control the position of the load, a feedback mechanism is used to sense the displacement and the force. For practical reasons, the direct sensing of the engine position is not feasible, so the position feedback usually uses the piston position measurement (9.3.6). This feedback can be sensed mechanically or electronically. The second component, the load feedback, depends on the differential pressure ΔP developed in the actuator. The differential pressure ΔP is equal to the actuator force (9.3.5) divided by the piston area;

$$\Delta P = \frac{1}{A_p} f_a. \tag{9.3.9}$$

Direct feedback of a signal proportional to the load force would result in an unacceptable steady-state positioning error; a detailed proof of this assertion is not offered here. The solution commonly used in actuator designs is to filter the load feedback (e.g., hydraulic pressure) through an electronic or hydraulic high-pass filter, which has the dynamics

$$\dot{u}_p = -\frac{1}{\tau_p} u_p + \Delta P. \tag{9.3.10}$$

This approach was first developed for hydraulic positioning systems in the late 1950's. The quantity u_p is the state of the load feedback filter, and the output that is used for feedback is the state derivative $\dot{u}_p$. The value τ_p is the time constant associated with the load feedback dynamics, and the input is the differential pressure ΔP. Combining (9.3.10), (9.3.9), and (9.3.2) gives the equation for the pressure feedback filter;

$$\dot{u}_p = -\frac{1}{\tau_p} u_p + \frac{1}{A_p} K_T \left(x_i - d_{arm} \beta_E \right). \tag{9.3.11}$$

The engine slew control is performed by modulating the flow Q, using both piston position feedback and pressure feedback. Combining these elements with an external command input gives the expression for the flow

$$Q = -k_v k_u \dot{u}_p - k_v k_x \left\{ \left(1 - \frac{K_T}{K_L} \right) d_{arm} \beta_E + \frac{K_T}{K_L} x_i \right\} + k_v k_q \beta_c \tag{9.3.12}$$

where k_u is the pressure feedback gain, k_x is the position feedback gain, and k_q is the command input gain. The valve flow gain is k_v. A substitution for

$\dot{u}_p$ from Eq. (9.3.11) can be made to yield the complete flow expression

$$Q = k_v k_q \beta_c + k_v k_u \frac{1}{\tau_p} u_p - \left(k_v K_T \left(\frac{k_u}{A_p} + \frac{k_x}{K_L} \right) \right) x_i$$
$$k_v d_{arm} \left(\frac{k_u}{A_p} K_T - k_x \left(1 - \frac{K_T}{K_L} \right) \right) \beta_E. \quad (9.3.13)$$

Eqs. (9.3.7), (9.3.11), and (9.3.13) can be incorporated into a state-space representation $\dot{\mathbf{x}} = \mathbf{A}\mathbf{x} + \mathbf{b}\beta_c$, $\mathbf{y} = \mathbf{C}\mathbf{x}$ using the state vector

$$\mathbf{x} = \begin{bmatrix} x_i & u_p & \beta_E & \dot{\beta}_E \end{bmatrix}^T \quad (9.3.14)$$

and the output

$$\mathbf{y} = \begin{bmatrix} \beta_E & \ddot{\beta}_E \end{bmatrix}^T. \quad (9.3.15)$$

The gimbal angular acceleration is included for the prescribed motion model of TWD effects. The state and output matrices are

$$\mathbf{A} = \begin{bmatrix} -\frac{k_v}{A_p} K_T \left(\frac{k_u}{A_p} + \frac{k_x}{K_L} \right) & \frac{k_v k_u}{\tau_p A_p} & \frac{k_v}{A_p} d_{arm} \left(\frac{k_u}{A_p} K_T - k_x \left(1 - \frac{K_T}{K_L} \right) \right) & 0 \\ \frac{1}{A_p} K_T & -\frac{1}{\tau_p} & -\frac{d_{arm}}{A_p} K_T & 0 \\ 0 & 0 & 0 & 1 \\ \frac{K_T d_{arm}}{I_E} & 0 & -\frac{\left(K_E + K_T d_{arm}^2 \right)}{I_E} & -\frac{C_E}{I_E} \end{bmatrix} \quad (9.3.16)$$

$$\mathbf{C} = \begin{bmatrix} 0 & 0 & 1 & 0 \\ \frac{K_T d_{arm}}{I_E} & 0 & -\frac{\left(K_E + K_T d_{arm}^2 \right)}{I_E} & -\frac{C_E}{I_E} \end{bmatrix}. \quad (9.3.17)$$

Since the input consists only of the flow command, the input matrix is simply

$$\mathbf{b} = \begin{bmatrix} \frac{k_v k_q}{A_p} & 0 & 0 & 0 \end{bmatrix}^T. \quad (9.3.18)$$

Parameters of a hypothetical TVC actuator for the vehicle described in Section 9.2.4 are given in Table 9.2. While the feedback gains k_x, k_u and the pressure time constant τ_p are usually fixed in the actuator hardware, they have been numerically tuned in this example to achieve the best step response.

The time and frequency response of the TVC actuator model is shown in Fig. 9.12. For a given engine mass, the achievable response time and bandwidth of a TVC actuator is limited by the compliance of the structure and the oil, regardless of the actuator power. For a typical large booster

Table 9.2 Parameters for hydraulic TVC model.

Parameter	Name	Value
k_v	Valve gain	$1000\ ^{in^3}/_{s}/in \cdot lbf$
k_x	Position feedback gain	$0.4\ ^{in \cdot lbf}/_{in}$
k_u	Pressure feedback gain	$1.0 \times 10^{-4}\ ^{in \cdot lbf}/_{psi}$
k_q	Input command gain	$9.617\ ^{in \cdot lbf}/_{rad}$
A_p	Piston area	$30\ in^2$
K_L	Load stiffness	$500,000\ ^{lbf}/_{in}$
K_o	Oil stiffness	$1,000,000\ ^{lbf}/_{in}$
K_T	Total stiffness	$333,333\ ^{lbf}/_{in}$
d_{arm}	Actuator moment arm	24 in
τ_p	Pressure time constant	0.4 s
I_E	Engine inertia	$120,000\ in \cdot lbf \cdot s^2$
C_E	Engine gimbal damping	$500,000\ ^{in \cdot lbf}/_{(rad/s)}$
K_E	Engine gimbal stiffness	$500,000\ ^{in \cdot lbf}/_{rad)}$

engine, the TVC bandwidth is usually on the order of 27 $^{rad}/_{s}$ or about 4 Hz.

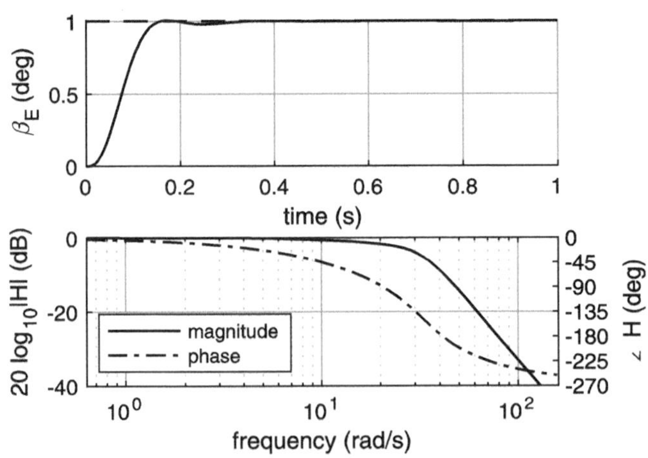

Figure 9.12 Response of typical TVC actuator model.

9.4 Stability analysis

The preceding sections have introduced several techniques based on time-domain analysis that are helpful in preliminary analysis. While simplified transfer functions and transfer function matrices provide insight, practical design problems involving just two or three sloshing or bending modes are essentially intractable without resorting to numerical methods. Fortunately, modern matrix computing tools are able to easily digest a high-order linear model such as that presented in Chapter 7. The primary tool for assessing the stability characteristics of the *closed-loop* dynamics is based on the frequency response of the *open-loop* linear system.

Elementary factors and frequency response

The frequency response of a transfer function is a complex function $H(j\omega)$ where $\sigma = 0$ and ω is a real frequency in radians per second. It is helpful to consider $H(j\omega)$ as simply a complex function $H : \mathbb{R} \rightarrow \mathbb{C}$, that is, a mapping of real angular frequency ω onto a complex number $H = re^{j\theta}$ which has a magnitude r and a phase θ. These quantities are related to H by

$$r = |H| = \sqrt{HH^\star} \tag{9.4.1}$$

$$\theta = \angle H = \arctan \frac{Im(H)}{Re(H)}, \tag{9.4.2}$$

where H^* is the complex conjugate of H.

A transfer function reflects the fact that for any sinusoidal input $u_n(t) = A_n \cos(\omega_n t - \phi_n)$, as $t \rightarrow \infty$ a stable linear system will produce an output $y_n(t)$ that is also a sinusoid at the same frequency ω_n,

$$y_n(t) = A'_n \cos(\omega_n t - \phi'_n) \tag{9.4.3}$$

where the gain $r = A'_n/A_n$ and phase $\theta = \phi'_n - \phi_n$ are determined by the transfer function $H(j\omega_n)$ using Eqs. (9.4.1) and (9.4.2). On the interval T any periodic function $u(t)$ may be approximated as a Fourier series of the form

$$u(t) \approx A_0 + \sum_{n=1}^{N} A_n \cos(n\omega_f t - \phi_n) \tag{9.4.4}$$

where A_0 is the average value of $u(t)$ and $\omega_f = 2\pi/T$ is the *fundamental frequency*. This implies that a transfer function can be used to evaluate the response of a linear system to arbitrary signals.

The graph of H as real frequency ω is varied is called the *frequency response* of H. Since H is complex valued, it is common to depict either the log magnitude $20\log_{10} r$ and phase θ on separate graphs (a Bode plot), the phase θ and magnitude r in polar coordinates (a Nyquist diagram), or the

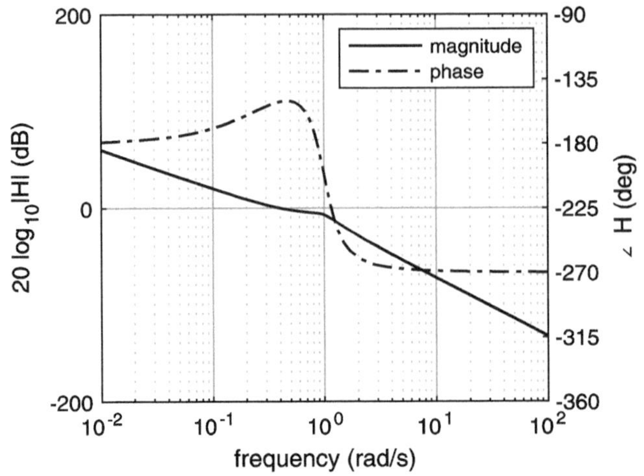

Figure 9.13 Example of a Bode plot.

log magnitude $20\log_{10} r$ and phase θ in Cartesian coordinates (a Nichols chart). A Bode plot of the system

$$H(s) = \frac{b_1 s + b_0}{s^2\left(s^2 + 2\zeta\omega_0 s + \omega_0^2\right)} \tag{9.4.5}$$

with $\omega_0 = 1$, $\zeta = 0.3$, and $\{b_0, b_1\} = \{0.1, 0.25\}$ is shown in Fig. 9.13.

In a Bode plot, the log magnitude is given in units of decibels (dB) versus log frequency. The response of the same system is shown in Fig. 9.14. The use of Nichols charts is widespread in the classical control design for flexible structures, particularly for rockets. This arises in part due to the ease of graphically determining stability margins, which will be discussed later.

In general, transfer function polynomials can be decomposed into a product of elementary factors such that $H = \prod N_i/D_i$, where each factor is either (1) a constant k, (2) a pure integrator or differentiator $s^{\pm 1}$, (3) a first-order lead, (4) a first order lag, or (5) a second-order factor. For linear systems, the magnitude and phase responses of elementary factors are asymptotic and additive on a logarithmic scale. "Asymptotic" means the response $H(j\omega)$ is evaluated far from the singularities in H (the poles and zeros), and the system gain response converges to a constant slope. This property can be used to construct frequency response charts without the aid of a computer, or more commonly, to deduce model properties from

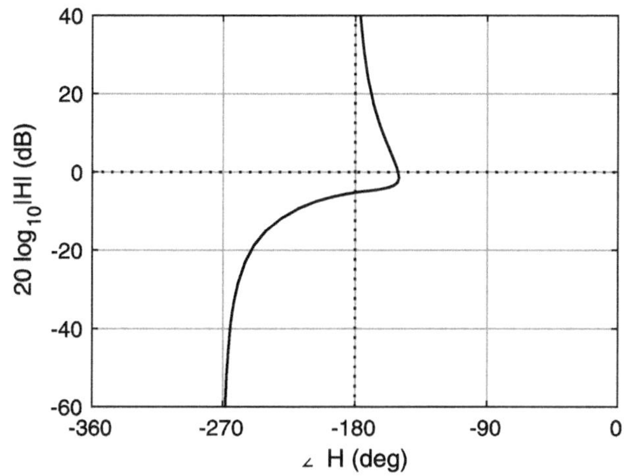

Figure 9.14 Example of a Nichols chart.

frequency response test data. Asymptotic gain behavior in a logarithmic plot is described in terms of *decades*, or powers of frequency, having units of dB/dec.

The factors N_i, D_i from which a transfer function model can be constructed have the following properties:

1. A gain k has a constant magnitude and zero phase for all ω.
2. A differentiator (integrator) $ks^{\pm 1}$ has a gain of k at $\omega = 1$ and a slope of ± 20 dB/dec with constant phase $\pm \frac{\pi}{2}$, with $\lim_{\omega \to 0} |H| = 0$, $\lim_{\omega \to \infty} |H| = \infty$ for a differentiator (ks) and $\lim_{\omega \to 0} |H| = \infty$, $\lim_{\omega \to \infty} |H| = 0$ for an integrator (ks^{-1}).
3. A lead factor $k(s + \omega_1)$ has a gain of $\lim_{\omega \to 0} H(j\omega) = k\omega_1$ at $\omega = 0$, a slope of zero at $\omega \ll \omega_1$, and a slope of $+20$ dB/dec at $\omega \gg \omega_1$. The phase is zero at $\omega \ll \omega_1$, $\pi/4$ at $\omega = \omega_1$, and $\pi/2$ at $\omega \gg \omega_1$. The gain is very close to $+3$ dB at $\omega = \omega_1$.
4. A lag factor $k(s + \omega_1)^{-1}$ has a gain of $\lim_{\omega \to 0} H(j\omega) = \frac{k}{\omega_1}$ at $\omega = 0$, a slope of zero at $\omega \ll \omega_1$, and a slope of -20 dB/dec at $\omega \gg \omega_1$. The phase is zero at $\omega \ll \omega_1$, $-\pi/4$ at $\omega = \omega_1$, and $-\pi/2$ at $\omega \gg \omega_1$. The gain is very close to -3 dB at $\omega = \omega_1$.

These frequency responses are shown in Fig. 9.15.

The fifth elementary factor is the quadratic factor

$$k\omega_0^2 \left(s^2 + 2\zeta \omega_0 s + \omega_0^2\right)^{\pm 1} \tag{9.4.6}$$

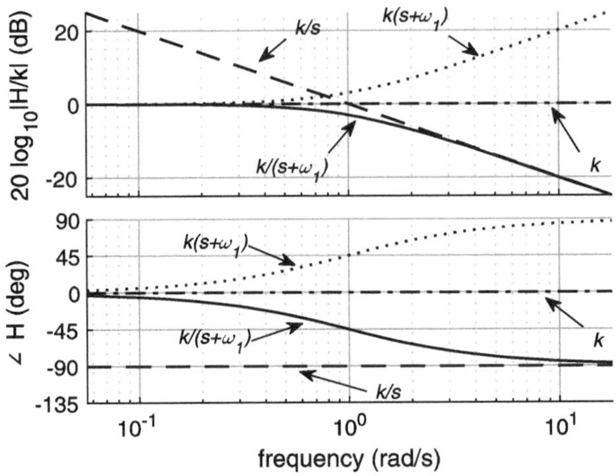

Figure 9.15 Elementary factors frequency response components.

whose time response was introduced in Section 9.2.1. The frequency response of this system for the underdamped case ($0 < \zeta < 1$) is an important model for many dynamic phenomena.

The frequency response of an underdamped quadratic pole, i.e., Eq. (9.4.6) with exponent -1, is shown in Fig. 9.16. It exhibits a peak near $\omega = \omega_0$ whose magnitude depends on the damping ratio. The frequency response of a quadratic pole is given by

$$|H| = k\left[\left(1 - \frac{\omega^2}{\omega_0^2}\right)^2 + \left(\frac{2\zeta\omega}{\omega_0}\right)^2\right]^{-\frac{1}{2}} \quad \angle H = -\tan^{-1}\frac{2\zeta\omega/\omega_0}{1-\omega^2/\omega_0^2}. \tag{9.4.7}$$

A quadratic factor has a gain of $\lim_{\omega\to 0} H(j\omega) = 0$ at $\omega = 0$, a slope of zero at $\omega \ll \omega_0$, and a slope of -40 dB/dec at $\omega \gg \omega_0$. The phase is zero at $\omega \ll \omega_0$, $-\pi/2$ at $\omega = \omega_0$, and $-\pi$ at $\omega \gg \omega_0$. The peak response has the absolute magnitude

$$r_m = \frac{k}{2\zeta\sqrt{1-\zeta^2}} \tag{9.4.8}$$

which occurs at the frequency

$$\omega_m = \omega_0\sqrt{1-2\zeta^2}. \tag{9.4.9}$$

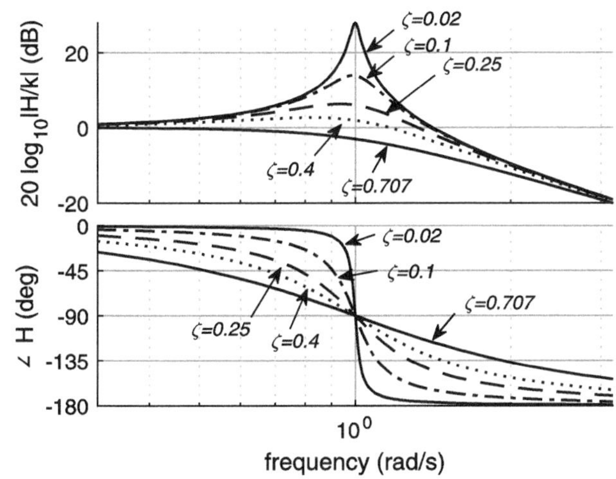

Figure 9.16 Frequency response of quadratic pole.

The value ω_m is called the *resonant frequency*. If $\zeta = \frac{\sqrt{2}}{2}$, $r_m = 1$ and ω_m is undefined. It is important to note that the system has no resonant peak in the frequency domain, but the time response is still underdamped.

The Nyquist criterion

The most important relationship in linear, frequency-domain control analysis is the Nyquist criterion. For linear, constant-coefficient systems of ordinary differential equations in a unity-gain single-input, single-output feedback configuration, the Nyquist criterion gives *necessary and sufficient* conditions for stability of the *closed loop* system based on two characteristics of the *open loop* system:

1. The frequency response $G(j\omega)$.
2. The number of poles of $G(s)$ in the complex right-half plane (RHP).

The frequency response of the system, combined with the Nyquist criterion, can be used to derive linear *stability margins*. The concept of a gain margin was previously introduced in Section 9.2.3, and this is only one of the margins that can be elucidated from the frequency response. Stability margins provide quantitative information about the robustness of a design to independent variations in the loop gain and phase. All of this information is available without explicitly solving for the eigenvalues of the closed-loop system.

The Nyquist criterion is based on Cauchy's residue theorem. The method is elegant but the details are omitted here; the reader is referred to the excellent derivation appearing in Reference [38].

Consider the system $G(s)$ in a unity feedback configuration. Using Eq. (9.2.14), the closed-loop transfer function is

$$G_{CL}(s) = \frac{G(s)}{1 + G(s)} = \frac{G(s)}{B(s)} \tag{9.4.10}$$

A necessary and sufficient condition for stability is that all of the solutions of $B(s) = 1 + G(s) = 0$ must be in the LHP. Recall from Section 9.2.1 that the *zeros* of the return difference function $B(s) = 1 + G(s)$, that is, the values of s that satisfy

$$B(s) = 0 \tag{9.4.11}$$

are the *poles* of the closed-loop system. Furthermore, the poles of $B(s)$ and the poles of $G(s)$ are the same. The Nyquist criterion is therefore concerned with finding the zeros of $B(s)$, or more precisely, the number of zeros of $B(s)$ that are in the RHP.

The zeros of $B(s)$ are related to the frequency response of $G(s)$ over $s \in [-j\infty, j\infty]$. It is helpful to think of the $G(j\omega)$ as simply a complex-valued function that varies as ω ranges over all frequencies. The relationships that lead to the Nyquist criterion can be summarized as follows:

- If $B(s)$ intersects the origin at some frequency $s = j\omega_c$, there is at least one pole of the closed-loop system on the imaginary axis.
- The number of clockwise encirclements of $B(s)$ about the origin is equal to the number of zeros of $B(s)$ in the RHP.
- The number of counterclockwise encirclements of $B(s)$ about the origin is equal to the number of poles of $B(s)$ in the RHP.

Therefore, the total number of counterclockwise encirclements N of $B(s)$ about the origin is equal to $P_R - Z_R$, where P_R and Z_R are the number of RHP poles and zeros, respectively, of $B(s)$:

$$N = P_R - Z_R. \tag{9.4.12}$$

The Nyquist criterion then establishes the following: if P_R is known and Z_R must be zero for stability, then *the number of counterclockwise encirclements N must equal the number of open-loop poles P_R*. Since $G(s) = B(s) - 1$, it is equivalent to use the frequency response of $G(s)$ with the point $-1 + j0$ instead of the origin of the complex plane. This point is called the *critical point*.

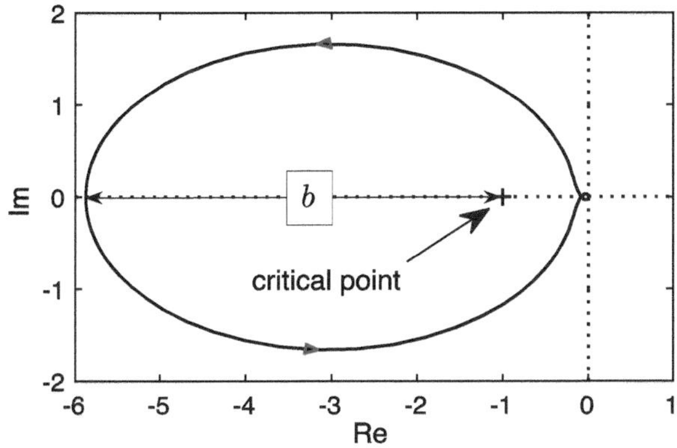

Figure 9.17 Nyquist plot for idealized vehicle pitch dynamics.

Consider the example given in Section 9.2.3 for the idealized open-loop system with proportional-feedback control (Eq. (9.2.29)) and the actuator dynamics given in Section 9.3. The open-loop transfer function is

$$G(s) = G_c(s)\,G_p(s)\,G_a(s), \tag{9.4.13}$$

where

$$G_a(s) = \frac{\beta_{Ey}(s)}{\beta_c(s)} \tag{9.4.14}$$

is the actuator transfer function computed from Eqs. (9.3.16), (9.3.17), and (9.3.18), omitting the engine acceleration (TWD) effect.

This system has real poles at $\lambda_i = \pm\sqrt{C_{N\alpha}\bar{q}S_{ref}X_{cp}}$, so the number of open-loop poles in the RHP is $P_R = 1$. The Nyquist diagram of this open-loop transfer function with the parameters in Table 9.1 is shown in Fig. 9.17. The feedback gains are given by Eqs. (9.2.37) and (9.2.38) for $\omega_c = 1.26$ and $\zeta_c = 0.814$. This Nyquist diagram has one counterclockwise encirclement of the critical point, so $N = P_R$ and the closed-loop system is stable.

The Nyquist criterion establishes stability for a broad class of linear systems, including those with infinite order (e.g., time delays). More importantly, once a system is shown to be stable using the Nyquist criterion, the degree of stability can be determined from the relationship of the frequency response to the critical point. The same system from Fig. 9.17 is shown in Fig. 9.18 in the vicinity of the critical point. The *phase margins* (PM) are the angles of rotation, or phase, such that the frequency response $G(j\omega)$ in-

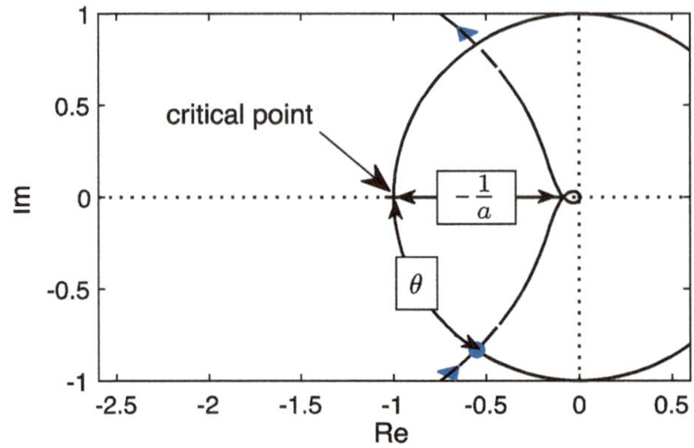

Figure 9.18 Nyquist plot with gain and phase margins.

tersects the critical point. Since the critical point lies on a circle of radius 1, each phase margin is an arc on the unit circle between the critical point and the frequency response. The *gain margins* (GM) are defined as the perturbations in the gain of $G(j\omega)$ such that the frequency response intersects the critical point. The gain or phase margin frequencies are those points on the trajectory of $G(j\omega)$ where the frequency response traverses the real line or the boundary of the unit circle, respectively. In Fig. 9.18, the phase margin is θ and the high-frequency gain margin is $20\log_{10} a$. The complementary (low-frequency) gain margin $20\log_{10} b$ is shown in Fig. 9.17.

It is important to recognize that there can be multiple gain and/or phase margins at multiple frequencies. In some cases, only the smallest of these need be considered. However, it is typical for rockets to exhibit *conditional stability*, meaning that there is a finite range of gain and phase perturbations where the system is stable. For example, an aerodynamically unstable rocket having an actuator of order 2 or higher will have at least two gain margins; one occurs at low frequency and one occurs at high frequency. It was already shown that the low frequency gain margin is approximately

$$\mathrm{GM}_a \approx 20\log_{10}\frac{X_G F_R k_P}{C_{N\alpha}\bar{q}S_{ref}X_{cp}}. \tag{9.4.15}$$

This is a negative value, and is the amount of gain *decrease* that can be tolerated before the system becomes unstable. Since the low frequency gain margin is usually defined by the aerodynamic parameters, it is called the aerodynamic gain margin (GM_a). Conversely, the high-frequency gain margin is called the *rigid-body gain margin* (GM_r) since it is usually associated with the coupled dynamics of the rigid body and the actuator. If the rocket is flexible, the control system will have multiple gain margins at frequencies above the rigid-body gain margin frequency.

The *forward gain* for the open-loop plant dynamics $G_p(s)$ is

$$k_0 = \frac{-X_G F_R}{I_{yy}}. \tag{9.4.16}$$

With the aerodynamic, actuator, and control gain parameters held constant, the gain margins are a measure of the robustness of the control loop to uncertainties in the CG location (X_G), the thrust (F_R) and the moment of inertia (I_{yy}). When an actuator is included, additional values are included in k_0, such as the gimbal moment arm d_{arm}, the input command gain, and so on.

Nichols charts

As the dynamics of the system become more complex, the depiction of multiple gain and phase margins on a Nyquist diagram becomes impractical. It is more common for the analysis of rocket control systems to use the Nichols chart, as was introduced in Fig. 9.14. The same open-loop system shown in Fig. 9.17 is depicted in a Nichols chart in Fig. 9.19. In this format, the gain and phase margins are readily deduced from the separation between the trajectory of $G(j\omega)$ and the critical point at $(-180°, 0$ dB$)$. Changes in the forward gain or the phase of the open-loop response appear as a vertical or lateral translation, respectively, of the frequency trajectory.

Frequency-domain design requirements such as gain and phase margin can be derived from industry experience, empirical guidelines, and more advanced stability theories like the circle criterion [39]. Since multiple margins may occur at different frequencies, the requirements may differ depending on the dynamics of the mode in question. For rigid-body margins as depicted in Fig. 9.19, typical values are ±6 dB for gain margin and a minimum of 30° of phase margin. Due to increased modeling uncertainties and a sensitivity to latency effects at high frequency, typical margins for bending dynamics are 12 dB of gain margin and 45° of phase margin. Phase margin requirements for sloshing propellant modes are sometimes relaxed, since in some cases lightly-damped modes can be benign even when marginally stable.

The value of the Nichols chart can be illustrated by considering a more complex, integrated system frequency response. In this example, the vehicle parameters of Table 9.1 are combined with the actuator dynamics

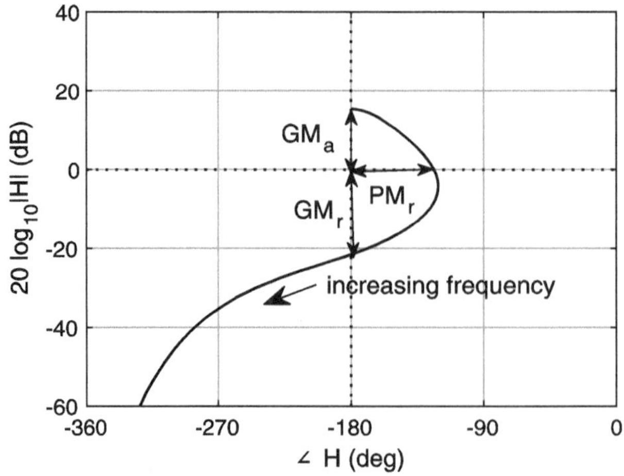

Figure 9.19 Nichols plot with gain and phase margins.

Table 9.3 Vehicle sloshing and bending parameters.

Parameter	Name	Value
m_{s1}	Oxidizer sloshing mass	1000 slug
m_{s2}	Fuel sloshing mass	100 slug
ω_{s1}	Oxidizer slosh frequency	1.56 rad/s
ω_{s2}	Fuel slosh frequency	1.25 rad/s
X_{s1}	Oxidizer slosh mass location	+80 ft
X_{s2}	Fuel slosh mass location	+100 ft
$\zeta_{s1} = \zeta_{s2}$	Slosh damping ratio	0.5%
ω_{B1}	First bending mode frequency	6.32 rad/s
ω_{B2}	First bending mode frequency	14.1 rad/s
ω_{B3}	First bending mode frequency	20 rad/s
ω_{B4}	First bending mode frequency	31.6 rad/s
ζ_B	Bending damping ratio	0.5%

given by Eqs. (9.3.16), (9.3.17), and (9.3.18) using the linear equations of Chapter 7. All inertial coupling (TWD) effects are retained. In addition, the vehicle model includes representative bending and sloshing modes as shown in Table 9.3. The details of the bending modal data are omitted from the table, but are included in the analysis.

The open-loop response of the vehicle from the gimbal command β_c^- to the rate gyro output $\hat{\dot{\phi}}_y$ is shown in Fig. 9.20. The Bode response is useful for identifying the primary modes of interest, such as the bending

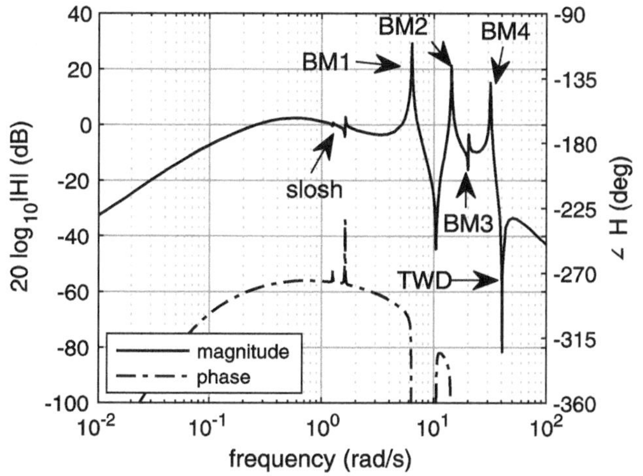

Figure 9.20 Bode response of a complex launch vehicle.

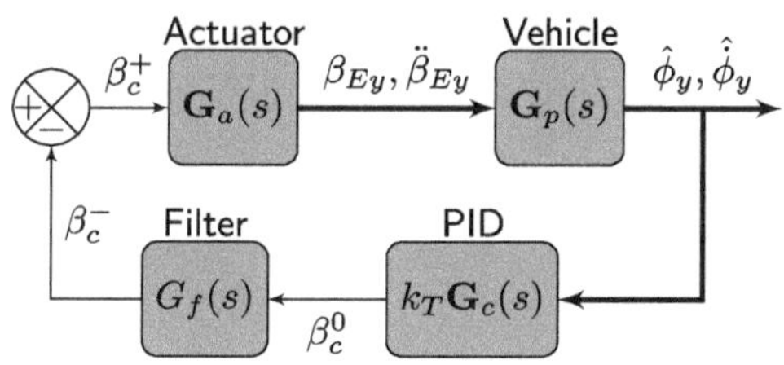

Figure 9.21 Feedback control structure with bending filter.

modes, sloshing modes, and the TWD zero. It should be noted that the TWD zero is shifted down due to interaction with the bending dynamics.

The flight control design for this example requires a *bending filter* to stabilize the bending and sloshing dynamics, and maintain sufficient phase margin that the autopilot feedback is stable. The structure of the system with feedback control is shown in Fig. 9.21. Although the design of such filters is not treated in this book, the filter structure is simple and can be represented as a transfer function that is applied to an intermediate PID controller output β_c^0. The bending filter transfer function is given by

$$G_f(s) = \left(\frac{\tau_1 s + 1}{\tau_2 s + 1}\right) \frac{k_L \omega_f^2}{s^2 + 2\zeta_f \omega_f s + \omega_f^2} \tag{9.4.17}$$

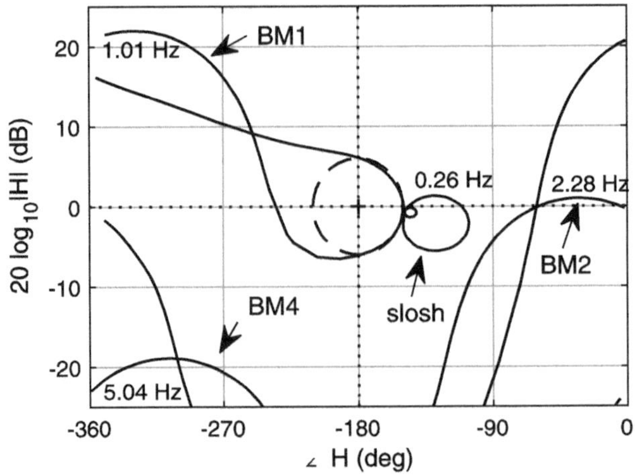

Figure 9.22 Nichols response of a complex launch vehicle.

with $k_L = 0.86$, $\tau_1 = 5.4$ s, $\tau_2 = 4$ s, $\omega_f = 5$ rad/s, and $\zeta_f = 0.707$. This is a simple low-pass filter with a phase lead, constructed using the relationships depicted in Figs. 9.15 and 9.16. The PID control law shown in Fig. 9.21 is designed using Eqs. (9.2.58) and (9.2.59) with $\omega_c = 0.63$ rad/s. An additional gain k_T has been included in the forward path and is set to a nominal value of $k_T = 1.12$.

The Nichols response associated with the complete open-loop system

$$G(s) = k_T G_f(s)\mathbf{G}_c(s)\mathbf{G}_p(s)\mathbf{G}_a(s) \tag{9.4.18}$$

is shown in Fig. 9.22, along with a *disc margin* in the magnitude-phase plane. The disc margin can be thought of as a "keep-out" zone near the critical point. As depicted, the disc margin has extents of ±30° and ±6 dB, consistent with classical stability margin guidelines for control systems. The response of bending modes in the frequency domain appears as the sum of several quadratic factors (Eq. (9.4.6)). Each mode has a peak amplitude that depends on the modal and system parameters, and the frequency response changes by −180° at the bending mode resonant frequencies. In order to depict several modes in the same diagram, the phase in Fig. 9.22 is depicted in modulo-2π form. This means that the phase is wrapped to always lie between 0 and −360°. For complex systems, it is helpful to annotate the mode frequencies as shown.

It is clear from this frequency response that there are multiple phase margins, or points where the frequency response is equal to unity gain. In particular, the first bending mode (BM1), second bending mode (BM2), and sloshing dynamics have a gain that exceeds 0 dB. These modes are said to be *phase stable*. It is worth mentioning that under certain circumstances, phase stabilization of bending will bring about a small increase in the closed-loop damping of the bending mode, which can be advantageous for reducing gust loads.

In contrast, the fourth bending mode (BM4) appears in the open-loop frequency response but is attenuated well below the critical value of 0 dB. Regardless of the relative phasing of this mode, it will still be stable so long as its gain does not increase substantially. A bending or sloshing mode that is attenuated in this fashion is said to be *gain stable*. A gain stable design is desirable as it is robust to phase uncertainty. High-frequency dynamics are more sensitive to latency, and the parameters of high-frequency bending modes are inherently difficult to accurately predict. For this reason, a gain-stable design is preferred for most rockets for all but the first or second bending modes.

Conclusions of Chapter 9

The present chapter has introduced several fundamental concepts that are used in the design and analysis of flight control systems for large rockets, and can assist the modeling and simulation engineer in assessing the stability and performance of preliminary designs. The general topic of launch vehicle flight control is vast, and many important topics for production systems have not been discussed here. These include, but are not limited to, digital effects including latencies and sampling errors, actuator nonlinearity, sensor dynamics, control allocation, blending of multiple rate gyros, and the design of optimal bending filters.

Excepting specialized configurations that require advanced or adaptive control schemes, however, the basic design and analysis techniques presented herein are the core of many launch vehicle designs, and experience shows that simplified analyses using these methods are a useful complement to the full-scale linear and nonlinear simulation approaches of the preceding chapters.

CHAPTER 10

Implementation and analysis

10.1 Numerical integration

The rocket dynamics analysis problem is distinguished by the fact that the mass properties are continuously changing. The analysis of Chapter 2 allows a fixed point to be selected as the origin of the coordinate system. Relative to this fixed point, the first and second moments of inertia must be recomputed continuously as the fuel is expended. The slosh mass locations and frequencies must also be recomputed.

In this discussion, we will consider as an example the equations of motion for the case where the engine dynamics are prescribed and the total vehicle mass is included in the FEM. The equations of motion have the form

$$\mathcal{M}\ddot{\mathbf{x}} = \mathcal{F} \tag{10.1.1}$$

The acceleration vector from Eq. (2.5.17) is given as

$$\ddot{\mathbf{x}} = \begin{bmatrix} \mathbf{a}_b^T & \dot{\boldsymbol{\omega}}^T & \ddot{\boldsymbol{\delta}}_{s1}^T & \ddot{\boldsymbol{\delta}}_{s2}^T & \ldots & \ddot{\boldsymbol{\delta}}_{sN}^T & \ddot{\eta}_1 & \ddot{\eta}_2 & \ldots & \ddot{\eta}_M \end{bmatrix}^T \tag{10.1.2}$$

with the mass matrix $\mathcal{M}$ from Eq. (2.5.19).

The most convenient way to compute the dynamic states $\mathbf{x}(t)$ is to recast these equations in first-order form. Let the quantity $\mathbf{r}_I$ represent the position vector of the body origin in inertial coordinates, and let $\mathbf{q}_{Ib}$ represent a body-to-inertial quaternion. (A helpful discussion on the use of quaternions in simulation is given in Reference [47].) A quaternion $\mathbf{q}$ is a four-element array having a scalar part q_0 and a vector part $\mathbf{q}_v$ such that

$$\mathbf{q} = \begin{bmatrix} q_0 \\ \mathbf{q}_v \end{bmatrix} \tag{10.1.3}$$

where $\mathbf{q}_v = \begin{bmatrix} q_1 & q_2 & q_3 \end{bmatrix}^T$. Using standard relationships for quaternion kinematics, a transformation from body to inertial coordinates can be com-

Dynamics and Simulation of Flexible Rockets
https://doi.org/10.1016/B978-0-12-819994-7.00015-7

puted from the associated quaternion;

$$\mathbf{C}_{Ib}(\mathbf{q}) = \begin{bmatrix} q_0^2 + q_1^2 - q_2^2 - q_3^2 & 2(q_1q_2 + q_0q_3) & 2(q_1q_3 - q_0q_2) \\ 2(q_1q_2 - q_0q_3) & q_0^2 - q_1^2 + q_2^2 - q_3^2 & 2(q_2q_3 + q_0q_1) \\ 2(q_1q_3 + q_0q_2) & 2(q_2q_3 - q_0q_1) & q_0^2 - q_1^2 - q_2^2 + q_3^2 \end{bmatrix}. \tag{10.1.4}$$

The velocity of the body origin in the inertial frame is obtained by integrating its acceleration, expressed in the inertial frame;

$$\dot{\mathbf{v}}_I = \mathbf{C}_{Ib}\mathbf{a}_b. \tag{10.1.5}$$

If the sloshing and elastic generalized displacement states are arranged in a column vector such that

$$\boldsymbol{\delta}_s = \mathrm{col}(\boldsymbol{\delta}_{s1}, \boldsymbol{\delta}_{s2}, \ldots \boldsymbol{\delta}_{sN}) \tag{10.1.6}$$

and

$$\eta = \mathrm{col}(\eta_1, \eta_2, \ldots \eta_M), \tag{10.1.7}$$

it follows that we can write the acceleration vector in Eq. (10.1.2) in the compact form

$$\ddot{\mathbf{x}} = \mathrm{col}\begin{pmatrix} \mathbf{a}_b & \dot{\boldsymbol{\omega}} & \ddot{\boldsymbol{\delta}}_s & \ddot{\eta} \end{pmatrix}. \tag{10.1.8}$$

We also define another acceleration vector that is nearly identical to (10.1.8), given by

$$\ddot{\mathbf{x}}^\star = \mathrm{col}\begin{pmatrix} \dot{\mathbf{v}}_I & \dot{\boldsymbol{\omega}} & \ddot{\boldsymbol{\delta}}_s & \ddot{\eta} \end{pmatrix}. \tag{10.1.9}$$

The former is obtained by inverting the mass matrix in Eq. (10.1.1). The first three elements of Eq. (10.1.9) are transformed to inertial coordinates using

$$\ddot{\mathbf{x}}^\star_{[1\cdots3]} = \mathbf{C}_{Ib}\ddot{\mathbf{x}}_{[1\cdots3]}. \tag{10.1.10}$$

The remaining elements of $\ddot{\mathbf{x}}^\star$ do not require any change. In order to actually integrate the equations numerically, we define a new state vector

$$\begin{aligned} \mathbf{y} &= \mathrm{col}\begin{pmatrix} \mathbf{r}_I & \mathbf{q}_I & \boldsymbol{\delta}_s & \eta & ; & \mathbf{v}_I & \boldsymbol{\omega} & \dot{\boldsymbol{\delta}}_s & \dot{\eta} \end{pmatrix} \\ &= \mathrm{col}\begin{pmatrix} \mathbf{r}_I & \mathbf{q}_I & \boldsymbol{\delta}_s & \eta & ; & \dot{\mathbf{x}}^\star \end{pmatrix}. \end{aligned} \tag{10.1.11}$$

The semicolons in these expressions divide the "top half", consisting of generalized coordinates, from the "bottom half" consisting of $\dot{\mathbf{x}}^\star$. This new state vector is twice as long as $\ddot{\mathbf{x}}$ plus one additional element, since $\mathbf{q}_{Ib}$ has four elements whereas $\boldsymbol{\omega}$ has only three. The transformation matrix $\mathbf{C}_{Ib}$ can be constructed from the quaternion using Eq. (10.1.4), or the transformation from body acceleration to inertial acceleration in Eq. (10.1.5) can be computed directly from the quaternion $\mathbf{q}_{Ib}$. Quaternion rates can be obtained from the body rates by an equation of the form

$$\dot{\mathbf{q}}_{Ib} = \boldsymbol{\Lambda}\left(\mathbf{q}_{Ib}\right)\boldsymbol{\omega} \tag{10.1.12}$$

where $\boldsymbol{\Lambda}$ is the matrix converting body rates to quaternion rates, given by

$$\boldsymbol{\Lambda}\left(\mathbf{q}_{Ib}\right) = \frac{1}{2}\begin{bmatrix} q_1 & q_2 & q_3 \\ -q_0 & -q_3 & q_2 \\ q_3 & -q_0 & -q_1 \\ -q_2 & q_1 & -q_0 \end{bmatrix}. \tag{10.1.13}$$

It is the normal practice that a time domain simulation of a rocket will utilize a quaternion library containing a function corresponding to $\boldsymbol{\Lambda}$ plus other functions such as quaternion multiplication, transpose, etc.

The state vector can be integrated from

$$\frac{d}{dt}\mathbf{y}(t) = \mathrm{col}\begin{pmatrix} \mathbf{v}_I & \dot{\mathbf{q}}_{Ib} & \dot{\boldsymbol{\delta}}_s & \dot{\boldsymbol{\eta}} & ; & \ddot{\mathbf{x}}^\star \end{pmatrix} \tag{10.1.14}$$

using a numerical scheme such as a Runge-Kutta method, which gives an explicit approximation of the integral

$$\Delta\mathbf{y} = \int_{t_0}^{t_0+\Delta T} \dot{\mathbf{y}}(t)\,dt \tag{10.1.15}$$

over some finite time interval ΔT (the simulation time step). This approach allows the computation of position and attitude from the integration of one long state vector.

The quaternion uses four elements to represent three angular degrees of freedom. For a valid representation of the attitude kinematics, the quaternion constraint

$$q_0^2(t) + q_1^2(t) + q_2^2(t) + q_3^2(t) = 1 \tag{10.1.16}$$

must be satisfied for all time. While the equations for quaternion integration (10.1.12) are exact, small inaccuracies in the numerical integration process will cause the quaternion to deviate from the ideal unit norm. One of the reasons for the popularity of quaternions is that it is not necessary to apply an algebraic constraint simultaneously during the integration of the equations. Thus one can first integrate (10.1.14) and subsequently correct $\mathbf{q}_{Ib}$ in order to enforce (10.1.16). Two methods have proven to provide good accuracy and numerical stability; the first is to simply renormalize a quaternion at each timestep using

$$\mathbf{q}_{k+1} = \mathbf{q}_k / \left\| \mathbf{q}_k \right\|, \tag{10.1.17}$$

which is usually adequate but can introduce small numerical artifacts. Another solution is to modify the computation of the quaternion rates in Eq. (10.1.12) as

$$\dot{\mathbf{q}}_{Ib} = \mathbf{\Lambda}\left(\mathbf{q}_{Ib}\right)\omega + k\left(1 - \mathbf{q}_{Ib}^T\mathbf{q}_{Ib}\right)\mathbf{q}_{Ib} \tag{10.1.18}$$

where $k > 0$ is a constant. This method projects the square error in the quaternion normalization onto the quaternion rates with a first-order convergence rate proportional to k, which tends to yield a smoother numerical response. A value of $k = 0.5$ usually gives acceptable results.

10.2 Constraints

The simulation analyst is sometimes confronted with a difficult problem that attempts to answer a simple question: how should one go about simulating the rocket's interactions with the launch structure? Most of the discussion of simulation is concerned with the rocket when it is in flight, but it is often necessary to understand the dynamics of the rocket when it has not yet left the ground. For the purposes of verifying normal operation prior to liftoff, some large rockets start their engines on the launch pad and burn for several seconds before committing to flight. During this time, important dynamic effects are still changing in time and can be computed in simulation, including the mass flow rates, sloshing parameters, and so on. One must also consider the response of the vehicle to thrust preload; that is, the structural potential energy that is accumulated by a propulsion system attempting to accelerate a vehicle that is being held down to a launch structure by pyrotechnic bolts or other supports.

A simple approach is to apply a force equal and opposite the product of the rocket's mass and gravity until its thrust to weight ratio equals one, at which time the force can be removed. However, this does not properly and consistently reproduce the dynamics of the actual vehicle. If the vehicle is flexible, its elastic degrees of freedom are also constrained at specific points, and it will act like a cantilever structure in response to disturbances such as winds. The simple method of applying a force equal to the rocket's weight does not account for these effects.

Witkin [48] presented a simple method for incorporating constraints into a simulation of a system of particles by introducing Lagrange multipliers. Here, we extend that approach to the specific case of a rotating flexible body constrained to a rigid structure which is attached to the earth. Since a body fixed to the earth is not stationary in an inertial frame, some additional mechanisms must be incorporated to properly approximate the constraints. For the present problem, each constraint representing an interface of the vehicle to the launch structure at a particular point can be expressed as a *point coincident constraint*. A point coincident constraint satisfies

$$\mathbf{r}_{IE_i}(t) - \mathbf{r}_{Ip_i}(t) = \mathbf{0} \tag{10.2.1}$$

for all time, where $\mathbf{r}_{IE_i}(t)$ is the position of the point on the earth, and $\mathbf{r}_{Ip_i}(t)$ is the position of the point on the body. The geometry of the constraints is shown in Fig. 10.1. The location of the body point $\mathbf{r}_{Ip_i}(t)$ is given by

$$\mathbf{r}_{Ip_i} = \mathbf{r}_I + \mathbf{C}_{Ib}\left(\mathbf{r}_{p_i} + \Delta\mathbf{r}_{p_i}\right) \tag{10.2.2}$$

where $\mathbf{r}_{p_i}$ is the location of the point with respect to the origin of the body frame, $\Delta\mathbf{r}_{p_i}$ is a perturbation in its position due to flexibility, and $\mathbf{r}_I$ is the location of the body frame origin in the inertial frame.

Let the global constraint function to be satisfied be given by

$$\mathfrak{C}(\mathbf{x}, t) = \mathbf{0} \tag{10.2.3}$$

where $\mathfrak{C}$ is a function with several elements, each in the form of the LHS of (10.2.1), and $\mathbf{x}$ is the state corresponding to the acceleration vector given in Eq. (10.1.2). In the process of integrating the accelerations, one encounters the same issue as in the previous section – quantities expressed in the body frame must first be transformed to an inertial frame before being integrated. This is accomplished indirectly in the following.

A general expression for the time derivatives of Eq. (10.2.3) follows from the chain rule;

$$\dot{\mathfrak{C}} = \frac{\partial \mathfrak{C}}{\partial \mathbf{x}} \frac{\partial \mathbf{x}}{\partial t} + \frac{\partial \mathfrak{C}}{\partial t} \tag{10.2.4}$$

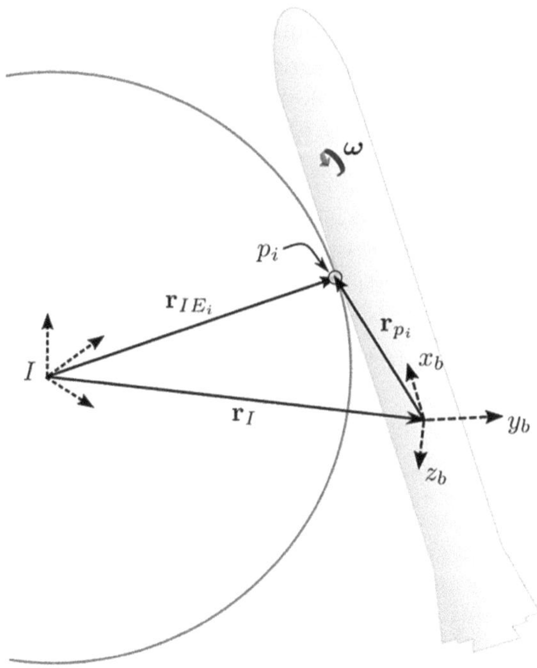

Figure 10.1 Geometry of a vehicle constrained to the earth.

and can be more compactly expressed as

$$\dot{\mathfrak{C}} = \mathbf{J}\dot{\mathbf{x}} + \mathbf{d}, \tag{10.2.5}$$

where $\mathbf{J}$ is the *constraint Jacobian*. The second time derivative of $\mathfrak{C}$ is

$$\ddot{\mathfrak{C}} = \dot{\mathbf{J}}\dot{\mathbf{x}} + \mathbf{J}\ddot{\mathbf{x}} + \dot{\mathbf{d}}. \tag{10.2.6}$$

If the dynamics are expressed by Eq. (10.1.1), suppose that we introduce a *constraint force* $\hat{\mathcal{F}}$ such that

$$\mathcal{M}\ddot{\mathbf{x}} = \mathcal{F} + \hat{\mathcal{F}} \tag{10.2.7}$$

where $\hat{\mathcal{F}}$ is designed to satisfy Eq. (10.2.3). A substitution for $\ddot{\mathbf{x}}$ can be made in Eq. (10.2.6) to yield

$$\ddot{\mathfrak{C}} = \dot{\mathbf{J}}\dot{\mathbf{x}} + \mathbf{J}\mathcal{M}^{-1}\left(\mathcal{F} + \hat{\mathcal{F}}\right) + \dot{\mathbf{d}}. \tag{10.2.8}$$

If Eq. (10.2.3) is to be satisfied, both $\dot{\mathfrak{C}}$ and $\ddot{\mathfrak{C}}$ must be identically zero. Thus,

$$\mathbf{J}\mathcal{M}^{-1}\hat{\mathcal{F}} = -\dot{\mathbf{J}}\dot{\mathbf{x}} - \mathbf{J}\mathcal{M}^{-1}\mathcal{F} - \dot{\mathbf{d}}. \tag{10.2.9}$$

It can be shown via the principle of virtual work that the constraint forces $\hat{\mathcal{F}}$ must be constructed from the vectors comprising the null space complement of $\mathbf{J}$ [48] and can be written as

$$\hat{\mathcal{F}} = \mathbf{J}^T\boldsymbol{\lambda} \tag{10.2.10}$$

where $\boldsymbol{\lambda}$ is a Lagrange multiplier vector. Substituting Eq. (10.2.10) into Eq. (10.2.9) yields

$$\left(\mathbf{J}\mathcal{M}^{-1}\mathbf{J}^T\right)\boldsymbol{\lambda} = -\dot{\mathbf{J}}\dot{\mathbf{x}} - \mathbf{J}\mathcal{M}^{-1}\mathcal{F} - \dot{\mathbf{d}}, \tag{10.2.11}$$

where the quantity in parentheses emphasizes the fact that it is possible to solve Eq. (10.2.11) for $\boldsymbol{\lambda}$. The rank of the matrix $\mathbf{J}\mathcal{M}^{-1}\mathbf{J}^T$ depends on the number of constraints. If the degrees of freedom equal the number of constraints, the system will have a unique solution. Otherwise, the system will either be overconstrained or underconstrained. In general, a numerical solution of Eq. (10.2.11) is possible using a pseudoinverse. Using this $\boldsymbol{\lambda}$ and Eq. (10.2.10), the augmented system (10.2.7) will satisfy the constraint for all time.

In order to implement this scheme, we must first construct the quantities $\mathbf{J}$ and $\dot{\mathbf{J}}$ in Eq. (10.2.11). As it turns out, these can be derived explicitly for the case of a point coincident constraint on a flexible rocket. Consider using (10.2.1) and (10.2.2) to represent the constraint function $\mathfrak{C}$ for a single point. It follows that

$$\mathfrak{C} = -\left[\mathbf{r}_I + \mathbf{C}_{Ib}\left(\mathbf{r}_{p_i} + \boldsymbol{\Psi}_{p_i}\boldsymbol{\eta}\right)\right] + \mathbf{r}_{IE_i}(t) \tag{10.2.12}$$

where the elastic displacement of a point p_i on the rocket is given by

$$\Delta\mathbf{r}_{p_i} = \boldsymbol{\Psi}_{p_i}\boldsymbol{\eta}. \tag{10.2.13}$$

The quantity $\boldsymbol{\Psi}_{p_i}$ is the familiar mode shape function at the point p_i. The first derivative of the constraint function is

$$\dot{\mathfrak{C}} = \dot{\mathbf{r}}_{IE_i}(t) - \left[\mathbf{v}_I + \mathbf{C}_{Ib}\boldsymbol{\Psi}_{p_i}\dot{\boldsymbol{\eta}} + \mathbf{C}_{Ib}\boldsymbol{\omega}^{\times}\left(\mathbf{r}_{p_i} + \boldsymbol{\Psi}_{p_i}\boldsymbol{\eta}\right)\right] \tag{10.2.14}$$

and its second derivative is given by

$$\begin{aligned}\ddot{\mathfrak{C}} = \ddot{\mathbf{r}}_{IE_i}(t) - [\mathbf{C}_{Ib}\mathbf{a}_b & \\ + \mathbf{C}_{Ib}\boldsymbol{\Psi}_{p_i}\ddot{\boldsymbol{\eta}} + \mathbf{C}_{Ib}2\boldsymbol{\omega}^{\times}\boldsymbol{\Psi}_{p_i}\dot{\boldsymbol{\eta}} + \mathbf{C}_{Ib}\dot{\boldsymbol{\omega}}^{\times}\left(\mathbf{r}_{p_i} + \boldsymbol{\Psi}_{p_i}\boldsymbol{\eta}\right) & \\ -\mathbf{C}_{Ib}\boldsymbol{\omega}^{\times}\left(\mathbf{r}_{p_i} + \boldsymbol{\Psi}_{p_i}\boldsymbol{\eta}\right)^{\times}\boldsymbol{\omega}\big] &\end{aligned} \tag{10.2.15}$$

where Eq. (10.1.5) has been used to write the inertial acceleration in terms of the body acceleration, $\mathbf{a}_b$. We can group terms and write these expressions in matrix form as

$$\dot{\mathfrak{C}} = -\mathbf{C}_{Ib}\begin{bmatrix} \mathbf{1} & -\left(\mathbf{r}_{p_i} + \boldsymbol{\Psi}_{p_i}\boldsymbol{\eta}\right)^{\times} & \mathbf{O}_{3\times N} & \boldsymbol{\Psi}_{p_i} \end{bmatrix}\begin{bmatrix} \mathbf{C}_{bI}\mathbf{v}_I \\ \boldsymbol{\omega} \\ \dot{\boldsymbol{\delta}}_s \\ \dot{\boldsymbol{\eta}} \end{bmatrix} + \dot{\mathbf{r}}_{IE_i}(t) \tag{10.2.16}$$

and

$$\begin{aligned}\ddot{\mathfrak{C}} = & \left\{-\mathbf{C}_{Ib}\boldsymbol{\omega}^{\times}\begin{bmatrix} \mathbf{O} & -\left(\mathbf{r}_{p_i} + \boldsymbol{\Psi}_{p_i}\boldsymbol{\eta}\right)^{\times} & \mathbf{O}_{3\times N} & 2\boldsymbol{\Psi}_{p_i} \end{bmatrix}\right\}\begin{bmatrix} \mathbf{C}_{bI}\mathbf{v}_I \\ \boldsymbol{\omega} \\ \dot{\boldsymbol{\delta}}_s \\ \dot{\boldsymbol{\eta}} \end{bmatrix} \\ & + \left\{-\mathbf{C}_{Ib}\begin{bmatrix} \mathbf{1} & -\left(\mathbf{r}_{p_i} + \boldsymbol{\Psi}_{p_i}\boldsymbol{\eta}\right)^{\times} & \mathbf{O}_{3\times N} & \boldsymbol{\Psi}_{p_i} \end{bmatrix}\right\}\begin{bmatrix} \mathbf{a}_b \\ \dot{\boldsymbol{\omega}} \\ \ddot{\boldsymbol{\delta}}_s \\ \ddot{\boldsymbol{\eta}} \end{bmatrix} + \ddot{\mathbf{r}}_{IE_i}(t).\end{aligned} \tag{10.2.17}$$

Comparing these expressions with (10.2.5) and (10.2.6), it is clear that for this point,

$$\mathbf{J}_i = -\mathbf{C}_{Ib}\begin{bmatrix} \mathbf{1} & -\left(\mathbf{r}_{p_i} + \boldsymbol{\Psi}_{p_i}\boldsymbol{\eta}\right)^{\times} & \mathbf{O}_{3\times N} & \boldsymbol{\Psi}_{p_i} \end{bmatrix} \tag{10.2.18}$$

$$\dot{\mathbf{J}}_i = -\mathbf{C}_{Ib}\boldsymbol{\omega}^{\times}\begin{bmatrix} \mathbf{O} & -\left(\mathbf{r}_{p_i} + \boldsymbol{\Psi}_{p_i}\boldsymbol{\eta}\right)^{\times} & \mathbf{O}_{3\times N} & 2\boldsymbol{\Psi}_{p_i} \end{bmatrix}, \tag{10.2.19}$$

with the body frame velocity state vector

$$\dot{\mathbf{x}} = \mathrm{col}\begin{pmatrix} \mathbf{C}_{bI}\mathbf{v}_I & \boldsymbol{\omega} & \dot{\boldsymbol{\delta}}_s & \dot{\boldsymbol{\eta}} \end{pmatrix}. \tag{10.2.20}$$

It should be noted that while the notation $\dot{\mathbf{J}}_i$ is convenient, this quantity should not be interpreted as a simple derivative since $\mathbf{J}_i$ is not a function of time alone. It is more appropriate to consider $\dot{\mathbf{J}}_i$ as a quantity that must be obtained by differentiating the constraint function.

The remaining quantities $\dot{\mathbf{r}}_{IE_i}(t)$ and $\ddot{\mathbf{r}}_{IE_i}(t)$ are simply the velocity and acceleration of a point fixed to the earth, which are

$$\dot{\mathbf{r}}_{IE_i}(t) = \mathbf{d} = \boldsymbol{\omega}_{IE}^{\times}\mathbf{r}_{IE_i}(t) \tag{10.2.21}$$

$$\ddot{\mathbf{r}}_{IE_i}(t) = \dot{\mathbf{d}} = \boldsymbol{\omega}_{IE}^{\times}\boldsymbol{\omega}_{IE}^{\times}\mathbf{r}_{IE_i}(t) \tag{10.2.22}$$

where $\boldsymbol{\omega}_{IE}$ is the earth angular rate expressed in the inertial frame.

In order to implement this scheme, we first determine the inertial time history of $\mathbf{r}_{IE_i}(t)$, which for a point on the earth, can be computed from its initial position using standard geodetic transformations [26]. Next, we construct the constraint Jacobian and its time derivative

$$\mathbf{J} = \begin{bmatrix} \mathbf{J}_1 \\ \mathbf{J}_2 \\ \vdots \end{bmatrix} \dot{\mathbf{J}} = \begin{bmatrix} \dot{\mathbf{J}}_1 \\ \dot{\mathbf{J}}_2 \\ \vdots \end{bmatrix} \tag{10.2.23}$$

and use the unconstrained force vector $\mathcal{F}$ to solve Eq. (10.2.11) for the Lagrange multiplier vector $\boldsymbol{\lambda}$. The total force vector is then substituted into the dynamic equations (10.2.7). The constraints can be arbitrarily added and deleted by removing their elements $\mathbf{J}_i$ from the Jacobian in Eq. (10.2.23).

A few practical issues arise in the implementation. First, the solution of (10.2.11) enforces constraints at the acceleration level, and so numerical errors will eventually cause the constraints to drift. Witkin suggested augmenting (10.2.11) with a proportional and derivative correction factor, such that we instead solve the equation

$$(\mathbf{J}\mathcal{M}^{-1}\mathbf{J}^T)\boldsymbol{\lambda} = -\dot{\mathbf{J}}\dot{\mathbf{x}} - \mathbf{J}\mathcal{M}^{-1}\mathcal{F} - \dot{\mathbf{d}} - k_0\mathfrak{C} - k_1\dot{\mathfrak{C}}. \tag{10.2.24}$$

The corrector coefficients k_0, k_1 can be chosen somewhat arbitrarily, but it is reasonable to choose k_0 empirically and then let

$$k_1 = 2\sqrt{k_0} \tag{10.2.25}$$

which gives the off-constraint dynamics an approximately critically-damped response. Secondly, the practical computation of the quantity $(\mathbf{J}\mathcal{M}^{-1}\mathbf{J}^T)^{-1}$ requires a numerically robust pseudoinverse routine, particularly one that discards near-zero singular values and handles variations in rank as constraints are added and removed.

10.3 Monte Carlo analysis

The general approach of Monte Carlo analysis in simulation is to vary the input parameters according to specific probability distributions, run the simulation many times, and then compute the statistics of the output. This process can be used to statistically verify requirements, in the sense that a single simulation *run* is treated as a member of a *sample* of runs. Since the number of runs is finite, and therefore the sample size is limited, any estimates using the output variables must take into account the uncertainty introduced by the finite sample size.

The formal theory of Monte Carlo analysis is based on acceptance sampling, originally developed for production quality assurance applications. These theories developed the formal definitions of *producer risk* and *consumer risk*. Producer risk is the probability of incorrectly rejecting a design which actually meets requirements, whereas consumer risk is the probability of incorrectly accepting a design which does not meet requirements.

The consumer risk is the same as Type II error probability in statistical hypothesis testing and is denoted by a probability β. The null hypothesis H_0, as given in Table 10.1, is that a rocket design meets requirements. The analyst of a rocket is mostly concerned with the reducing β. That is, the cost of rejecting a good design is ultimately less than accepting a bad one, in the sense that a rocket failure is more costly than an excessively safe one. For this reason, in this section the "confidence" $(1 - \beta)$ in the estimator will be used to describe the complement of β, rather than the complement of the Type I error probability α. A detailed discussion of the theory is given in Reference [33].

Table 10.1 Hypothesis test for Monte Carlo simulation.

	H_0 **true (design meets requirements)**	H_0 **false (design does not meet requirements)**
Design accepted	Correct decision	Type II error (β)
Design rejected	Type I error (α)	Correct decision

Consumer risk is a quantifiable value that depends on the sample size, or the population being tested. Suppose a small sounding rocket for lofting science payloads is manufactured by a producer and flight tested by its customer. The customer might require that a certain number of test flights, say ten, be conducted during an "acceptance testing" period before the rocket is deemed reliable enough to be declared "operational." Since the rocket is being produced continuously, it can be assumed for simplicity that the ten vehicles selected from the production line are a random sample of an infinite population. In this example, suppose that nine out of ten launches are successful, so the rocket manufacturer proudly concludes that the system has 90% reliability. This estimate has significant uncertainty, with two possibilities to be considered:

1. The rocket reliability is much better than 90%, and the customer just happened to have an unlikely failure (producer risk);
2. The rocket reliability is much worse than 90%, and the customer just happened to get 9 good rockets (consumer risk).

The astute customer is likely to adjust his or her estimates of the reliability to a specified consumer risk; that is, the customer would like the estimate to have a probability of error (underestimating the failure rate) of no greater than, say, 5%. This is analogous to saying that if the estimated reliability is $\hat{p}_s$ (for probability of success) and the consumer risk is β, subsequent samples (sets of test flights) have no greater than a β chance of exhibiting a failure rate greater than $\hat{p}_f = 1 - \hat{p}_s$.

For the numerical example at hand, the sample size is $N = 10$ and the number of failures is $k = 1$. Let us introduce the *naive estimator*

$$\hat{p}_{s?} = 1 - \frac{k}{N} \tag{10.3.1}$$

which gives $\hat{p}_{s?} = 0.9$. The "?" symbol indicates that the consumer risk is uncontrolled and varies with the sample size. Eq. (10.3.1) is an estimator of the actual success rate p_s, which can never be known unless the entire population is sampled. As it turns out, the *best* consumer risk for the naive estimator (as $N \rightarrow \infty$) is 50%! In this example, the estimated reliability at the 95% confidence level ($\beta = 0.05$) is

$$\hat{p}_{s95} = 0.606. \tag{10.3.2}$$

A test using only ten flights tells the customer very little about the actual overall reliability. In fact, the naive estimator is grossly under-conservative for "small" N (<100,000). The subject of acceptance sampling theory is concerned, in part, with developing better estimators that provide results having small β for reasonably-sized samples.

In the realm of Monte Carlo analysis, these principles are used to make predictions about the success or failure of a mission or a set of requirements based on a sample of simulation runs. The most common metric is a binomial metric, that is, some output of the simulation that can assume only a value of "pass" or "fail". An example would be using a Monte Carlo simulation to determine the proportion of runs which were able to achieve orbit.

The problems that arise in using the naive estimator are shown in Fig. 10.2. Clearly one desires an estimate with high confidence for a modest number of Monte Carlo simulations. This issue can be resolved fairly easily using basic binomial statistics.

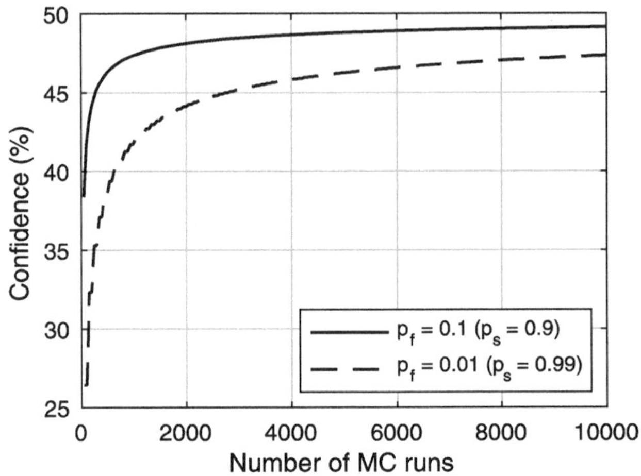

Figure 10.2 Confidence in Monte Carlo analysis for naive binomial estimator.

A slightly more complicated metric is a quantile estimate, which involves extending the methods developed for binomial statistics to continuous, nonparametric output distributions using the method of order statistics. The term nonparametric means that quantiles can be computed without requiring that the output variables be fit to some distribution, e.g., a normal distribution. An example of such a quantile estimate would be the value of the dynamic pressure the majority (say, 99%) of the actual rockets flown will not exceed. This type of value can be estimated using Monte Carlo analysis, and anchored to a value of acceptable consumer risk. The use of quantile estimators using nonparametric statistics is common in rocket designs, and is discussed extensively in Reference [33].

In practice, a rocket is developed by integrating subsystems across disciplines, and each subsystem must meet a set of requirements. These requirements are statistical in nature. It is the job of the systems engineer to allocate each subsystem's probability of meeting requirements with the acceptable consumer risk in the estimates so that the overall integrated system has acceptable performance. Increasing the reliability or performance requirements of a system and verifying them with very little risk rapidly becomes costly. For simulation analysis, this means a large number of Monte Carlo runs, and for hardware, a very large number of tests.

The methods of acceptance sampling are based on the binomial distribution,

$$P_{\mathrm{BIN}}\left(k, p_f, N\right) = \binom{N}{k} p_f^k \left(1 - p_f\right)^{N-k}. \tag{10.3.3}$$

The binomial distribution is the probability that a sample of size N will have exactly k failures when the actual population failure rate (which, in practice, is unknown) is p_f. The cumulative distribution

$$F_{\mathrm{BIN}}\left(k, p_f, N\right) = \sum_{j=0}^{k} \binom{N}{j} p_f^j \left(1 - p_f\right)^{N-j} \tag{10.3.4}$$

is the probability that k or *less* failures are observed in sample of size N when the underlying failure rate is p_f. Here, N could be considered a number of Monte Carlo runs and k is the observed number of failures. The consumer risk is the probability that the actual failure rate is equal to or greater than the failure rate predicted by the sample. For k, N, and β fixed, Eq. (10.3.4) can be solved numerically to yield the estimated failure probability $\hat{p}_f^\star$:

$$\hat{p}_f^\star \mid F_{\mathrm{BIN}}\left(k, \hat{p}_f^\star, N\right) \leq \beta \tag{10.3.5}$$

The inequality (10.3.5) defines the relationship between consumer risk, the characteristics of the sample, and the failure rate. It thus forms the basis of finding *sampling plans*, which help the analyst to determine the number of acceptable failures k in a Monte Carlo set of N runs. For example, one solution for verifying a requirement that a vehicle design is 99% successful ($p_f = 0.01$) with 90% confidence ($\beta = 0.1$) is that $N = 531$ and $k \leq 2$. A few examples of sampling plans are given in Table 10.2. A comprehensive set of tables is provided in Reference [33].

Sampling plans provide the guidance as to how many simulations are necessary to verify a given requirement. It is helpful that for $k = 0$,

$$\beta = F_{\mathrm{BIN}}\left(0, p_f, N\right) = p_s^N \tag{10.3.6}$$

Table 10.2 Example sampling plans for Monte Carlo analysis.

N	k	p_s	β (%)
52	2	0.9	<10%
105	2	0.95	<10%
531	2	0.99	<10%
852	0	0.9973	<10%
1705	0	0.99865	<10%

and the minimum number of Monte Carlo runs needed to verify a requirement at the level of p_s with a consumer risk β is

$$N_{min} = \left\lceil \frac{\ln \beta}{\ln p_s} \right\rceil \tag{10.3.7}$$

Likewise, the maximum verifiable requirement for a given number of Monte Carlo runs is

$$p_{s\max} = \beta^{\frac{1}{N}}. \tag{10.3.8}$$

Intuitively, verifying very high performance or high-reliability systems at a high level of confidence increases cost. For example, a requirement of $p_s =$ 99.999% verified via Monte Carlo simulation at $\beta = 0.1$ (90% confidence) requires 230,258 runs with no failures.

The reader is cautioned that the statistical methods of Monte Carlo simulation address the problem of *aleatory uncertainty*, which can be thought of as the result of stochastic variation in the model parameters under the assumption that the model is correct. In essence, it assumes that the model is representative of reality, and a simulated flight and actual flight are equivalent. Obviously, a simulation is not a perfect model. Such "unknowable" differences between the simulation and the actual flight are classified as *epistemic uncertainty*, which must be accounted for separately through the allocation of margins and a careful design of simulation experiments. This topic is discussed in detail in Reference [33].

APPENDIX A

List of symbols and acronyms

Lower case bold (vectors)

$\mathbf{a}_b$	acceleration of the origin expressed in the body frame
$\mathbf{a}_{b0}$	nominal sensed acceleration in body frame, $\begin{bmatrix} \frac{F-D}{m_T} & 0 & 0 \end{bmatrix}^T$
$\mathbf{a}'_b$	$\mathbf{a}_b - \mathbf{a}_{b0}$
$\mathbf{b}_1$	location of hinge point H for body 1 (Fig. 4.1)
$\mathbf{b}_j$	location of hinge point H for body j
$\mathbf{b}_{sj}$	equilibrium position of the jth slosh mass (Fig. 2.1)
$\mathbf{b}_{Hj}$	location of hinge point H for slosh pendulum j, Eq. (4.3.2)
$\mathbf{c}_{EFi}$	engine/flex coupling vector, Eq. (2.4.22)
$\mathbf{d}$	system first moment of inertia about the origin, Eq. (4.1.4). Same as $\mathbf{s}_{TD}$.
$\mathbf{d}$	Chapter 10, velocity of earth constraint point.
$\mathbf{f}$	sum of external forces (thrust, gravity, RCS, and aerodynamic)
$\mathbf{f}_0$	external force acting on body 0 (part of $\mathbf{f}$)
$\mathbf{f}_1$	external force acting on body 1 (part of $\mathbf{f}$)
$\mathbf{f}_{NL}$	nonlinear force
$\mathbf{f}_{presc}$	force due to prescribed engine motion
$\mathbf{f}_R$	gimbaled thrust vector
$\mathbf{f}_{sj}$	force acting on slosh mass j
$\mathbf{g}$	sum of external torques about the origin (thrust, gravity, aerodynamic, and fuel offset)
$\mathbf{g}_1$	external torque acting on body 1
$\mathbf{g}_E$	total torque acting on the engine (TVC, gravity, aerodynamic) in engine frame
$\mathbf{g}_{Eb}$	total torque acting on the engine in rocket body frame. See Eq. (2.4.18).
$\mathbf{g}_{H1}$	torque about hinge from body 0 acting on body 1 (interbody torque)
$\mathbf{g}_{NL}$	nonlinear torque about the rocket origin

$\mathbf{g}_{\text{offset}}$	torque due to fuel offset, Section 5.2.
$\mathbf{g}_{\text{presc}}$	torque due to prescribed engine motion
$\mathbf{g}_{\text{RCS}}$	total moment about origin due to RCS thrusters
$\mathbf{h}$	system angular momentum
$\mathbf{h}_0$	angular momentum of body 0
$\mathbf{h}_1$	angular momentum of body 1
$\mathbf{p}$	system linear momentum
$\mathbf{p}_0$	linear momentum of body 0
$\mathbf{p}_1$	linear momentum of body 1
$\mathbf{q}_{Ib}$	quaternion transforming from body to inertial coordinates
$\mathbf{r}$	Location of a mass element with the rocket (Fig. 2.1).
$\mathbf{r}_1$	Chapter 2, $\mathbf{r} - \mathbf{r}_G$, location of engine mass element relative to gimbal
$\mathbf{r}_1$	Chapter 4, pendulum vector for body 1 (Fig. 4.1)
$\mathbf{r}_{cm}$	location of center of mass
$\mathbf{r}_G$	location of the gimbal point (Fig. 2.1).
$\mathbf{r}_I$	location of the body frame origin in an inertial frame
$\mathbf{r}_j$	pendulum vector for the jth slosh mass
$\mathbf{r}_{sj}$	location of the jth slosh mass (Fig. 2.1).
$\mathbf{r}_{\text{RCS}k}$	vector from origin to RCS thruster k
$\mathbf{s}_E$	first moment of inertia of the engine about its gimbal, expressed in engine frame
$\mathbf{s}_{Eb}$	first moment of inertia of the engine about its gimbal, in body frame
$\mathbf{s}_T$	first moment of inertia of the total vehicle about its origin, with slosh and engine masses in their equilibrium positions.
$\mathbf{s}_{TD}$	dynamic first moment of inertia, slosh and engine masses in actual positions
$\mathbf{u}$	input to state space model in the form of Eq. (7.0.1)
$\mathbf{u}_{\text{RCS}}$	RCS on-off vector, Eq. (5.1.72)
$\mathbf{u}_v$	unit vector in the downward vertical direction expressed in body frame
$\mathbf{v}$	velocity of the body frame origin in body coordinates
$\mathbf{v}_H$	absolute velocity of hinge point, Table 4.1
$\mathbf{v}_I$	velocity of the body frame origin in inertial coordinates
$\mathbf{v}_{\text{rel}}$	Velocity of body frame origin with respect to atmosphere, including wind
$\mathbf{v}_{sj}$	velocity of slosh mass j relative to the attach point
$\boldsymbol{\delta}_{sj}$	perturbation in the position of the jth slosh mass (Fig. 2.1)

$\boldsymbol{\phi}$	(no subscript) vector of body axis rotations, $\begin{bmatrix} \phi_x & \phi_y & \phi_z \end{bmatrix}^T$
$\boldsymbol{\phi}_i$	eigenvector of mode i of the structural dynamic analysis
$\boldsymbol{\psi}_i(\mathbf{r})$	translational displacement vector for flex mode i, Eq. (2.2.28)
$\boldsymbol{\psi}_{ji}$	translational displacement vector for slosh mass j and mode i Eq. (2.3.13)
$\boldsymbol{\psi}_{\beta i}$	translational displacement vector for flex mode i at the gimbal point
$\boldsymbol{\sigma}_i(\mathbf{r})$	vector of rotations for flex mode i at location $\mathbf{r}$
$\boldsymbol{\sigma}_{\beta i}$	vector of rotations for flex mode i at the gimbal
$\boldsymbol{\beta}_E$	vector of engine angular displacements, $\begin{bmatrix} 0 & \beta_{Ey} & \beta_{Ez} \end{bmatrix}^T$
$\vec{\rho}$	vector from the origin to the location of mass element dm (Fig. 3.1)
$\rho(\mathbf{r})$	density of rocket at location $\mathbf{r}$
$\boldsymbol{\omega}$	absolute angular velocity of the body coordinate frame
$\boldsymbol{\omega}_1$	absolute angular velocity of body 1
$\boldsymbol{\omega}_j$	absolute angular velocity of slosh pendulum j
$\boldsymbol{\omega}_{rj}$	angular velocity of pendulum j relative to the rocket body frame
$\boldsymbol{\omega}_E$	relative angular velocity of the engine expressed in the engine frame
$\boldsymbol{\omega}_{Eb}$	relative angular velocity of the engine expressed in the rocket body frame
$\boldsymbol{\chi}_{Ei}$	engine flex displacement factor for mode i, Eq. (2.6.2)

Upper case bold (matrices)

$\mathbf{1}$	identity matrix. (3×3 identity matrix for all but Section 2.2)
$\mathbf{C}_{01}$	transformation matrix from body 1 to body 0 coordinates
$\mathbf{C}_{10}$	transformation matrix from body 0 to body 1 coordinates
$\mathbf{C}_{0I}$	transformation matrix from inertial to body 0 coordinates
$\mathbf{C}_{I0}$	transformation matrix from body 0 to inertial coordinates
$\mathbf{C}_{Ib}$	transformation matrix from integrated body frame to inertial coordinates
$\mathbf{I}_T$	inertia matrix of the total vehicle about the origin, slosh masses in equilibrium position
$\mathbf{I}_{TD}$	inertia matrix of the total vehicle about the origin, slosh masses in actual positions
$\mathbf{I}_E$	inertia matrix of the engine about the gimbal, expressed in engine frame

$\mathbf{I}_{Eb}$	inertia matrix of the engine about the gimbal, expressed in rocket body frame
$\mathbf{I}_{TWD}$	Tail Wags Dog inertia, Eq. (2.4.23)
$\mathbf{J}$	system inertia matrix (same as $\mathbf{I}_{TD}$)
$\mathbf{J}_1$	inertia matrix of body 1 in body 1 frame.
$\mathbf{J}_{01}$	Eq. (4.2.19)
$\mathbf{J}_{10}$	Eq. (4.2.15)
$\mathbf{K}_B$	stiffness matrix of the structural dynamic model
$\mathbf{M}_B$	mass matrix of the structural dynamic model
$\mathbf{O}$	null matrix
$\mathbf{0}$	null vector
$\mathbf{\Omega}_B$	diagonal mode frequency matrix
$\mathbf{\Omega}_{sj}$	diagonal slosh frequency matrix

Italic (scalars)

Units shown assume each generalized modal mass = 1

a_{bx}, a_{by}, a_{bz}	components of $\mathbf{a}_b$	ft/sec^2
S_{ref}	aerodynamic reference area	ft^2
C_A	Chapter 6, actuator damping coefficient	ft · lbf · sec/rad
$C_{N\alpha}$	rigid-body aerodynamic normal force coefficient slope (negative)	(-)
$C_{Y\beta}$	rigid-body aerodynamic side force coefficient slope (positive)	(-)
$C_{y\eta i}$	wind bending Y force coefficient, Eq. (5.3.25)	$1/(\text{rad}\sqrt{\text{slug}})$
$C_{z\eta i}$	wind bending Z force coefficient, Eq. (5.3.26)	$1/(\text{rad}\sqrt{\text{slug}})$
D	drag force	lbf
$D(s)$	Chapter 9, denominator polynomial	-
F	thrust of all engines	lbf
F_0	thrust of fixed (ungimbaled) engines	lbf
F_R	thrust of gimbaled engines	lbf
g	acceleration due to gravity	ft/sec^2
$\bar{g}$	sensed acceleration $(F-D)/m_T$	ft/sec^2
I_{EG}	y or z component, moment of inertia of engine about gimbal	slug-ft^2
I_{yy}, I_{zz}	pitch, yaw moment of inertia (including engines and sloshing fluid) about origin	slug-ft^2

k_A	Chapter 6, actuator torsional spring constant	ft · lbf/rad
K	Chapter 5, number of RCS thrusters	(-)
K_A	Chapter 6, actuator linear spring constant	lbf/ft
K_C	Chapter 6, actuator attach point spring constant	lbf/ft
l_E	distance from gimbal to engine CM (negative aft of gimbal), Fig. 6.5	ft
l_{pj}	length of slosh pendulum	ft
L	Chapter 7, dimension of X vector, $6 + 2N + M$	(-)
m_T	total vehicle mass (including engines and sloshing fluid)	slugs
m_E	mass of gimbaled engines	slugs
m_{Bi}	generalized mass, i^{th} bending mode.	(-)
m_{sj}	slosh mass, j^{th} tank	slugs
M	number of flexible modes	(-)
N	number of tanks.	(-)
$N(s)$	Chapter 9, numerator polynomial	
$\bar{q}$	dynamic pressure	lbf/ft^2
$Q_{\eta i}$	excitation of the ith mode, Eq. (5.3.29)	$\text{lbf}/(\text{rad}\sqrt{\text{slug}})$
S_{Ex}	x-component of $\mathbf{s}_E$	slug-ft
$\ddot{x}_b, \ddot{y}_b, \ddot{z}_b$	accelerations of body-frame origin	ft/sec^2
x_{sj}, y_{sj}, z_{sj}	displacements of slosh mass j	ft
X_1	location of beginning of solid propellant for jet damping analysis, Fig. 5.3	
X_{cp}	center of pressure x location measured from origin	ft
X_{ex}	location of propellant exit plane for jet damping analysis, Fig. 5.2	ft
X_G	gimbal x coordinate (negative number), Fig. 5.1	
X_{liq}	location of liquid surface for jet damping analysis, Fig. 5.2	
X_{sj}	slosh mass equilibrium x location measured from origin	ft
α_y, α_z	angle of attack of undeformed vehicle	rad
β_{Ey}	engine angle about the pitch axis	rad
β_{Ez}	engine angle about the yaw axis	rad
η_i	amplitude of the ith flexible mode	$\text{ft}\sqrt{\text{slug}}$
ϕ_y	pitch angle of undeformed centerline relative to trajectory frame	rad

ϕ_z	yaw angle of undeformed centerline relative to trajectory frame	rad
ψ_{yji}	y-bending mode shape of the ith mode at the jth slosh location	$1/\sqrt{slug}$
ψ_{zji}	z-bending mode shape of the ith mode at the jth slosh location	$1/\sqrt{slug}$
$\psi_{y\beta i}$	y-bending mode shape of the ith mode at the gimbal	$1/\sqrt{slug}$
$\psi_{z\beta i}$	z-bending mode shape of the ith mode at the gimbal	$1/\sqrt{slug}$
$\sigma_{y\beta i}$	y-rotation of the ith mode at the gimbal (pitch)	$rad/(foot\sqrt{slug})$
$\sigma_{z\beta i}$	z-rotation of the ith mode at the gimbal (yaw)	$rad/(foot\sqrt{slug})$
ζ_{Bi}	damping factor, i^{th} bending mode, (same for y and z)	(-)
ζ_{sj}	slosh damping factor, j^{th} tank, (same for y and z)	(-)
ω_A	Chapter 6, actuator natural frequency	rad/sec
Ω_{Bi}	natural frequency, i^{th} bending mode	rad/sec
Ω_{sj}	slosh natural frequency, j^{th} tank	rad/sec

Acronyms

DOF	Degree of Freedom
DWT	Dog Wags Tail
FEM	Finite Element Model
FVT	Final Value Theorem
GSVD	Generalized Singular Value Decomposition
ITAE	Integral product of Time and Absolute Error
LHS	Left Hand Side
LSQI	Least Squares Quadratic Inequality
RHS	Right Hand Side
SISO	Single Input Single Output
TVC	Thrust Vector Control
TWD	Tail Wags Dog

APPENDIX B

Quadruple vector product

This appendix is concerned with an apparent contradiction in the expressions for centrifugal torque. Assuming a coordinate frame rotating at an angular rate $\boldsymbol{\omega}$, and a point mass m fixed at a location $\mathbf{r}$ in this frame, the centrifugal force on the mass is given by $m\boldsymbol{\omega}^{\times}\mathbf{r}^{\times}\boldsymbol{\omega}$. The centrifugal torque about the origin is thus

$$\mathbf{g}_c = m\mathbf{r}^{\times}\boldsymbol{\omega}^{\times}\mathbf{r}^{\times}\boldsymbol{\omega} \tag{B.0.1}$$

For the case of a body, the Euler coupling term $-\boldsymbol{\omega}^{\times}\mathbf{I}\boldsymbol{\omega}$ (that is, the gyroscopic torque) represents the integrated effect of all the centrifugal torques from each point of the body. In the present case, we are only considering one such point, so the "inertia matrix" from (2.1.21) can be written as

$$\mathbf{I} = -m\mathbf{r}^{\times}\mathbf{r}^{\times} \tag{B.0.2}$$

Thus

$$\mathbf{g}_c = -\boldsymbol{\omega}^{\times}\mathbf{I}\boldsymbol{\omega} = m\boldsymbol{\omega}^{\times}\mathbf{r}^{\times}\mathbf{r}^{\times}\boldsymbol{\omega} \tag{B.0.3}$$

The apparent contradiction is that the two forms (B.0.1) and (B.0.3) appear to differ. The objective of this appendix is to demonstrate that the two expressions are in fact equivalent.

We have the following vector identity:

$$\vec{\mathbf{a}} \times \left(\vec{\mathbf{b}} \times \vec{\mathbf{c}}\right) = (\vec{\mathbf{a}} \cdot \vec{\mathbf{c}})\,\vec{\mathbf{b}} - \left(\vec{\mathbf{a}} \cdot \vec{\mathbf{b}}\right)\vec{\mathbf{c}} \tag{B.0.4}$$

If $\vec{\mathbf{a}} = \vec{\mathbf{b}}$, then

$$\vec{\mathbf{c}} \times (\vec{\mathbf{a}} \times (\vec{\mathbf{a}} \times \vec{\mathbf{c}})) = \vec{\mathbf{c}} \times ((\vec{\mathbf{a}} \cdot \vec{\mathbf{c}})\,\vec{\mathbf{a}} - (\vec{\mathbf{a}} \cdot \vec{\mathbf{a}})\,\vec{\mathbf{c}}) = \vec{\mathbf{c}} \times (\vec{\mathbf{a}} \cdot \vec{\mathbf{c}})\,\vec{\mathbf{a}} = (\vec{\mathbf{a}} \cdot \vec{\mathbf{c}})\,\vec{\mathbf{c}} \times \vec{\mathbf{a}} \tag{B.0.5}$$

Reversing the roles of $\vec{\mathbf{c}}$ and $\vec{\mathbf{a}}$, we obtain

$$\vec{\mathbf{a}} \times (\vec{\mathbf{c}} \times (\vec{\mathbf{c}} \times \vec{\mathbf{a}})) = (\vec{\mathbf{a}} \cdot \vec{\mathbf{c}})\,\vec{\mathbf{a}} \times \vec{\mathbf{c}} = -\vec{\mathbf{c}} \times (\vec{\mathbf{a}} \times (\vec{\mathbf{a}} \times \vec{\mathbf{c}})) \tag{B.0.6}$$

In matrix form this becomes

$$\mathbf{a}^{\times}\mathbf{c}^{\times}\mathbf{c}^{\times}\mathbf{a} = -\mathbf{c}^{\times}\mathbf{a}^{\times}\mathbf{a}^{\times}\mathbf{c} \tag{B.0.7}$$

Using $\boldsymbol{\omega}$ and $\mathbf{r}$ rather than $\mathbf{a}$ and $\mathbf{c}$ gives

$$\boldsymbol{\omega}^{\times}\mathbf{r}^{\times}\mathbf{r}^{\times}\boldsymbol{\omega} = -\mathbf{r}^{\times}\boldsymbol{\omega}^{\times}\boldsymbol{\omega}^{\times}\mathbf{r} \tag{B.0.8}$$

Interchanging the last two terms on the RHS

$$\boldsymbol{\omega}^{\times}\mathbf{r}^{\times}\mathbf{r}^{\times}\boldsymbol{\omega} = \mathbf{r}^{\times}\boldsymbol{\omega}^{\times}\mathbf{r}^{\times}\boldsymbol{\omega} \tag{B.0.9}$$

This result may seem a bit strange, since the order of the first two matrices is reversed, and it is readily shown that $\boldsymbol{\omega}^{\times}\mathbf{r}^{\times} \neq \mathbf{r}^{\times}\boldsymbol{\omega}^{\times}$. However, in this particular instance this reversal is permitted.

APPENDIX C

Finite element model unit conversions

Consider the matrix defined in Eq. (2.2.5),

$$\mathbf{\Phi} \equiv \text{Eigenvector matrix from the FEM}$$

This must be converted from the FEM axis system to the body axis system. Typically this involves a 180 degree rotation about the y axis. A second issue arises if the FEM has been created using units of inches, pounds, and slinches, but the dynamic model uses feet, pounds, and slugs. These units are common in the US aerospace industry. This problem is compounded if the mode shapes have been mass normalized. The mass normalization procedure uses

$$\mathbf{\Phi}^T \mathbf{M}_B \mathbf{\Phi} = \mathbf{1}$$

where $\mathbf{1}$ is the identity matrix (dimensionless). Write the eigenvector of mode i and node n as

$$\boldsymbol{\phi}_{ni} = \begin{bmatrix} \bar{\psi}_x & \bar{\psi}_y & \bar{\psi}_z & \bar{\sigma}_x & \bar{\sigma}_y & \bar{\sigma}_z \end{bmatrix}^T$$

The overbar represents a variable in the inch system. Let the symbol $\propto$ mean "has units of." Since $\mathbf{M}_B \propto$ slinches and slinch-inch^2, this results in

$$\begin{aligned} \bar{\psi} &\propto \frac{1}{\sqrt{\text{slinch}}} \\ \bar{\sigma} &\propto \frac{1}{\text{inch}\sqrt{\text{slinch}}} \\ \bar{\eta} &\propto \text{inch}\sqrt{\text{slinch}} \end{aligned}$$

where $\bar{\eta}$ is the modal amplitude. To convert these outputs into units that are compatible with slugs and feet, define

$$\begin{aligned} \psi &\equiv \frac{1}{\sqrt{12}}\bar{\psi} \\ \sigma &\equiv \sqrt{12}\bar{\sigma} \end{aligned}$$

$$\eta \equiv \frac{1}{\sqrt{12}}\bar{\eta}$$

Thus

$$\psi \propto \sqrt{\frac{1}{\text{slug}}}$$

$$\sigma \propto \frac{1}{\text{feet}}\sqrt{\frac{1}{\text{slug}}}$$

$$\eta \propto \text{feet}\sqrt{\text{slug}}$$

In this manner the product $\psi\eta$ =feet, and the product $\sigma\eta$ is dimensionless (radians).

APPENDIX D

Second-order coordinate transformation

The authors are indebted to Richard Metzinger of Draper, who wrote the memo upon which the following material is based.

Eq. (2.1.2) provides a first-order transformation of the following form

$$\mathbf{v}_b = \left(\mathbf{1} - \boldsymbol{\theta}^{\times}\right)\mathbf{v}_a \tag{D.0.1}$$

where $\mathbf{v}_a$ is a vector expressed in Frame a, $\mathbf{v}_b$ is the same vector expressed in Frame b, $\boldsymbol{\theta} = \left[\begin{array}{ccc} \theta_1 & \theta_2 & \theta_3 \end{array}\right]^T$, and θ_1, θ_2, and θ_3 represent small angle rotations about the x, y and z axes respectively, that are necessary to get from Frame a to Frame b. In (2.1.2), Frame b is the body frame and Frame a is an inertial frame, but this same transformation can be used to represent small angular misalignments between an engine frame, for example, and the body frame. It is proposed that a second-order version of this can be written in the form

$$\mathbf{v}_b = \left(\mathbf{1} - \boldsymbol{\theta}^{\times} + \frac{1}{2}\boldsymbol{\theta}^{\times}\boldsymbol{\theta}^{\times}\right)\mathbf{v}_a \tag{D.0.2}$$

The purpose of this appendix is to analyze this expression and form an understanding of its correctness. First we perform the indicated squaring operation:

$$\boldsymbol{\theta}^{\times}\boldsymbol{\theta}^{\times} = \left[\begin{array}{ccc} 0 & -\theta_3 & \theta_2 \\ \theta_3 & 0 & -\theta_1 \\ -\theta_2 & \theta_1 & 0 \end{array}\right]\left[\begin{array}{ccc} 0 & -\theta_3 & \theta_2 \\ \theta_3 & 0 & -\theta_1 \\ -\theta_2 & \theta_1 & 0 \end{array}\right] \tag{D.0.3}$$

$$\boldsymbol{\theta}^{\times}\boldsymbol{\theta}^{\times} = \left[\begin{array}{ccc} -\left(\theta_2^2 + \theta_3^2\right) & \theta_1\theta_2 & \theta_1\theta_3 \\ \theta_1\theta_2 & -\left(\theta_1^2 + \theta_3^2\right) & \theta_2\theta_3 \\ \theta_1\theta_3 & \theta_2\theta_3 & -\left(\theta_1^2 + \theta_2^2\right) \end{array}\right] \tag{D.0.4}$$

The complete transformation is thus

$$\mathbf{1}-\boldsymbol{\theta}^{\times}+\frac{1}{2}\boldsymbol{\theta}^{\times}\boldsymbol{\theta}^{\times}=\begin{bmatrix} 1-\frac{(\theta_2^2+\theta_3^2)}{2} & +\theta_3+\frac{\theta_1\theta_2}{2} & -\theta_2+\frac{\theta_1\theta_3}{2} \\ -\theta_3+\frac{\theta_1\theta_2}{2} & 1-\frac{(\theta_1^2+\theta_3^2)}{2} & \theta_1+\frac{\theta_2\theta_3}{2} \\ \theta_2+\frac{\theta_1\theta_3}{2} & -\theta_1+\frac{\theta_2\theta_3}{2} & 1-\frac{(\theta_1^2+\theta_2^2)}{2} \end{bmatrix} \tag{D.0.5}$$

Next we develop a "truth" model by multiplying rotation matrices for three consecutive angle rotations together. We use the notation that c_1 denotes $\cos\theta_1$, s_1 denotes $\sin\theta_1$, etc. There are six ways to perform this depending on the sequence of rotations. Sequences that have two rotations about the same axis are not included. The first one is written out in detail and the results for the other five are simply stated. The six sequences are tabulated below. The matrix order in this tabulation results from the fact that matrix multiplications proceed from right to left. The first one, labeled 3-2-1, has the first rotation about z, the second about y and the final rotation about x. The six possibilities are:

Euler Sequence	Matrix order
3-2-1	x-y-z
2-3-1	x-z-y
1-2-3	z-y-x
2-1-3	z-x-y
1-3-2	y-z-x
3-1-2	y-x-z

$$\mathbf{T}_{321}=\begin{bmatrix} 1 & 0 & 0 \\ 0 & c_1 & s_1 \\ 0 & -s_1 & c_1 \end{bmatrix}\begin{bmatrix} c_2 & 0 & -s_2 \\ 0 & 1 & 0 \\ s_2 & 0 & c_2 \end{bmatrix}\begin{bmatrix} c_3 & s_3 & 0 \\ -s_3 & c_3 & 0 \\ 0 & 0 & 1 \end{bmatrix} \tag{D.0.6}$$

$$\mathbf{T}_{321}=\begin{bmatrix} 1 & 0 & 0 \\ 0 & c_1 & s_1 \\ 0 & -s_1 & c_1 \end{bmatrix}\begin{bmatrix} c_2c_3 & c_2s_3 & -s_2 \\ -s_3 & c_3 & 0 \\ s_2c_3 & s_2s_3 & c_2 \end{bmatrix} \tag{D.0.7}$$

$$\mathbf{T}_{321}=\begin{bmatrix} c_2c_3 & c_2s_3 & -s_2 \\ (-s_3c_1+s_2c_3s_1) & (c_3c_1+s_2s_3s_1) & s_1c_2 \\ (s_1s_2+s_2c_3c_1) & (-s_1c_3+s_2s_3c_1) & c_1c_2 \end{bmatrix} \tag{D.0.8}$$

In like manner the results for the other possibilities are:

$$\mathbf{T}_{231} = \begin{bmatrix} c_2c_3 & s_3 & -s_2c_3 \\ (-c_1c_2s_3 + s_2s_1) & c_3c_1 & (s_2s_3c_1 + s_1c_2) \\ (c_2s_3s_1 + s_2c_1) & -s_1c_3 & (-s_1s_2s_3 + c_2c_1) \end{bmatrix} \tag{D.0.9}$$

$$\mathbf{T}_{123} = \begin{bmatrix} c_2c_3 & (s_2c_3s_1 + c_1s_3) & (-s_2c_3c_1 + s_3s_1) \\ -c_2s_3 & (c_3c_1 - s_1s_2s_3) & (s_2s_3c_1 + s_1c_3) \\ s_2 & -c_2s_1 & c_2c_1 \end{bmatrix} \tag{D.0.10}$$

$$\mathbf{T}_{213} = \begin{bmatrix} (c_3c_1 + s_1s_2s_3) & c_1s_3 & (-s_2c_3 + c_2s_3s_1) \\ (-c_2s_3 - s_1s_2c_3) & c_3c_1 & (s_2s_3 + s_1c_2c_3) \\ c_1s_2 & -s_1 & c_1c_2 \end{bmatrix} \tag{D.0.11}$$

$$\mathbf{T}_{132} = \begin{bmatrix} c_2c_3 & (c_1c_2s_3 + s_1s_2) & (-s_2c_1 + c_2s_3s_1) \\ s_3 & c_3c_1 & c_3s_1 \\ s_2c_3 & (-c_2s_1 + c_1s_2s_3) & (s_1s_2s_3 + c_1c_2) \end{bmatrix} \tag{D.0.12}$$

$$\mathbf{T}_{312} = \begin{bmatrix} (c_2c_3 - s_1s_2s_3) & (c_2s_3 + s_1s_2c_3) & -s_2c_1 \\ -c_1s_3 & c_3c_1 & s_1 \\ (s_2c_3 + c_2s_1s_3) & (s_2s_3 - s_1c_2c_3) & c_1c_2 \end{bmatrix} \tag{D.0.13}$$

The approach will be to substitute the second order Taylor series expansion for the sine and cosine in (D.0.8) through (D.0.13), and then only retain the first and second order terms. The Taylor Series expansions for sine and cosine are

$$\sin\theta = \theta \tag{D.0.14}$$

$$\cos\theta = 1 - \frac{\theta^2}{2} \tag{D.0.15}$$

We find that all the transformation matrices have one diagonal term of the form:

$$\cos(a)\cos(b) \pm \sin(a)\sin(b)\sin(c) \tag{D.0.16}$$

Substituting in the series expansions we get, for example

$$c_2c_3 \pm s_1s_2s_3 = \left(1 - \frac{\theta_2^2}{2}\right)\left(1 - \frac{\theta_3^2}{2}\right) \pm \theta_1\theta_2\theta_3 \tag{D.0.17}$$

$$c_2c_3 \pm s_1s_2s_3 = \left(1 - \frac{(\theta_2^2 + \theta_3^2)}{2} - \frac{\theta_2^2\theta_3^2}{4}\right) \pm \theta_1\theta_2\theta_3 \tag{D.0.18}$$

Retaining only terms to second order we find:

$$c_2c_3 \pm s_1s_2s_3 = c_2c_3 = 1 - \frac{(\theta_2^2 + \theta_3^2)}{2} \tag{D.0.19}$$

All the diagonal terms reduce to the same form as (D.0.19). Examining the off-diagonal terms we find that they come in four forms (each with permutations of the indices):

$$\begin{aligned} &s_2 \\ &c_2s_3 \\ &s_3c_1 + s_2c_3s_1 \\ &s_3s_1 + s_2c_3c_1 \end{aligned} \tag{D.0.20}$$

Inserting the series expansion for sine and cosine and retaining only terms to 2nd order we find:

$$\begin{aligned} s_2 &= \theta_2 \\ c_2s_3 &= \theta_3 \\ s_3c_1 + s_2c_3s_1 &= \theta_3 + \theta_1\theta_2 \\ s_3s_1 + s_2c_3c_1 &= \theta_3\theta_1 + \theta_2 \end{aligned} \tag{D.0.21}$$

Making these substitutions into the transformation matrices (D.0.8) through (D.0.13) we find that:

$$\mathbf{T} = \begin{bmatrix} 1 - \frac{(\theta_2^2+\theta_3^2)}{2} & \begin{pmatrix} \theta_3 \\ \text{or} \\ \theta_3 + \theta_1\theta_2 \end{pmatrix} & \begin{pmatrix} -\theta_2 \\ \text{or} \\ -\theta_2 + \theta_1\theta_3 \end{pmatrix} \\ \begin{pmatrix} -\theta_3 \\ \text{or} \\ -\theta_3 + \theta_1\theta_2 \end{pmatrix} & 1 - \frac{(\theta_1^2+\theta_3^2)}{2} & \begin{pmatrix} \theta_1 \\ \text{or} \\ \theta_1 + \theta_2\theta_3 \end{pmatrix} \\ \begin{pmatrix} \theta_2 \\ \text{or} \\ \theta_2 + \theta_1\theta_3 \end{pmatrix} & \begin{pmatrix} -\theta_1 \\ \text{or} \\ -\theta_1 + \theta_2\theta_3 \end{pmatrix} & 1 - \frac{(\theta_1^2+\theta_2^2)}{2} \end{bmatrix} \tag{D.0.22}$$

The off-diagonal terms have two alternate embodiments, and each embodiment occurs three times. So since we really don't know which of the six possible rotation sequences to use, we take the average. For example, the

(1, 2) term has the form of θ_3 for three versions, and the form $\theta_3 + \theta_1\theta_2$ for three versions. The average value is thus:

$$\theta_3 + \frac{\theta_1\theta_2}{2} \tag{D.0.23}$$

Using this rule we determine that the average small angle rotation matrix using a second order Taylor series expansion for the sines and cosines is:

$$\mathbf{T} = \begin{bmatrix} 1 - \frac{(\theta_2^2+\theta_3^2)}{2} & \theta_3 + \frac{\theta_1\theta_2}{2} & -\theta_2 + \frac{\theta_1\theta_3}{2} \\ -\theta_3 + \frac{\theta_1\theta_2}{2} & 1 - \frac{(\theta_1^2+\theta_3^2)}{2} & \theta_1 + \frac{\theta_2\theta_3}{2} \\ \theta_2 + \frac{\theta_1\theta_3}{2} & -\theta_1 + \frac{\theta_2\theta_3}{2} & 1 - \frac{(\theta_1^2+\theta_2^2)}{2} \end{bmatrix} \tag{D.0.24}$$

This is identical to the expression on the right hand side of (D.0.5). We have thus shown that the left side of (D.0.5) is a transformation that represents a good average of the six possible Euler sequences.

As a side note, any attempt to compare the above Euler transformations with similar transformations from other sources will require careful attention to how the indices are defined. Table 2.1 of Hughes [6] is a set of direction cosine matrices (DCM's) with the indices based on the order in which the rotation angle occurs, rather than the axis about which the rotation occurs. Regan and Anandrakrishnan [25], Appendix C, present a table that is identical to that of Hughes. They do a much better job of explaining that a backwards indexing scheme is used, i.e. they have $\theta_3 =$ angle of first rotation, $\theta_2 =$ angle of second rotation, $\theta_1 =$ angle of third rotation. Schaub and Junkins [49] also provide a table of DCM's in their Appendix B, but they fail to provide any explanation of their indexing. It turns out that they too base their indices on the order of rotation, but they use forward indexing. Thus to get their results to match those of Hughes, it is necessary to swap θ_3 and θ_1. To match to the axis-based indexing in this Appendix, a different mapping is required for each Euler sequence. Axis-based indexing is preferred as long as the same axis is not used for two rotations.

APPENDIX E

Angular momentum of free-free modes

The purpose of this appendix is to demonstrate that the angular momentum integral in (2.2.30) is zero for any origin of the coordinate system, i.e. it is not necessary to locate the origin at the center of mass. Overbars are used to indicate variables written in such a barycentric coordinate system. Thus $\bar{\mathbf{r}}$ is a vector from the center of mass to mass element dm. Using a different origin, we can write

$$\mathbf{r} = \bar{\mathbf{r}} + \mathbf{r}_{cm} \tag{E.0.1}$$

where $\mathbf{r}_{cm}$ is the location of the cm in the non-overbar coordinate system.

The term "free-free" vibrations originates from analysis of beams and corresponds to the condition that neither end of the beam has a constraint such as being pinned or cantilevered. This same terminology is used for rockets. As discussed in Chapter 2, no linear momentum is generated by such vibrations. This statement becomes

$$\int \bar{\boldsymbol{\psi}}_i(\bar{\mathbf{r}})\, \mathrm{d}m = 0, \text{ all } i \tag{E.0.2}$$

Let $\boldsymbol{\psi}_i$ be the mode shape in the non-overbar coordinate frame. $\boldsymbol{\psi}_i$ and $\bar{\boldsymbol{\psi}}_i$ are just two ways of writing the same mode shape. At corresponding points, they have the same value.

$$\bar{\boldsymbol{\psi}}_i(\bar{\mathbf{r}}) = \boldsymbol{\psi}_i(\mathbf{r}) \tag{E.0.3}$$

Thus

$$\int \boldsymbol{\psi}_i(\mathbf{r})\, \mathrm{d}m = 0, \text{ all } i \tag{E.0.4}$$

It is also true that no net angular momentum about the center of mass is generated. We write this as

$$\int \bar{\mathbf{r}}^{\times} \bar{\boldsymbol{\psi}}_i(\bar{\mathbf{r}})\, \mathrm{d}m = 0, \text{ all } i \tag{E.0.5}$$

Substituting (E.0.1) into (E.0.5) gives

$$\int (\mathbf{r} - \mathbf{r}_{cm})^{\times} \bar{\boldsymbol{\psi}}_i(\bar{\mathbf{r}})\, \mathrm{d}m = \int \mathbf{r}^{\times} \bar{\boldsymbol{\psi}}_i(\bar{\mathbf{r}})\, \mathrm{d}m - \mathbf{r}_{cm}^{\times} \int \bar{\boldsymbol{\psi}}_i(\bar{\mathbf{r}})\, \mathrm{d}m = 0 \tag{E.0.6}$$

Using (E.0.3), this becomes

$$\int \mathbf{r}^{\times} \boldsymbol{\psi}_i(\mathbf{r})\, \mathrm{d}m - \mathbf{r}_{cm}^{\times} \int \bar{\boldsymbol{\psi}}_i(\bar{\mathbf{r}})\, \mathrm{d}m = 0 \tag{E.0.7}$$

From (E.0.2), the second integral is zero. Thus

$$\int \mathbf{r}^{\times} \boldsymbol{\psi}_i(\mathbf{r})\, \mathrm{d}m = 0 \tag{E.0.8}$$

Bibliography

[1] A. Greensite, Analysis and Design of Space Vehicle Flight Control Systems, Volume I - Short Period Dynamics, Tech. Rep. CR-820, NASA, 1967.

[2] M. Rheinfurth, Control-Feedback Stability Analysis, Tech. Rep. DA-TR-2-60, U.S. Army Ballistic Missile Agency, 1960.

[3] D. Garner, Control Theory Handbook, Tech. Rep. TM X-53036, NASA, Apr. 1964.

[4] J. Frosch, D. Vallely, Saturn AS-501/S-IC Flight Control System Design, J. Spacecr. 4 (8) (1967) 1003–1009.

[5] H. Bauer, Vehicle Stability and Control, Tech. Rep. SP-106, The Dynamic Behavior of Liquids in Moving Containers, NASA, 1966, p. 225, Ch. 7.

[6] P. Hughes, Spacecraft Attitude Dynamics, Dover, 2004.

[7] H. Bauer, Fluid Oscillations in Containers of a Space Vehicle and Their Influence on Stability, Tech. Rep. TR-R-187, NASA, Feb. 1964.

[8] F. Dodge, D. Kana, Moment of Inertia and Damping of Liquids in Baffled Cylindrical Tanks, J. Spacecr. Rockets 3 (1966) 153–155.

[9] H. Bauer, Comments on 'Moment of Inertia and Damping of Liquids in Baffled Cylindrical Tanks', J. Spacecr. Rockets (Jun. 1966).

[10] R. Craig, A. Kurdila, Fundamentals of Structural Dynamics, Wiley, 2006.

[11] G. Strang, Linear Algebra and Its Applications, Academic Press, 1976.

[12] W. Hurty, Dynamics of Structures, Prentice-Hall, 1964.

[13] H. Abramson, The Dynamic Behavior of Liquids in Moving Containers, Tech. Rep. SP-106, NASA, 1966.

[14] F. Dodge, The New Dynamic Behavior of Liquids in Moving Containers, Southwest Research Institute, San Antonio, TX, 2000.

[15] H. Lamb, Hydrodynamics, 6th edition, Dover, 1945.

[16] D. Lomen, Liquid Propellant Sloshing in Mobile Tanks of Arbitrary Shape, Tech. Rep. CR-230, NASA, 1965.

[17] D. Lomen, Digital Analysis of Liquid Propellant Sloshing in Mobile Tanks With Rotational Symmetry, Tech. Rep. CR-222, NASA, 1965.

[18] D. Kana, Validated Spherical Pendulum Model for Rotary Liquid Slosh, J. Spacecr. 26 (3) (1989) 188–195.

[19] H. Bauer, Nonlinear Mechanical Model for the Description of Propellant Sloshing, AIAA J. 4 (9) (1966) 1662–1668.

[20] J. Roberts, E. Basurto, P. Chen, Slosh Design Handbook I, Tech. Rep. CR-406, NASA, May 1966.

[21] I. Summer, Experimentally Determined Pendulum Analogy of Liquid Sloshing in Spherical and Oblate-Spheroidal Tanks, Tech. Rep. TN D-2737, NASA, Apr. 1965.

[22] W. Thomson, Introduction to Space Dynamics, Dover, 1961.

[23] J.R. Wertz (Ed.), Spacecraft Attitude Determination and Control, Kluwer Academic Publishers, Dordrecht, Netherlands, 1978.

[24] R. Stengel, Flight Dynamics, Princeton University Press, 2004.

[25] F. Regan, S. Anandrakrishnan, Dynamics of Atmospheric Re-Entry, AIAA Education Series, 1993.

[26] D. Vallado, Fundamentals of Astrodynamics and Applications, McGraw-Hill, 1997.

[27] R. Bisplinghoff, H. Ashley, R. Halfman, Aeroelasticity, Addison-Wesley, 1955.

[28] D. Lukens, A. Schmitt, G. Broucck, Approximate Transfer Functions for Flexible-Booster-and-Autopilot Analysis, Tech. Rep. WADD-TR-61-93, Convair Astronautics Division, General Dynamics Corp., 1961.

[29] J. Nielsen, Missile Aerodynamics, American Institute of Aeronautics and Astronautics, 1988.
[30] J. Orr, M. Johnson, J. Wetherbee, J. McDuffie, State Space Implementation of Linear Perturbation Dynamics Equations for Flexible Launch Vehicles, AIAA Guidance, Navigation, and Control Conference, Chicago, Illinois, AIAA-2009-5962, 2009.
[31] R. Bartels, P. Chwalowski, S. Massey, J. Heeg, Computational Aeroelastic Analysis of the Ares I Crew Launch Vehicle During Ascent, J. Spacecr. Rockets 49 (4) (2012) 651–658.
[32] J. Orr, A Coupled Aeroelastic Model for Launch Vehicle Stability Analysis, AIAA Atmospheric Flight Mechanics Conference, Toronto, CA, AIAA-2010-7642, 2010.
[33] J. Hanson, B. Beard, Applying Monte Carlo Simulation to Launch Vehicle Design and Requirements Analysis, Tech. Rep. TP-2010-216447, NASA, Sep. 2010.
[34] G. Golub, C. Van Loan, Matrix Computations, 3rd ed., Johns Hopkins University Press, 1996.
[35] T. Chan, J. Olkin, D. Cooley, Solving Quadratically Constrained Least Squares Using Black Box Solvers, BIT Numer. Math. 32 (1992) 481–494.
[36] P. Hansen, Regularization Tools: A MATLAB Package for Analysis and Solution of Discrete Ill-Posed Problems, Numer. Algorithms 6 (1994) 1–35.
[37] R. Jurenko, T. Bush, J. Ottander, Elastic Model Transitions: a Hybrid Approach Utilizing Quadratic Inequality Constrained Least Squares (LSQI) and Direct Shape Mapping (DSM), American Astronautical Society (AAS) Guidance, Navigation, and Control Conference, 2014.
[38] J. D'Azzo, C. Houpis, Feedback Control System Analysis and Synthesis, McGraw-Hill, 1960.
[39] J. Hsu, A. Meyer, Modern Control Principles and Applications, McGraw-Hill, 1968.
[40] F. Graham, R. Lathrop, Synthesis of Optimum Transient Response - Criteria and Standard Forms, Tech. Rep. 53-66, USAF Wright Air Development Center, Aug. 1953.
[41] Y. Cao, Correcting the Minimum ITAE Standard Forms of Zero-Displacement-Error Systems, J. Zhejiang Univ. Sci. 23 (4) (1989) 550–559.
[42] J. Orr, J. Wall, T. VanZwieten, C. Hall, Space Launch System Ascent Flight Control Design, American Astronautical Society Guidance, Navigation, and Control Conference, AAS 14-036, 2014.
[43] B. Starr, I. Yunis, A. Olds, Use of Flexible Body Coupled Loads in Assessment of Day of Launch Flight Loads, AIAA Atmospheric Flight Mechanics Conference, AIAA 2011-6465, 2011.
[44] R. Hoelker, Theory of Artificial Stabilization of Missiles and Space Vehicles With Exposition of Four Control Principles, Tech. Rep. D-555, NASA Marshall Space Flight Center, Jun. 1961.
[45] Z. Thompson, J. Hung, A Mathematical Model for a Space Vehicle Thrust Vector Control System, Tech. Rep. R-15, University of Tennessee, Control Theory Group, Aug. 1967.
[46] Z. Thompson, J. Hung, An Apparent Adaptive Notch Filter for Thrust Vector Control, in: M.E.A. Lunc (Ed.), Propulsion Re-Entry Physics, International Aeronautical Congress, 1970, pp. 101–107.
[47] J. Kuipers, Quaternions and Rotation Sequences: A Primer With Applications to Orbits, Aerospace, and Virtual Reality, Princeton University Press, Princeton, N.J., 1999.
[48] A. Witkin, An Introduction to Physically Based Modeling: Constrained Dynamics, Tech. rep., Carnegie Mellon University Robotics Institute, 1997.
[49] H. Schaub, J. Junkins, Analytical Mechanics of Space Systems, AIAA, 2018.

Index

B

C

D

T

U

V

FSC
www.fsc.org
MIX
Papier | Fördert
gute Waldnutzung
FSC® C083411

Kolibri 360 GmbH
Ferdinand-Jühlke-Straße 7
99095 Erfurt, Deutschland
produktsicherheit@kolibri360.de